AN INTRODUCTION TO THE ATOMIC AND RADIATION PHYSICS OF PLASMAS

Plasmas comprise more than 99% of the observable universe. They are important in many technologies and are potential sources for fusion power. Atomic and radiation physics is critical for the diagnosis, observation and simulation of astrophysical and laboratory plasmas, and plasma physicists working in a range of areas from astrophysics, magnetic fusion and inertial fusion utilise atomic and radiation physics to interpret measurements. This book develops the physics of emission, absorption and interaction of light in astrophysics and in laboratory plasmas from first principles, using the physics of various fields of study, including quantum mechanics, electricity and magnetism, and statistical physics. Linking undergraduate-level atomic and radiation physics with the advanced material required for postgraduate study and research, the text adopts a highly pedagogical approach and includes numerous exercises within each chapter to reinforce students' understanding of key concepts.

G. J. TALLENTS is Professor in Physics at the York Plasma Institute at the University of York. His current research centres on the effects of high plasma density on spectroscopy and the interaction of extreme ultraviolet lasers with solid targets.

AN INTRODUCTION TO THE ATOMIC AND RADIATION PHYSICS OF PLASMAS

G. J. TALLENTS
University of York

CAMBRIDGE
UNIVERSITY PRESS

University Printing House, Cambridge CB2 8BS, United Kingdom

One Liberty Plaza, 20th Floor, New York, NY 10006, USA

477 Williamstown Road, Port Melbourne, VIC 3207, Australia

314–321, 3rd Floor, Plot 3, Splendor Forum, Jasola District Centre, New Delhi – 110025, India

79 Anson Road, #06–04/06, Singapore 079906

Cambridge University Press is part of the University of Cambridge.

It furthers the University's mission by disseminating knowledge in the pursuit of education, learning, and research at the highest international levels of excellence.

www.cambridge.org
Information on this title: www.cambridge.org/9781108419543
DOI: 10.1017/9781108303538

First published 2018

A catalogue record for this publication is available from the British Library.

Library of Congress Cataloging-in-Publication Data
Names: Tallents, G. J., author.
Title: An introduction to the atomic and radiation physics of plasmas / G.J. Tallents, University of York.
Description: Cambridge, United Kingdom ; New York, NY : Cambridge University Press, 2018. | Includes bibliographical references and index.
Identifiers: LCCN 2017042303 | ISBN 9781108419543 (hardback) | ISBN 1108419542 (hardback)
Subjects: LCSH: Plasma radiation. | Radiation–Measurement.
Classification: LCC QC718.5.R3 T35 2018 | DDC 530.4/4–dc23 LC record available at https://lccn.loc.gov/2017042303

ISBN 978-1-108-41954-3 Hardback

Contents

Preface

This book provides an introduction to the physics of emission, absorption and interaction of light in astrophysics and in laboratory plasmas. Such study necessarily requires a wide range of modern physics understanding involving electricity and magnetism, relativity, atomic structure, quantum mechanics, particle collision theory, statistical physics and more. Indeed, the analysis of light emission and collisional processes relevant to plasmas has provided much of the experimental evidence for quantum mechanics. The atomic and radiation physics of plasmas is, consequently, an ideal subject for study as an extension to material taught to physics undergraduates. The book combines undergraduate-level studies of the quantum mechanics of ions/atoms with the atomic and radiation physics of plasmas, though non-quantum models are used extensively. Atomic and radiation physics is presented at a level aimed at undergraduates in their final two years through to graduate students and researchers. Material needed for research in plasma physics and astrophysics is derived.

Plasma physicists working in a range of areas from astrophysics, magnetic fusion and inertial fusion to low-temperature plasmas of technological significance utilise atomic and radiation physics to interpret measurements. Plasma physics is a growing research area with the construction of the ITER tokamak, new laser-plasma facilities and the development of new methods of creating plasma, such as with free-electron lasers. Atomic and radiation physics is also an essential component in the theoretical development and simulation of astrophysical and laboratory plasmas. One aim of this book is to emphasise the overlap of atomic/radiation physics between astrophysical and laboratory plasmas, an imbrication exploited in the expanding field of laboratory astrophysics where physical scenarios relevant to astrophysics are simulated in the laboratory.

Due to the range of understanding required for research in the atomic and radiation physics of plasmas, the underlying physics is often not developed in research publications in astrophysics and plasma spectroscopy. An aim of this book has been to start with the knowledge obtained by physics graduates before they begin to specialise and to develop formulae and explain techniques used in plasma spectroscopy. The areas of plasma research utilising aspects of atomic and radiation physics are briefly introduced before spectroscopic applications are covered, but this book concentrates on the underlying atomic and radiation physics.

As this is a textbook, rather than a monograph, some presented treatments are not the most comprehensively complete available, but illustrate the way to standard formulae and techniqes. Similarly, the citations presented are representative and do not give a full coverage of the development of topics. I offer my apologies to those whose contribution to knowledge is described but not cited.

Exercises are included at the end of each chapter and form an integral component of the text. Where a numerical answer is required, this is added in brackets, sometimes along with comment indicating, for example, wider implications of the exercise. Material is presented using the International System of Units (SI) unless explicitly defined otherwise. The convention common in laboratory plasma work to define temperatures (T) in units of energy (k_BT) using electron volts (eV) is widely used in the text. Here k_B is Boltzmann's constant. In SI units, $1\ \text{eV} = 1.6 \times 10^{-19}$ J and corresponds to 11 605°K. While formulas are developed in SI units, research areas often use centimeter-gram-second (CGS) units, so some expressions are converted from SI where numerical values are presented.

Much of the content presented here has been developed for courses taught at the University of York. The treatment of the atomic physics of the hydrogen and multi-electron atoms has been taught to third-year students for several years, while other material has featured in lecture courses presented to MSc and PhD students of fusion energy. I am grateful to the University of York for the opportunity to develop some of these lecture courses into the present book and also thank many students for their questions, comments and corrections. I am grateful to Professor Geoff Pert FRS for his comments on a draft of the manuscript and to Dr Erik Wagenaars for providing lecture material.

1
Plasma and Atomic Physics

A plasma is created by adding energy to a gas so that electrons are removed from atoms, producing free electrons and ions. Electric and magnetic fields interact strongly with the charged electrons and ions in plasmas (unlike solids, liquids and gases) and, consequently, plasmas behave differently to imposed electric and magnetic fields and modify electromagnetic waves in different ways to solids, liquids and gases. The different behaviour of plasmas has caused them to be regarded as a fourth fundamental state of matter in addition to solids, liquids and gases.

More than 99% of the observable universe is plasma. For example, the Sun is a plasma and has mass comprising 99.85% of the solar system, so the fraction of plasma in the solar system is slightly higher once interplanetary plasma is included. Present understanding of the universe has been enabled by the detection of electromagnetic radiation emitted by or passing through plasma material. To understand the universe, we need to understand plasmas, and, in particular, we need to understand the processes of light emission and propagation in plasmas.[1]

Plasmas have many realised and potential applications. The fusion of isotopes of hydrogen in plasmas confined using magnetic fields or confined by inertia before a dense plasma can expand should provide a new source of energy production to replace the burning of fossil fuels, though the exact physics and many technical issues are not yet resolved [35]. The fuel for a fusion reactor (the deuterium isotope of hydrogen) is abundant in seawater (at concentration 33 mg/litre). Large-scale experiments are under way to make fusion reactors because of the enormous potential impact of the development of a fusion power plant [79, 67].

Plasmas are used in many technological applications, including semiconductor etching and thin-film coating [15]. Plasma is created during the welding of

[1] In astrophysics, plasma material is sometimes referred to as an 'ionised gas', while in laboratory plasma work involving partial ionisation of atoms, the term 'gaseous electronics' has been employed to denote the physics of 'low-temperature' plasmas. The use of the word 'plasma' to describe both 'ionised gases' and 'gaseous electronics', however, is now almost ubiquitous.

solid material and is under study for biological and medical applications such as bacterial sterilisation. The emission of light from plasmas has many applications, ranging from fluorescent tubes to the use of extreme ultra-violet light emitted from laser plasmas for the lithography of semiconductors [105]. Many different lasers utilising plasmas have been developed, including argon ion lasers and an extensive array of plasma lasers designed to operate at short wavelengths [108, 91], with the record for saturated lasing achieved at wavelength 5.9 nm [125, 100]. A road map for plasma applications shows the range of applicability of plasmas in technology [97].

A plasma can be defined as a collection of ions and free electrons where the charged ions and electrons produce collective responses to electric and magnetic fields, but the net charge density averages to zero over longer-length scales. Similar definitions have often been used to define material in the plasma state (see [17, 35, 5]). Our given definition of plasma leads to the concept of the plasma frequency, which is a minimum frequency for an oscillating field to exist in a plasma, and to the concept of the Debye length, which is the distance over which electron and ion charges average to zero. We start our examination of plasmas by considering the plasma frequency and Debye length in Section 1.1. The plasma frequency is particularly important for the physics of the propagation of electromagnetic radiation in a plasma.

To ionise material so that free electrons and ions are present to form a plasma, elevated temperatures are required, causing plasmas to emit electromagnetic radiation, depending on the temperature of the plasma, in, typically, the infra-red to X-ray spectral range, though the spectrum of emission can extend to longer wavelength microwaves and radio waves, and to high-photon-energy gamma rays. In plasmas, electrons often occupy the excited bound quantum states of the ions and the free unbound quantum states. Such excitation and ionisation lead to radiation emission. The atomic physics producing electromagnetic waves in plasmas, and the subsequent propagation and absorption of electromagnetic radiation in plasmas, are the main subjects of this book.

For the relatively low-density but hot plasmas found in the laboratory, atoms and ions can be regarded as having an atomic physics structure close to that of an isolated atom or ion, but with quantum-state populations far from equilibrium. Free electrons, photons, ions and atoms have ‘collisions’ with the ions, causing excitation and ionisation. Astrophysical and space plasmas span energy-density ranges from extremely low (interstellar space) to extremely high (e.g. dwarf stars), and are associated with long timescales, often with equilibrium population and radiation fields.

The atomic and radiation physics of plasmas covers a wide range of modern physics understanding involving electricity and magnetism, relativity, atomic structure, quantum mechanics, particle collision theory, statistical physics and more. The analysis of light emission and collisional processes relevant to plasmas has provided much of the experimental evidence for quantum mechanics.

Analysis of the emission and absorption of light is an effective and non-invasive method to measure plasma conditions, such as density and temperature. Analysing spectral emission and absorption is the sole diagnostic technique applicable to astrophysical plasmas and is essential for diagnosing conditions in magnetic and inertial fusion plasmas. To determine plasma conditions, light probing involving scattering, absorption and radiation phase measurements (interferometry) can be used in laboratory plasmas. An understanding of radiation interaction in plasmas allows the interpretation of such probing.

There are books which concentrate on the diagnosis of plasma conditions using radiation emission – a subject known as plasma spectroscopy [51, 38]. Comprehensive research-level treatments of the atomic [95] and radiation [93] physics of plasmas, as well as an introduction to astronomical spectroscopy [111] and a graduate-level text emphasising atomic physics of relevance to astrophysics [86] are available. Codes and databases relevant to the atomic physics of plasmas include the Atomic Data and Analysis Structure (ADAS), (see, for example, Guzman et al. [41]) the FLYCHK code [18], the Astrophysical Plasma Emission Code (APEC) [101], and the National Institute of Standards and Technology (NIST) Atomic Spectra [63].

An understanding of the atomic physics of plasmas is needed for plasma simulation. The emission and absorption of light in a plasma can affect the plasma dynamics by, for example, transporting energy. However, as well as affecting plasma dynamics, the atomic and radiation physics of plasmas enables simulations of plasma density to be 'closed'. Fluid codes require a relationship between material density and pressure which, in turn, requires a knowledge of the degree of ionisation. Simulation particle codes similarly need a measure of the degree of ionisation for closure. This requirement for closure is explored in Section 1.1. We also show how atomic physics affects the velocity of sound in a plasma. This first chapter then presents an introduction to some radiation and atomic physics which is important in plasmas. The equilibrium relationship for ionisation (the Saha–Boltzmann equation), the distribution of speeds and energies of the particles (the Maxwellian distribution) and the Bohr model for atomic and ionic energy levels are introduced.

1.1 Plasma Physics

Some fundamental aspects of plasma physics are encapsulated in the definition of a plasma given above: 'A plasma is a collection of ions and free electrons where the charged ions and electrons produce collective responses to electric and magnetic fields with the net charge density over longer-length scales averaging to zero.' The concept of a collective response and the idea of the charge density averaging to zero lead to the concept of the plasma frequency and Debye length.

Plasma Frequency

Consider a uniform plasma of free electrons and ions which is neutrally charged and occupying a defined space. An imposed electric field can cause the centre of mass of the lighter electrons to be displaced by a distance x relative to the more massive ions. To deduce the necessary electric field, we can use the integral form of Gauss' law (see Appendix A.2) given by

$$\int_S \mathbf{E} \cdot d\mathbf{A} = \frac{1}{\epsilon_0} \int_V \rho_c dV.$$

Assume a cubic volume V extends into the plasma with one surface of area A perpendicular to the electric field and parallel to the plane of the plasma edge. The total charge enclosed by the volume is $-n_e e A x$, where n_e is the electron number density and $-e$ is the charge of the electron. The electric field across the area is then

$$E = -\frac{n_e e x}{\epsilon_0} \tag{1.1}$$

after cancelling the area A from both sides of the expression for Gauss's law. If the imposed electric field is switched off, there is a force eE on the electrons in the opposite direction to the force arising from the imposed electric field and an equation of motion of the electrons such that

$$m_0 \frac{d^2 x}{dt^2} = eE = -\frac{n_e e^2}{\epsilon_0} x \tag{1.2}$$

where m_0 is the mass of an electron. Solutions of this equation are of form

$$x(t) = x(0) \exp(-i\omega_p t)$$

where, upon substitution into Equation 1.2, we have

$$\omega_p^2 = \frac{n_e e^2}{m_0 \epsilon_0}. \tag{1.3}$$

The frequency ω_p is known as the plasma frequency. It represents the natural, collective oscillation frequency of the electrons relative to the ions and, we shall later see, defines a minimum frequency of light that can propagate in a plasma in the absence of a magnetic field.

The Debye Length

The characteristic distance for charge neutrality in a plasma can be found by considering a situation with two parallel plates separated in the x-direction by a distance $2a$. The plates are assumed to be at earth potential with electrons of density n_e filling the space between the plates. The electrostatic potential V_P is related to charge density ρ at positions x by Poisson's equation (see Appendix A.2). We have

$$\nabla^2 V_P = \frac{d^2 V_P}{dx^2} = -\frac{\rho_e}{\epsilon_0} = \frac{n_e e}{\epsilon_0}.$$

At the midway point between the plates (distance $x = a$), the potential is given by

$$V_P = \frac{n_e e a^2}{2\epsilon_0}$$

and the energy required to move another electron to the midway point between the plates is $V_P e = n_e e^2 a^2/(2\epsilon_0)$. In one direction, the average kinetic energy of an electron at temperature T_e is $(1/2)k_B T_e$ (see Exercise 1.2). We can equate this kinetic energy to the energy required to move an electron to the midway distance between the two plates:

$$\frac{1}{2} k_B T_e = \frac{n_e e^2 a^2}{2\epsilon_0}.$$

The distance a, where the electron kinetic energy is equal to the energy required to move an electron to the midway points between the plates, is the distance over which the ground potential of the plates stops influencing the 'average' electron. The distance is given by

$$a = \lambda_D = \left(\frac{\epsilon_0 k_B T_e}{n_e e^2}\right)^{1/2}. \tag{1.4}$$

This distance is known as the Debye length λ_D.

The number of electrons in a sphere of radius λ_D reflects the number of electrons likely to move 'collectively' together during, for example, light scattering from a plasma. We can write for the number of electrons in a Debye sphere

$$N_D = n_e \frac{4}{3}\pi \lambda_D^3 = \left(\frac{4}{3}\right)\pi \left(\frac{\epsilon_0 k_B T}{n_e e^2}\right)^{3/2} n_e. \tag{1.5}$$

Plasma Pressure and the Speed of Sound in a Plasma

In a plasma, pressure P is related to the mass density ρ and temperature T by adding up the electron, ion and atom pressure given by Boyle's law. We can write for a plasma where all particles have the same temperature and behave as ideal gases that

$$P = n_i k_B T_i + n_e k_B T_e = \frac{\rho(1 + Z_{av}) k_B T}{A m_p} \tag{1.6}$$

where A is the average atomic mass, m_p is the mass of the proton and the average degree of ionisation $Z_{av} = n_e/n_i$, where n_e is the electron density and n_i is the ion-plus-atom density. The electron temperature T_e and ion temperature T_i are assumed equal to T for the last equality. Equation 1.6 is an example of an equation of state relationship between state variables in thermodynamic equilibrium.

Changes of mass density ρ and the velocity $\mathbf{u}$ of a plasma fluid can be related by the continuity equation (representing conservation of mass) and the equation of motion (representing a fluid version of Newton's law that force is equal to mass times acceleration). The equation of motion for a fluid is also known as the Navier-Stokes equation. We can use standard fluid treatments (e.g. [84]) and write for these two equations respectively

$$\frac{\partial \rho}{\partial t} + \nabla.\rho\mathbf{u} = 0, \tag{1.7}$$

$$\rho\left(\frac{\partial \mathbf{u}}{\partial t} + (\mathbf{u}.\nabla)\mathbf{u}\right) = -\nabla P. \tag{1.8}$$

The fluid equation of continuity and the equation of motion can be used to simulate plasmas if the pressure P and the density ρ can be related to each other by a known closure relationship, for example, Equation 1.6. This is not always straightforward. For example, the pressure/density relationship is affected strongly by the value of the degree of ionisation Z_{av} and it is often necessary to evaluate separate temperatures for the different ion, atomic and electron components of the plasma.

For variations in mass density, velocity and pressure in one dimension z, the continuity equation and equation of motion can be written such that

$$\frac{\partial \rho}{\partial t} + \frac{\partial \rho u}{\partial z} = 0,$$

$$\rho\left(\frac{\partial u}{\partial t} + u\frac{\partial u}{\partial z}\right) = -\frac{\partial P}{\partial z}.$$

Assuming a small time-varying deviation of density $\rho = \rho_0 + \rho_1$, velocity $u = u_0 + u_1$ and pressure $P = P_0 + P_1$ from steady-state values ρ_0, u_0 and P_0, it is possible to show that to a good approximation the one-dimensional continuity equations combine to give an equation for the propagation of the deviation of density such that

$$\frac{\partial^2 \rho_1}{\partial z^2} + \frac{1}{(\partial P/\partial \rho)}\frac{\partial^2 \rho_1}{\partial t^2} = 0.$$

This equation has the form of a wave equation where the speed c_s of the wave is given by

$$c_s^2 = \left(\frac{\partial P}{\partial \rho}\right). \tag{1.9}$$

The propagation of a disturbance in density is usually known as a sound wave, so this equation shows that the sound speed is given by the square root of the rate of change of pressure with density changes. (The partial derivative here means that other parameters such as entropy and energy density are held constant.) The sound speed determines the rate of expansion of a freely expanding plasma and the speed of shock waves and other disturbances propagating in plasmas.

Statistical mechanics tells us that for a gas (or other system) characterised by a temperature T, the average energy per degree of freedom per particle is equal to $(1/2)k_BT$ (see Exercise 1.2). A degree of freedom can be represented by translational motion in one direction (giving three degrees of freedom for a monatomic gas or a plasma species such as the electrons), but can also include, for example, vibrational degrees of freedom for polyatomic gases.

Rather than consider degrees of freedom in a plasma, it is often more convenient to define a parameter γ_{eos} using the relationship between pressure and energy density. We introduce the energy density per unit mass (ϵ_m) and write for the energy density per unit volume $U = \rho\epsilon_m$ that

$$U = \rho\epsilon_m = \frac{P}{\gamma_{eos} - 1} \tag{1.10}$$

which then defines γ_{eos}. Equation 1.10 illustrates that the pressure P and energy density per unit volume U are essentially the same thing, as for an ideal gas with n_d degrees of freedom, we can write that

$$\gamma_{eos} = 1 + \frac{2}{n_d}. \tag{1.11}$$

The energy density per unit volume of the electrons or ions is then given by

$$U_{e,i} = \frac{n_d}{2} n_{e,i} k_B T \tag{1.12}$$

where $n_{e,i}$ represents the electron or ion number density. The energy density per unit mass is consequently given by

$$\epsilon_m = \frac{3}{4}\frac{(1 + Z_{av})k_BT}{Am_p} \tag{1.13}$$

upon substituting into Equation 1.10 using our Boyle's law expression for the pressure (Equation 1.6) and setting $n_d = 3$. A particle such as an electron or ion has $n_d = 3$ degrees of freedom as there are three directions for the components of velocity. This energy per unit mass expression is equivalent to the one that can be obtained by counting $(1/2)k_BT$ energy per degree of freedom assuming that there are three degrees of freedom for both the electrons and ions.

Differentiating the equation that defines γ_{eos} (Equation 1.10), we get

$$\frac{\partial P}{\partial \rho} = (\gamma_{eos} - 1)\left(\epsilon + \rho\frac{\partial \epsilon_m}{\partial \rho}\right). \tag{1.14}$$

The change of energy content $\Delta\epsilon_m$ per unit mass of a gas is given by the summation of energy added (Δq), minus the work done by the gas due to volume changes ($-P\Delta V$), a statement often known as the first law of thermodynamics. We can write that

$$\Delta\epsilon = \Delta q - P\Delta\left(\frac{1}{\rho}\right) \tag{1.15}$$

as the volume change ΔV is equal to the change of $1/\rho$. As

$$\frac{d(1/\rho)}{d\rho} = -\frac{1}{\rho^2},$$

the partial derivative of the energy content per unit mass with respect to density can now be evaluated from Equation 1.15. We use a partial derivative, which means that quantities other than density are held constant (so the heat flow $\Delta q = 0$) and obtain

$$\frac{\partial \epsilon_m}{\partial \rho} = \frac{P}{\rho^2}.$$

Substituting into Equation 1.14 and using Equation 1.10 gives another expression for the sound speed

$$c_s^2 = \frac{\partial P}{\partial \rho} = \frac{\gamma_{eos}P}{\rho}. \tag{1.16}$$

Interestingly, we see that any factor that affects the relationship between energy density and pressure (Equation 1.10) will affect the speed of sound in the plasma. For example, the degree of ionisation in a plasma affects this relationship, so we have the seemingly perverse result that different ionisation can cause changes in the speed of sound in a plasma. The speed of sound determines the velocity of propagation of shock waves and rarefaction waves and the speed of expansion of an unconstrained plasma [5].

1.1.1 Adiabatic Condition

An adiabatic process is one that occurs without transfer of heat or matter between a thermodynamic system and its surroundings. The adiabatic condition for a plasma element means that no external energy is added so that any change in the internal energy per unit volume $dU_{e,i}$ of the electrons or ions is balanced by the work $P_{e,i}dV$ associated with a volume change dV. As the electron or ion pressure $P_{e,i} = n_{e,i}k_BT$ (see Equation 1.6), using Equation 1.12 we have

$$dU_{e,i} = \frac{n_d}{2}(P_{e,i}dV + VdP_{e,i}). \tag{1.17}$$

Equating $dU_{e,i}$ and $-PdV$ gives

$$\frac{n_d}{2}(P_{e,i}dV + VdP_{e,i}) = -P_{e,i}dV.$$

Rearranging, we have

$$\frac{dP_{e,i}}{P_{e,i}} = -\left(1 + \frac{2}{n_d}\right)\frac{dV}{V} = -\gamma_{eos}\frac{dV}{V}$$

using Equation 1.11. Integrating the pressure from P_0 to $P_{e,i}$ and volume from V_0 to V gives

$$\ln\left(\frac{P_{e,i}}{P_0}\right) = -\gamma_{eos}\ln\left(\frac{V}{V_0}\right).$$

We can write that

$$P_{e,i}V^{\gamma_{eos}} = P_0V_0^{\gamma_{eos}}. \tag{1.18}$$

Equation 1.18 means that for an adiabatic element of plasma, $P_{e,i}V^{\gamma_{eos}}$ is constant.

Another way of stating this adiabatic condition for a plasma is found by recognising that the volume V of a plasma element is proportional to the inverse of the number density $1/n_{e,i}$ and that $P_{e,i} = n_{e,i}k_BT$. For a perfect gas, $\gamma_{eos} = 5/3$, so that the constant $P_{e,i}V^{\gamma_{eos}}$ is equivalent to a constant $n_{e,i}/(k_BT)^{3/2}$. A freely expanding plasma volume element is often adiabatic with constant value of $n_{e,i}/(k_BT)^{3/2}$.

1.2 Free Electron Speed and Energy Distributions

We discuss the division of particles into fermions and bosons in Section 8.1, but we can utilise here the main result of that discussion: that only one fermion can occupy a quantum state. For particles in a thermodynamic equilbrium, the probability $P(E)$ of occupancy by a particle of a quantum state of energy E is given by the proportionality

$$P(E) \propto \exp\left(\frac{N(\mu - E)}{k_BT}\right) \tag{1.19}$$

where N is the number of particles occupying the state with energy E, μ is the chemical potential and T is the temperature. The chemical potential is the energy required to add one more particle to the 'gas' of particles. If the state is not occupied by a particle $P(E) \propto 1$ as $N = 0$. As electrons are fermions, a state can only be occupied by one electron, or it can be unoccupied. If occupied, the probability relationship is $P(E) \propto \exp((\mu - E)/k_BT)$. The proportionality constants to turn the probabilities into absolute probabilities are the same for both occupied and not-occupied states, so the average occupancy $n(E)$ of a state of energy E is given by the ratio of the probabilities here for $P_{occupied}/(P_{occupied} + P_{notoccupied})$ giving

$$n(E) = \frac{\exp((\mu - E)/k_BT)}{\exp((\mu - E)/k_BT) + 1} = \frac{1}{1 + \exp((E - \mu)/k_BT)}. \tag{1.20}$$

This average occupancy of a quantum state can be immediately utilised to obtain an expression for the distribution of speeds of electrons. The number of electrons per unit volume $f_v(v)dv$ with speeds between v and $v + dv$ is given by the proportionality

$$f_v(v)dv \propto 4\pi v^2 n(E)dv$$

where $E = (1/2)m_0v^2$ is the electron energy for electron mass m_0. The factor $4\pi v^2 dv$ is the velocity space volume corresponding to the speed range v to $v + dv$ given by the volume of a shell of radius v and thickness dv. The expression for the electron distribution of speeds can then be written as

$$f_v(v)dv \propto 4\pi v^2 \frac{dv}{1 + \exp(((1/2)m_0v^2 - \mu)/k_BT)}.$$

To convert the proportionality constant here to an absolute value of the distribution of speeds requires normalisation. We choose to require that integrating over all possible speeds gives the total electron number density n_e per unit volume. We then have that

$$\int_0^\infty f_v(v)dv = n_e.$$

The probability distribution function with this normalisation gives the number of electrons per unit volume with speeds between v and $v + dv$. An alternative normalisation with $\int_0^\infty \hat{f}_v(v)dv = 1$ would give the probability of finding an electron with a speed in the range v to $v + dv$ (not the number of electrons) and is used in Chapter 12. Unfortunately, the integrations to do the normalisation are not straightforward, except in the limiting case where the chemical potential is large and negative corresponding to the thermodynamic state of a lower-density electron gas where the electron quantum states are not close to being fully occupied. We consider the chemical potential in Chapter 13. In the case of a lower-density gas, we have the Maxwellian distribution of speeds with

$$f_v(v)dv = n_e \left(\frac{m_0}{2\pi k_BT}\right)^{3/2} 4\pi v^2 \exp\left(-\frac{m_0v^2}{2k_BT}\right) dv. \tag{1.21}$$

By equating the number $f_E(E)dE$ of electrons with energy between E and $E + dE$ to the Maxwellian distribution of speeds using $f_v(v)dv = f_E(E)dE$ (and noting that $E = (1/2)m_0 v^2$ and $dE = m_0 v dv$), we can obtain an expression for the Maxwellian distribution of energies:

$$f_E(E)dE = n_e \frac{2}{\sqrt{\pi}} \left(\frac{1}{k_B T} \right)^{3/2} E^{1/2} \exp\left(-\frac{E}{k_B T} \right) dE. \tag{1.22}$$

Almost all familiar plasmas have equilibrium distributions given by the Maxwellian distribution. The distributions given by Equations 1.21 and 1.22 similarly apply to ions and atoms in a plasma, provided they are in thermal equilibrium (though the mass m_0 needs to change to the mass of the ion or atom in the Maxwellian distribution of speeds). In equilibrium the particle velocity and energy distributions are simply determined by the total number density and the temperature.

1.3 The Density of Quantum States for Free Electrons

Electrons bound by the central potential of the nuclear charge of an ion are clearly in discrete quantum state and there are large energy gaps between some of the quantum states. The energies of the quantum states of free electrons (not bound by an atomic or ionic potential) are close together, but still represent discrete quantum states.

Solving the time-independent Schrodinger equation shows that a free electron (i.e. an electron not in a potential energy field) has a sinusoidally oscillating wavefunction with a wavelength λ_{DB} given by h/p, where h is Planck's constant and p is the electron momentum (see Exercise 1.3). The wavelength λ_{DB} is known as the de Broglie wavelength. To calculate the density of quantum states for free electrons, we need to evaluate how many wavefunctions with a certain de Broglie wavelength can occupy a volume.

Consider a cubic volume with sides L. If electrons are confined within this volume, a steady-state requirement is that only a half-integer number of de Broglie wavelengths occurs between the walls of the cube, otherwise the oscillating nature of the free electron wavefunction will cause cancellation of wavefunction (in a similar manner to interference effects with light). The cube walls can be imagined to be infinitely high, infinitely steep potential barriers for this thought experiment.

The requirement that $m\lambda_{DB}/2 = L$ in the three directions between the cube walls, where m is an integer, means that the momentum associated with each m value is $mh/(2L)$ and the momentum spacing between $m + 1$ and m is $h/(2L)$. The volume in p-space (momentum space) occupied by one electron quantum state is thus $(h/(2L))^3$.

We can now evaluate the number $g(p)dp$ of free electron states per unit volume with momentum in the range p to $p + dp$. The volume of p-space to be considered is $(1/8)4\pi p^2 dp$ which is the volume of an octant of p-space with momentum in the range p to $p + dp$ corresponding to positive values of p in three Cartesian co-ordinate directions. Negative momentum components have no physical meaning here, so only the volume of an octant of p-space, where the p_x, p_y and p_z components in the directions x, y and z are all positive, should be considered. The number of free electron quantum states is then given by

$$g(p)dp = 2\frac{(1/8)4\pi p^2 dp}{(h/(2L))^3 L^3} \tag{1.23}$$

after allowing for the two possible electron spin states. Equation 1.23 divides the allowed p-state volume for free electrons by the volume in p-space occupied by one electron state and the volume L^3 of the cube. Simplifying Equation 1.23 gives the density $g(p)$ of free electron states per unit volume per unit of momentum. The number of free electron quantum states between momentum p and $p + dp$ is given by

$$g(p)dp = \frac{8\pi p^2}{h^3} dp. \tag{1.24}$$

The density of free electron quantum states per unit of momentum per unit volume given by Equation 1.24 is independent of the size L of the cube imagined for the thought experiment. Due to the cancellation of the cube size L, the density of states given by Equation 1.24 applies for all free electrons (i.e. all electrons in a zero potential field as assumed inside the cube).

Using $p = m_0 v$, $dp = m_0 dv$ and $g(p)dp = g(v)dv$, we can immediately write down an expression for the number $g(v)dv$ of free electron quantum states with speeds between v and $v + dv$:

$$g(v)dv = 8\pi v^2 \left(\frac{m_0}{h}\right)^3 dv. \tag{1.25}$$

Using $E = (1/2)m_0 v^2$ and $dE = m_0 v dv$ the number of free electron quantum states with energies between E and $E + dE$ can be similarly obtained:

$$g(E)dE = \frac{4}{\sqrt{\pi}} \left(\frac{2\pi m_0}{h^2}\right)^{3/2} E^{1/2} dE. \tag{1.26}$$

1.4 The Degree of Ionisation

We represented the degree of ionisation of a plasma by $Z_{av} = n_e/n_i$, where the electron density is n_e and the ion density is n_i. For many calculations, such as the evaluation of plasma pressure and the determination of the plasma sound speed, a single value of Z_{av} is sufficient. However, to calculate Z_{av} it is necessary to evaluate the populations of a range of ionisation stages. In lower-density plasmas, this is a major undertaking requiring a model of the collisional and radiative processes populating the discrete quantum states (see Chapter 12). For higher-density plasmas, it is often sufficient to assume that the populations are given by an equilibrium relationship known as the Saha–Boltzmann equation. This can be derived by extending the Boltzmann population ratio used to infer the equilibrium ratio of populations between quantum states within a single ionisation stage or atom.

At high densities, electron–ion collisional processes dominate over radiative processes in the populating and depopulating processes between quantum states. The electron–ion collisions cause a transfer of an electron from one bound state to another, or by collisional ionisation cause a transition of an electron from a bound state to a free electron quantum state. The inverse of this last process, where free electrons interact by colliding with an ion to cause one electron to move to a bound state, while another electron absorbs the energy of the recombining electron, is known as three-body recombination. The important issue is that each collisional excitation or ionisation process, for a system in thermal equilibrium, proceeds at the same rate as the inverse collisional de-excitation or three-body recombination process. This equality of rate for each process is known as the principle of detailed balance. Provided that all important processes and their inverses occur, this means that quantum state populations will be in equilbrium. At low densities, radiative decay rates become significant compared to the electron collisional decay rates, but the radiative process of spontaneous decay is often not balanced by the other detailed radiative processes of photo-absorption and stimulated emission. This can cause the populations for low-density plasmas to differ from the thermal equilibrium populations.

1.4.1 The Saha–Boltzmann Equation

Consider Z_i and $(Z_i + 1)$ charged ions with population densities n_{Z_i} and n_{Z_i+1} in two discrete energy states. We want to calculate the population ratio n_{Z_i+1}/n_{Z_i} for a plasma in equilibrium. We assume that the $Z_i + 1$ ion energy state is the ground state, so that the energy difference between the two states is equal to the ionisation energy E_{ion} of the Z_i charged ion. We assume the two energy states have degeneracies of g_{Z_i} and g_{Z_i+1}: i.e. the number of quantum states and hence the maximum number of electrons per ion that can exist at the energies associated with the two states are g_{Z_i} and g_{Z_i+1}. In this context, degeneracies are also known as 'statistical weights'.

The Boltzmann population ratio between two quantum states (upper u and lower l) in the same ion can be written as

$$\frac{n_u}{n_l} = \frac{g_u}{g_l} \exp\left(-\frac{\Delta E}{k_B T}\right) \tag{1.27}$$

where g_u and g_l represent the degeneracies of the upper and lower states respectively and ΔE is the energy difference between the quantum states. This ratio arises from Fermi–Dirac statistics (Equation 1.19) when the average occupancy N of a quantum state is such that $N \ll 1$.

In the ionisation process of converting a Z_i charged ion to a $(Z_i + 1)$ charged ion, a free electron with a speed (say v) is created. The Boltzmann ratio becomes

$$\frac{f_v(v)dv}{n_Z} = \frac{g_{Z_i+1}(1/n_{Z_i+1})g(v)dv}{g_{Z_i}} \exp\left(-\frac{E_{ion} + (1/2)m_0 v^2}{k_B T}\right) \tag{1.28}$$

where $g(v)dv$ is the density of free electron states (see Equation 1.25) and $f_v(v)dv$ is the number density of electrons with speeds in the range v to $v + dv$. The quantity $g_{Z_i+1}(1/n_{Z_i+1})g(v)dv$ represents the degeneracy associated with the upper energy state for electrons with speeds v to $v + dv$. It comprises the number of quantum states in the volume of one Z_i+1 ion ($(1/n_{Z_i+1})g(v)dv$) multiplied by the degeneracy of the $Z_i + 1$ energy state (g_{Z_i+1}). The density of speeds is given by the Maxwellian distribution (see Equation 1.21). Substituting Equations 1.21 and 1.25 into Equation 1.28, we obtain the Saha–Boltzmann equation for the population ratio n_{Z_i+1}/n_{Z_i} such that

$$\frac{n_{Z_i+1}}{n_{Z_i}} = \frac{g_{Z_i+1}}{g_{Z_i}} \frac{2}{n_e} \left(\frac{m_0}{h}\right)^3 \left(\frac{2\pi k_B T}{m_0}\right)^{3/2} \exp\left(-\frac{E_{ion}}{k_B T}\right). \tag{1.29}$$

The exponentiated electron kinetic energy $(1/2)m_0 v^2$ and a $v^2 dv$ term cancel on both sides. Interestingly, for an adiabatic element of plasma with $n_e/(k_B T)^{3/2}$ constant (see Section 1.1.1), Equation 1.29 predicts constant ionisation. Adiabatically expanding plasmas such as the plume of a laser-produced plasma are known to exhibit a frozen state of ionisation which does not change in time [106].

The Saha–Boltzmann equation is sometimes written in terms of the thermal de Broglie wavelength Λ defined by $\Lambda = h/\sqrt{2\pi m_0 k_B T}$ such that

$$\frac{n_{Z_i+1}}{n_{Z_i}} = \frac{g_{Z_i+1}}{g_{Z_i}} \frac{2}{n_e \Lambda^3} \exp\left(-\frac{E_{ion}}{k_B T}\right).$$

The quantity $2/(n_e \Lambda^3)$ can be regarded as representing the average degeneracy of the free electrons created in the ionisation process.

Equation 1.29 is usually referred to as the Saha–Boltzmann equation when an excited state is considered for the Z_i-charged ion. With appropriate modification of the ionisation energy E_{ion}, Equation 1.29 represents the population ratio between any bound quantum state of the Z_i-charged ion and the ground state of the $Z_i + 1$-charged ion. If the Z_i-charged ion quantum state in question is the ground state, Equation 1.29 is usually referred to simply as the Saha equation. When temperatures and densities in a plasma are not too high, the population of the ground state of ions significantly exceeds the populations of the excited states and the Equation 1.29 population ratio for the two ground states then represents the total ratio of population in all quantum states of the $Z_i + 1$-charged to Z_i-charged ions.

We find (see Section 12.7) that an equilibrium ratio of population densities is only achieved at high plasma densities, when there is significant radiation flux or for closely spaced quantum states. The Saha–Boltzmann equation is important in these limiting cases, but also enables the calculation of the rates of inverse collisional and radiative processes using the assumption of detailed balance. With detailed balance, a plasma in equilibrium is said to have equal rates of inverse processes. Examples of inverse processes include spontaneous and stimulated emission/photo-excitation (see Section 4.2), collisional excitation/collisional de-excitation (see Section 12.1) and collisional ionisation/three-body recombination (see Section 12.2).

1.5 The Bohr Energy Level Model for Atoms and Ions

The simplest atomic system is a hydrogen atom comprising an electron and a proton. It is possible to treat a slightly more complex situation of a hydrogen-like ion comprising a nucleus of Z protons and one electron with almost identical equations. Ions with the same number of electrons form an 'isoelectronic series' with identical electron configurations. The simplest isoelectronic series is the hydrogen-like series.

We first discuss an early model of the hydrogen atom which incorporates some of the first ideas about quantum mechanics (that an electron has a wave-like nature with an associated wavelength) and predicts the electron energies in atoms to good accuracy. The Bohr model to describe atoms and ions was developed before a full understanding of quantum mechanics, but it is remarkedly accurate in many predictions and so is still usually taught to students of atomic physics before they engage with more complete quantum-mechanical descriptions of atoms. The Bohr model of hydrogen and the hydrogen-like ion in the simplest form assumes that an electron orbits a point nucleus of charge Ze in a planetary circular orbit of radius, say r. It is assumed that around the circumference of the circular orbit there are an integer number n of the de Broglie wavelengths λ_{DB} of a free electron. The de Broglie wavelength is the wavelength associated with the wavefunction of a free electron of given velocity. As we saw in Section 1.3, the de Broglie wavelength is given by

$$\lambda_{DB} = \frac{h}{m_0 v}$$

where m_0 is the electron mass and v the electron orbital speed. This condition on the number of de Broglie wavelengths gives rise to a quantisation of the angular momentum $L = m_0 v r$ such that

$$L = n\hbar$$

where $\hbar = h/2\pi$.

Balancing the Coulomb force of the attraction between the electron of charge e and the nucleus of charge Ze and the centripetal force associated with an orbiting mass, we can write that

$$\frac{1}{4\pi\epsilon_0}\frac{Ze^2}{r^2} = \frac{m_0 v^2}{r}$$

and obtain an expression for the total kinetic and potential energy E_n of the electron using

$$E_n = \frac{1}{2}m_0 v^2 - \frac{Ze^2}{4\pi\epsilon_0 r}.$$

The potential energy of the electron is the work done by an electron in moving from infinity to a radius r from the nucleus and is obtained by integrating the electrostatic force between the nucleus and the electron from infinity to distance r. Manipulating the above equations, the total energy E_n and the radius r of the orbit are found to be

$$E_n = -\frac{Z^2 e^4}{(4\pi\epsilon_0)^2}\frac{m_0}{2\hbar^2}\frac{1}{n^2} = -\frac{1}{2}\frac{Ze^2}{4\pi\epsilon_0(a_0/Z)}\frac{1}{n^2} = -\frac{Z^2 R_d}{n^2}, \tag{1.30}$$

$$r = 4\pi\epsilon_0\frac{\hbar^2}{Ze^2}\frac{1}{m_0}n^2 = a_0\frac{n^2}{Z}. \tag{1.31}$$

It is apparent that the electron energies scale as Z^2/n^2 and the electron orbital radius as n^2/Z. The integer n introduced here is obviously important and consequently is referred to as the 'principal quantum number'. The ground state $n = 1$ of hydrogen $Z = 1$ has an energy known as the Rydberg energy with a numerical value $R_d = 13.6$ eV. The orbital radius of the ground state of hydrogen is numerically 0.529 angstrom (0.529×10^{-10} m) and is referred to as the Bohr radius. We use the Bohr radius later, so introduce a symbol for its designation (a_0).

The Bohr model is remarkably good at predicting the energies of electrons in hydrogen and hydrogen-like ions and the radius of the orbiting electron is equal to the peak of the radial wavefunction found using quantum mechanics for the lowest energy ground state where $n = 1$. However, unexplained features of atomic energy levels cannot be understood using the simple Bohr model. Spectral lines arising from radiative transitions between quantum states were seen, for example, as doublets, where a single line would be expected and the angular momentum $L = n\hbar$ in the Bohr model was found to depend on another quantum number and not on the principal quantum number n.

An extension to the Bohr model by the physicist Sommerfeld partly explained the angular-momentum issue and enabled an explanation of energy splitting by assuming the electron orbits are elliptical. The so-called Bohr–Sommerfeld model considered that the electron swung around the nucleus in rotating elliptical orbits, much as a comet orbits around the Sun. Relativistic effects became important as the electron speed approached the speed of light as it passed close to the nucleus. The degree of ellipticity described by another quantum number (not the principal quantum number n) thus caused small relativistic corrections to the electron energy. Assuming that the elliptical orbits rotate around the nucleus produces energy differences due to the generation of different magnetic fields and predicted the energy difference observed between, for example, the photon energies of doublet spectral lines.

The elliptical orbit treatment became obsolete with the development of quantum mechanics and the appreciation that the concept of a free electron 'particle' orbiting in either a circular or elliptical orbit is at best an approximation. Nevertheless, we should recall the 'correspondence principle' before moving on to consider quantum-mechanical solutions for atoms. This can be stated as follows: When the density of quantum states becomes high, quantum mechanical solutions approach the classical theory.

For high principal quantum numbers (large n), the energies of the orbiting electrons become close to each other – varying proportionally as $1/n^2$ according to the Bohr model (and as we shall see, quantum mechanics). These high n electron states start to behave increasingly like the Bohr model of the electron with an electron orbiting around the nucleus in a classical orbit. Typically, for an atom of atomic number Z, a high n electron sees a charge close to the nucleus of Z protons and $(Z - 1)$ electrons so that the central atom potential appears to the high n electron like a hydrogen nucleus. Atoms with an electron in high n orbits are now referred to as Rydberg atoms as they orbit classically as described by Bohr with their energies given by the Rydberg energy divided by n^2. Electron transitions between high n values give rise to radio-wave absorption or emission and such Rydberg atoms are important in radio astronomy.

The Bohr–Sommerfeld model with electrons tracing out elliptical orbits around the nucleus has a similarity to the quantum mechanical distribution of wavefunctions around the nucleus. Such a similarity is to be expected, particularly for higher n states, because of the correspondence principle. We should also expect a similarity because the Bohr–Sommerfeld model accurately predicts the energies of the atoms, including the fine-structure energy levels giving rise, for example, to spectral-line doublets, where the simple Bohr model predicts a single spectral line.

Exercises

1.1 In many plasmas, the electrons and ions have different temperatures. Using the expression for the sound speed $c_s^2 = (\gamma_{eos} P/\rho)$ (see Equation 1.16), show that for a plasma with average ionisation Z_{av} the sound speed is given by

$$c_s = \left(\frac{5(k_B T_i + Z_{av} k_B T_e)}{3 A m_p} \right)^{1/2}$$

where the electron temperature is T_e and the ion temperature is T_i. Here A is the atomic mass and m_p is the mass of a proton.

1.2 The average energy $\hat{E}$ associated with a degree of freedom of a gas or plasma constituent can be obtained by averaging the energy over the velocity distribution. We have

$$\hat{E} = \frac{\int_{-\infty}^{\infty} \exp(-E/k_B T) E \, dv}{\int_{-\infty}^{\infty} \exp(-E/k_B T) \, dv}.$$

Show that

$$\hat{E} = \frac{1}{2} k_B T.$$

[You need the values of the definite integrals $\int_{-\infty}^{\infty} \exp(-x^2) dx = \sqrt{\pi}$ and $\int_{-\infty}^{\infty} x^2 \exp(-x^2) dx = \sqrt{\pi}/2$.]

1.3 For a free electron, the Schrodinger equation can be written as

$$-\frac{\hbar^2}{2m_0} \nabla^2 \psi = -\frac{\hbar^2}{2m_0} \frac{\partial^2 \psi}{\partial z^2} = E\psi.$$

Show that $\psi = e^{ikz}$ is a solution with $k = 2\pi/\lambda_{DB}$, where $\lambda_{DB} = h/(m_0 v)$ and v is the electron velocity. The wavelength λ_{DB} is known as the de Broglie wavelength.

1.4 In low-temperature plasmas, there is often little ionisation so that only a small number of the neutral atoms are ionised. For an equilibrium low-temperature plasma, show that the Saha–Boltzmann equations can be approximated to give for the electron density

$$n_e^2 \approx \frac{\rho}{m_p} \frac{g_1}{g_0} \frac{1}{\Lambda^3} \exp\left(-\frac{E_{ion}}{k_B T} \right)$$

where g_0 and g_1 are the degeneracies of the neutral atom and ionised ion ground states respectively, ρ is the mass density, m_p is the mass of a proton and the other quantities are defined in Section 1.4.1.

1.5 Using the expression for the density of free electron quantum states (Equation 1.26), show that the number of states per cubic centimetre (i.e. in units of cm^{-3}) for an electron energy range from zero to an energy E is given by

$$\int_0^E g(E')dE' = 4.5 \times 10^{21} E^{3/2}$$

where E is measured in electron volts.

1.6 Considering the balance of the Coulomb force between electron and nucleus and the centripetal force, show for the ground state $n = 1$ in the Bohr model for hydrogen that the velocity of the orbiting electron is given by

$$v = \alpha c,$$

where c is the speed of light and

$$\alpha = \frac{e^2}{4\pi \epsilon_0 \hbar c} \cong \frac{1}{137}.$$

The quantity α is known as the fine structure constant.

1.7 Show that the potential energy of an electron in the ground state of the hydrogen atom in terms of the hydrogen ionisation energy defined as 1 Rydberg (1 R_d) is such that

$$\left| -\frac{e^2}{4\pi \epsilon_0 a_0} \right| = 2R_d$$

where a_0 is the Bohr radius.

1.8 Use the result of Exercise 1.7 to deduce the kinetic energy in Rydbergs of an electron in the ground state of the hydrogen atom. [$+1R_d$]

1.9 Use the results from Exercises 1.6 and 1.7 to show that

$$\hbar c \alpha = 2R_d a_0.$$

1.10 In a Rydberg atom, given that the energy levels are determined by $E_n = -13.6/n^2$ eV, show that the energy separation of adjacent energy levels associated with principal quantum number n and $n + 1$ is approximately $27.2/n^3$ eV.

2

The Propagation of Light

To achieve a plasma with free electrons requires elevated temperatures and hence light emission, propagation and absorption can be important. The propagation of light, unlike many other familiar waves, does not need a medium in which to oscillate.[1] Light propagates in plasma and the free electrons are driven to oscillate, but the electrons generally impede the wave oscillation rather than aid the process. It becomes impossible for light to travel through an unmagnetised plasma if the frequency of the radiation is less than the plasma's natural oscillation frequency: the plasma frequency is discussed in Section 1.1. At low frequencies, the electron oscillations relative to the ions dampen the electromagnetic oscillation of light. Light of all frequencies passes with no absorption or alteration in phase in vacuum, though the intensity from any finite-sized source ultimately falls proportionally to the inverse of the square of distance from the source.

In this chapter, we show how Maxwell's equations describing the relationships between electric and magnetic fields (and electric current and electric charge) are consistent with oscillating electric and magnetic fields propagating in vacuum at the velocity of light c. The oscillating fields are solutions of Maxwell's equations. We treat the electric currents generated in a plasma by light to show how the currents affect electromagnetic waves. The acceleration of any charge is shown to produce transverse electric field oscillations, thus providing a mechanism for the production of electromagnetic waves.

The electromagnetic spectrum of interest in plasma physics ranges from radio waves to X-rays and gamma rays (see Figure 2.1). The propagation of the different

[1] The aether was a construct in which light was supposed to propagate. Its existence was negated by the Michelson–Morley experiment in 1887 which showed that light always propagates at the same velocity in vacuum. Albert Michelson and Edward Morley, at what is now Case Western Reserve University in Cleveland, Ohio, undertook the experiment using an interferometer which Michelson invented. The measured constant velocity of light helped Einstein develop his theories of relativity based on the idea of the constant velocity of light: the theory of special relativity published in 1905 and the theory of general relativity published in 1916.

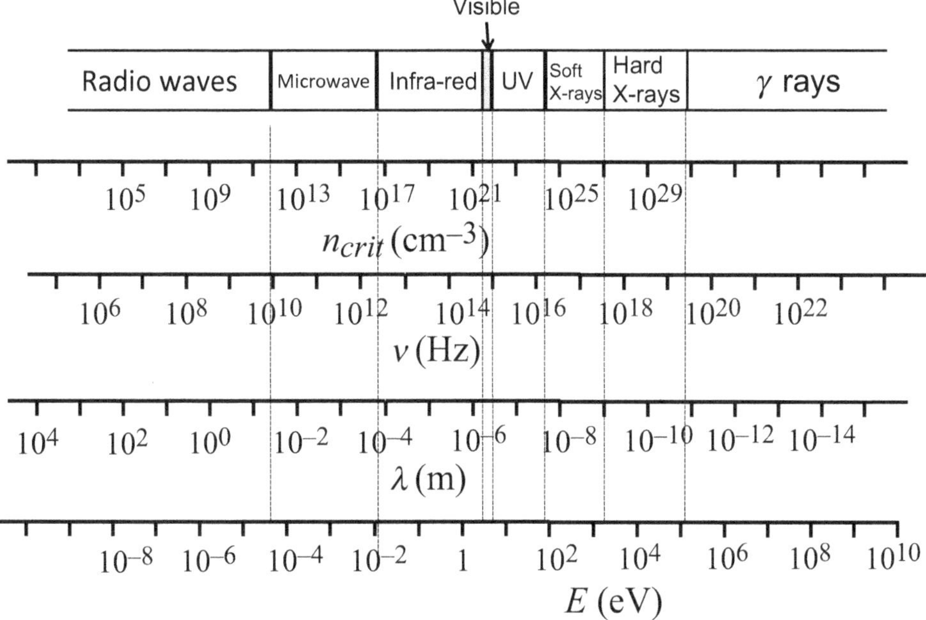

Figure 2.1 The regions of the electromagnetic spectrum showing the logarithmic variation of photon energy E in electron volts, wavelength λ in metres, frequency ν in Hz and the maximum density for electromagnetic propagation (the critical density n_{crit} in cm^{-3}). Apart from the visible light boundaries (λ between 400 and 700 nm), the designated ranges for the different types of electromagnetic radiation are flexible depending on context by up to an order of magnitude.

components of the electromagnetic spectrum involves identical physics with variations only occurring when the radiation interacts with matter. The highest-frequency, highest-energy gamma rays at photon energies above, say, 100 keV are not created typically by thermal processes, but can be important when fast 'superthermal' particles are created in plasmas. Radio-wave propagation can be important in low-density plasmas, such as the ionosphere.

2.1 Electromagnetic Waves in Plasmas

The propagation of radiation through a medium can be examined using Maxwell's equations. We need to utilise three of Maxwell's equations. We need Faraday's law

$$\nabla \times \mathbf{E} = -\frac{\partial \mathbf{B}}{\partial t} \tag{2.1}$$

and Ampere's law

$$\nabla \times \mathbf{B} = \mu_0 \mathbf{J} + \epsilon_0 \mu_0 \frac{\partial \mathbf{E}}{\partial t} \tag{2.2}$$

which describe how magnetic **B** and electric **E** fields can be produced by temporal changes of each field, and equivalently how a magnetic field can be created by a flow of current density **J** (see Appendix A.2 for a brief discussion of Maxwell's equations). Taking the curl ($\nabla\times$) of Faraday's law and substituting Ampere's law for $\nabla \times \mathbf{B}$ gives that

$$\nabla \times \nabla \times \mathbf{E} = -\frac{\partial}{\partial t}\left(\mu_0 \mathbf{J} + \epsilon_0 \mu_0 \frac{\partial \mathbf{E}}{\partial t}\right).$$

In a plasma, the density of charge ρ for length scales longer than the Debye length is zero. Gauss's law then gives that

$$\nabla.\mathbf{E} = \frac{\rho}{\epsilon_0} = 0. \tag{2.3}$$

Using the vector identity $\nabla \times \nabla \times \mathbf{E} = \nabla(\nabla.\mathbf{E}) - \nabla^2\mathbf{E}$ and our result for $\nabla.\mathbf{E} = 0$ gives an equation known as the wave equation. We have

$$\nabla^2\mathbf{E} = \mu_0 \frac{\partial \mathbf{J}}{\partial t} + \epsilon_0 \mu_0 \frac{\partial^2 \mathbf{E}}{\partial t^2}. \tag{2.4}$$

The last of Maxwell's equations ($\nabla.\mathbf{B} = 0$) is not used to derive the wave equation.

The wave equation has solutions for the electric and magnetic field which oscillate in time and space and which are known as electromagnetic radiation. Substituting in the wave equation (Equation 2.4) verifies that variations of electric field of the following form satisfy the wave equation at position **r**:

$$\mathbf{E} = \mathbf{E}_0 \exp[i(\mathbf{k}.\mathbf{r} - \omega t)] \tag{2.5}$$

where $\mathbf{E}_0$ is a field amplitude, **k** is a wavevector representing the rapid spatial variation of the field and ω is an angular frequency representing the rapid temporal variation of the field. The angular frequency is the phase change of the field in radians per second and is related to the frequency in cycles per second (i.e. Hertz) by $\omega = 2\pi\nu$. Taking the divergence of **E** in Equation 2.5 gives

$$\nabla \cdot \mathbf{E} = i\mathbf{k} \cdot \mathbf{E}$$

which from Equation 2.3 is identically equal to zero. With $\mathbf{k} \cdot \mathbf{E} = 0$, we have that the wavevector **k** is perpendicular to the electric field **E**.

The orientation of the electric field **E** in the plane perpendicular to **k** is the polarisation direction of the electromagnetic wave. Any electromagnetic wave can be decomposed into polarisation components in two orthogonal directions both perpendicular to the wavevector **k**. Unpolarised light can be regarded as the independent supposition of two polarised beams with polarisation directions at angle $\pi/2$ to each other.

If we assume that the **k** vector is directed in the z-direction with the electric field **E** directed in the x-direction, then we can write that the solutions of the wave equation have the form $E = E_0 \exp[i(kz-\omega t)]$. The velocity of electrons in a plasma is found by integrating the acceleration $(-eE/m_0)$ on the electrons to give

$$\frac{dx}{dt} = \frac{eE}{im_0\omega}.$$

The current density is consequently directed in the x-direction with amplitude given by

$$J = -n_e e\frac{dx}{dt} = i\frac{n_e e^2}{m_0\omega}E. \tag{2.6}$$

The wave equation then becomes

$$\frac{\partial^2 E}{\partial z^2} = \mu_0 i\frac{n_e e^2}{m_0\omega}\frac{\partial E}{\partial t} + \epsilon_0\mu_0\frac{\partial^2 E}{\partial t^2}. \tag{2.7}$$

In vacuum (with $n_e = 0$), the first term on the right-hand side here vanishes, and substituting Equation 2.5 gives that $\omega/k_0 = 1/\sqrt{\epsilon_0\mu_0}$, where k_0 is the vacuum wavenumber. For constant phase $k_0 z - \omega t$, we see that the speed of light in vacuum $c = dz/dt = \omega/k_0$. Consequently, we have that

$$\epsilon_0\mu_0 = \frac{1}{c^2}. \tag{2.8}$$

Substituting solutions $E = E_0 \exp[i(kz - \omega t)]$ into Equation 2.7 shows that

$$k^2 = \frac{\omega^2}{c^2} - \frac{n_e e^2}{\epsilon_0 m_0 c^2} = \frac{\omega^2 - \omega_p^2}{c^2} \tag{2.9}$$

where we have used the expression for the plasma frequency ω_p (see Equation 1.3) for the resonance oscillation of electrons relative to the ions in a plasma:

$$\omega_p = \sqrt{\frac{n_e e^2}{\epsilon_0 m_0}}. \tag{2.10}$$

For electromagnetic radiation to propagate in the original forward direction requires $k \geq 0$. Value of $k < 0$, can be interpreted as the radiation being reflected backwards. For example, radiation incident into an increasing density of plasma only propagates up to where the laser frequency $\omega = \omega_p$. The corresponding maximum electron density for light penetration can be found using Equation 2.10 and is known as the critical density. Re-arranging Equation 2.10 and setting $\omega = \omega_p$ shows that the critical electron density is given by

$$n_{crit} = \frac{\epsilon_0 m\omega^2}{e^2}. \tag{2.11}$$

Evaluating Equation 2.11 numerically, we find that for light of wavelength $\lambda = 1.06\,\mu$m, the critical electron density $n_{crit} = 10^{21}\,\text{cm}^{-3}$ and scales proportionally to $1/\lambda^2$ (see Figure 2.1).

The refractive index η is the ratio of the speed of light in vacuum to the phase velocity of light in a medium, so $\eta = kc/\omega$. Using Equation 2.9 we have

$$\eta = \sqrt{1 - \frac{\omega_p^2}{\omega^2}} = \sqrt{1 - \frac{n_e}{n_{crit}}}. \tag{2.12}$$

From Equation 2.12 we have that refractive indices in plasma are less than unity, while at the critical density the refractive index value decreases to zero. Radiation can no longer propagate at frequencies below the plasma frequency or at densities greater than the critical density.

2.1.1 The Magnetic Field and the Irradiance of an Electromagnetic Wave

We need to determine the corresponding oscillating magnetic field arising from the oscillating electric field for an electromagnetic wave in a plasma. Ampere's law (Equation 2.2) is given by

$$\nabla\times\mathbf{B} = \mu_0\mathbf{J} + \epsilon_0\mu_0\frac{\partial\mathbf{E}}{\partial t} = i\mu_0\frac{n_e e^2}{m_0\omega}\mathbf{E} - i\omega\epsilon_0\mu_0\mathbf{E} \tag{2.13}$$

upon differentiating $\mathbf{E}$ and using Equation 2.6. Equation 2.13 requires that the magnetic field varies as $\mathbf{B} = \mathbf{B}_0\exp[i(kz - \omega t)]$ in phase with the electric field variation. The definition of $\nabla\times$ in Cartesian co-ordinates determines that $\mathbf{B}$ is perpendicular to $\mathbf{E}$ and gives $-ikB = (i\mu_0 n_e e^2/(m_0\omega) - i\omega\epsilon_0\mu_0)E$. Using the equivalences $1/c^2 = \epsilon_0\mu_0$ and $c = \eta\omega/k$ and the expression for the refractive index (Equation 2.12), we have with a little manipulation that

$$\frac{E}{B} = \frac{c}{\eta}. \tag{2.14}$$

The classical energy density of an electric field is $((1/2)\epsilon E^2)$ and for a magnetic field is $((1/2)B^2/\mu_0)$, where the dielectric constant for a medium $\epsilon = \epsilon_0\eta^2$ (ϵ_0 is the dielectric constant in vacuum). Two exercises at the end of this chapter illustrate how both of these quantities for the energy density of the fields can be obtained by considering the charging of a capacitor (for the electric field) and the increase of current into a solenoid (for the magnetic field). The average of the variation of the electric and magnetic fields in time is given by the average of the real and imaginary components of $\exp[i(kz - \omega t)]$, that is

$$\frac{\int_{\omega t=0}^{2\pi} \cos^2(\omega t)d(\omega t)}{\int_{\omega t=0}^{2\pi} d(\omega t)} = \frac{\int_{\omega t=0}^{2\pi} \sin^2(\omega t)d(\omega t)}{\int_{\omega t=0}^{2\pi} d(\omega t)} = 1/2.$$

Adding up the energy in the electric and magnetic fields of an electromagnetic beam, we find that the electric and magnetic energies are equal and of magnitude $(1/4)\epsilon_0\eta^2E_0^2$. Imagine a unit cross-section volume of unit area and length c/η along the direction of propagation of the beam. In unit time, the energy from the volume of c/η will pass through the unit area. This energy passing per unit time per unit area normal to the direction of the beam is known as the intensity or irradiance I and is given by

$$I = \frac{1}{2}\epsilon_0 c\eta E_0^2. \tag{2.15}$$

2.1.2 The Group Velocity

The refractive indices of plasmas as given by Equation 2.12 are less than unity which means that the phase velocity of light is greater than the vacuum speed of light. The principle of relativity is not violated, however, as it is not possible to convey information at the phase velocity. The light has to be modulated or pulsed to send a signal.

Rather than using just a single frequency ω and wavevector $\mathbf{k}$, sending a signal requires a pulse or packet of slightly different frequencies and wavevectors. We can suppose that a pulse of radiation has a total electric field $E(z,t)$ at position z and time t given by

$$E(z,t) = \int_{-\infty}^{\infty} E_0(k)\exp(i(kz - \omega t))dk \tag{2.16}$$

where the integration is over different wavevectors which contribute an electric field $E_0(k)dk$ in the range k to $k + dk$. Assuming a small range of frequencies ω around a central frequency ω_m associated with a small range of wavevectors k centred on k_m, we can write that

$$\omega \approx \omega_m + (k - k_m)\frac{d\omega}{dk}. \tag{2.17}$$

Substituting Equation 2.17 into Equation 2.16, we have

$$E(z,t) = \exp(i(k_m z - \omega_m t)) \int_{-\infty}^{\infty} E_0(k)\exp\left(i\,(k - k_m)\left(z - \frac{d\omega}{dk}t\right)\right)dk. \tag{2.18}$$

Equation 2.18 represents a wave of frequency ω_m and wavevector k_m in a pulse with an amplitude varying with the values given by the integral. The value of the integral changes in space z and time t such that $z - (d\omega/dk)t$ is constant. Differentiating this constant value with respect to time gives

$$\frac{dz}{dt} = \frac{d\omega}{dk}. \tag{2.19}$$

The quantity $d\omega/dk$ is consequently the velocity of the amplitude variation of the pulse of electromagnetic wave and is known as the group velocity. It is the speed of propagation of the pulse of radiation and is the velocity that a signal can propagate.

Using Equation 2.9, we can find the group velocity v_g for a light pulse in an unmagnetised plasma. We have

$$v_g = \frac{d\omega}{dk} = c\left(1 - \frac{\omega_p^2}{\omega^2}\right)^{1/2} \tag{2.20}$$

which is less than the vacuum speed of light c.

An astrophysical application of Equation 2.20 for the group velocity is to pulsars. Pulsars are magnetised neutron or white dwarf stars that emit a rotating beam of electromagnetic radiation with a typically broad spectrum of radiation. A particular line of sight to Earth receives pulses of light separated by times from milliseconds to seconds when the rotating beam is aligned. For example, the first pulsar, observed in 1967 by Jocelyn Bell (later Bell Burnell) and Antony Hewish [45], exhibited pulses of radio waves separated by 1.33 seconds. The spectrum of light from pulsars is dispersed by interaction with interstellar plasma. Each frequency of the light travels at a slightly different group velocity and reaches Earth at slightly different times.

For a pulsar at distance z_p. the time t_p required for a pulse to reach Earth at frequency ω is given by an integration along the line of sight from the pulsar to Earth. We have

$$t_p = \int_0^{z_p} \frac{dz}{v_g}.$$

The inverse of the group velocity is given by

$$\frac{1}{v_g} = \frac{1}{c}\left(1 - \frac{\omega_p^2}{\omega^2}\right)^{-1/2} \approx \frac{1}{c}\left(1 + \frac{\omega_p^2}{2\omega^2}\right)$$

as the interstellar plasma frequency is low (typically 10^3 radians s^{-1}). We obtain for the time for the pulse to reach Earth:

$$t_p \approx \frac{z_p}{c} + \frac{1}{2c\omega^2} \int_0^{z_p} \omega_p^2 dz. \tag{2.21}$$

The first term is the time for a pulse to travel in vacuum, while the second term is a correction to the travel time due to the interstellar plasma. If the rate of change of the arrival time of the pulse as a function of frequency is measured, a measurement of the distance of the pulsar can be obtained. Differentiating Equation 2.21, we have

$$\frac{dt_p}{d\omega} = -\frac{e^2}{m_0 \epsilon_0 c \omega^3} \int_0^{z_p} n_e dz. \tag{2.22}$$

Using a typical value for electron density n_e in interstellar space ($n_e \approx 0.02\,\mathrm{cm}^{-3}$), a measurement of the distance of the pulsar can be obtained (see Exercise 2.5).

2.2 Electromagnetic Waves in a Magnetised Plasma

An electromagnetic wave traveling in a plasma with an imposed magnetic field $\mathbf{B}_0$ will be affected in different ways depending on the direction of travel. The equation of motion of a single electron is given by the Lorentz force arising from the electric field $\mathbf{E}$ of the light with

$$m_0 \frac{d^2\mathbf{r}}{dt^2} = -e\left(\mathbf{E} + \frac{d\mathbf{r}}{dt} \times \mathbf{B}_0\right).$$

As in Section 2.1, we assume that the electron velocity $\mathbf{v} = d\mathbf{r}/dt \propto \exp(-i\omega t)$, but now we must consider the effect of the static magnetic field. In order to simplify the algebra we take the z direction to be the direction of the magnetic field. The three components of the equation of motion can be written:

$$\begin{aligned} -m_0\, i\,\omega\, v_x &= -e\,E_x - e\,B_0\,v_y \\ -m_0\, i\,\omega\, v_y &= -e\,E_y + e\,B_0\,v_x \\ -m_0\, i\,\omega\, v_z &= -e\,E_z. \end{aligned} \tag{2.23}$$

We may solve for $\mathbf{v}$ in terms of E:

$$\begin{aligned} v_x &= -\frac{ie}{\omega m_0} \frac{1}{1 - \omega_c^2/\omega^2} \left(E_x - i\frac{\omega_c}{\omega} E_y\right) \\ v_y &= -\frac{ie}{\omega m_0} \frac{1}{1 - \omega_c^2/\omega^2} \left(i\frac{\omega_c}{\omega} E_x + E_y\right) \\ v_z &= -\frac{ie}{\omega m_0} E_z \end{aligned} \tag{2.24}$$

where $\omega_c = eB_0/m_0$ is known as the electron cyclotron frequency, gyrofrequency or Larmor frequency. It can be shown that ω_c is the frequency that an electron orbits around a magnetic field of given strength B_0. The current density, $\mathbf{J}$, is given by:

$$\mathbf{J} = -en_e\mathbf{v} = \sigma\mathbf{E} \tag{2.25}$$

where the conductivity tensor, σ, is given by:

$$\sigma = \frac{in_e e^2}{m_0\omega}\frac{1}{1-\omega_c^2/\omega^2}\begin{bmatrix} 1 & -i\omega_c/\omega & 0 \\ i\omega_c/\omega & 1 & 0 \\ 0 & 0 & 1-\omega_c^2/\omega^2 \end{bmatrix}. \tag{2.26}$$

We want to solve the wave equation (Equation 2.4) using the above expression for the plasma current density. The solutions are of the form

$$\mathbf{E} = \mathbf{E_0}\exp\left[i(\mathbf{k}.\mathbf{r} - \omega t)\right] \tag{2.27}$$

where $\mathbf{k}$ is the wavevector and $\mathbf{r} = (x, y, z)$ is the position is space. We may choose axes so that $k_x = 0$, that is

$$\mathbf{k} = (0, k\sin\theta, k\cos\theta) \tag{2.28}$$

where θ is the angle between $\mathbf{k}$ and $\mathbf{B}_0$. Substituting Equations 2.24 to 2.28 into Equation 2.4 results in a vector for the electric field in the coordinate directions

$$\begin{bmatrix} E_x \\ E_y \\ E_z \end{bmatrix}$$

such that

$$\begin{bmatrix} -\eta^2 + 1 - \frac{X}{1-Y^2} & \frac{iXY}{1-Y^2} & 0 \\ -\frac{iXY}{1-Y^2} & -n^2\cos^2\theta + 1 - \frac{X}{1-Y^2} & n^2\sin\theta\cos\theta \\ 0 & n^2\sin\theta\cos\theta & -n^2\sin^2\theta + 1 - X \end{bmatrix}\begin{bmatrix} E_x \\ E_y \\ E_z \end{bmatrix} \tag{2.29}$$

is equal to the zero vector

$$\begin{bmatrix} 0 \\ 0 \\ 0 \end{bmatrix}.$$

We have defined

$$X = \omega_p^2/\omega^2 \qquad Y = \omega_c/\omega \tag{2.30}$$

and used the refractive index $\eta = kc/\omega$ identity. Equation 2.29 is equal to the zero vector if the determinant matrix shown is equal to zero. We need:

$$\begin{vmatrix} -\eta^2 + 1 - \frac{X}{1-Y^2} & \frac{iXY}{1-Y^2} & 0 \\ -\frac{iXY}{1-Y^2} & -\eta^2 \cos^2\theta + 1 - \frac{X}{1-Y^2} & \eta^2 \sin\theta\cos\theta \\ 0 & \eta^2 \sin\theta\cos\theta & -\eta^2 \sin^2\theta + 1 - X \end{vmatrix} = 0 \quad (2.31)$$

This determinantal equation can be solved, though the algebra is heavy. The solutions for the refractive index are:

$$\eta^2 = 1 - \frac{X(1-X)}{1 - X - \frac{1}{2}Y^2\sin^2\theta \pm \left[\left(\frac{1}{2}Y^2\sin^2\theta\right)^2 + (1-X)^2Y^2\cos^2\theta\right]^{1/2}} \quad (2.32)$$

which is known as the Appleton–Hartree formula. Due to the $\pm$ terms for the expression for η^2 in Equation 2.32, there are at least three physically meaningful (positive η) refractive index solutions for some angles θ.

In the absence of a magnetic field $Y = \omega_c/\omega = eB_0/m_0\omega$ goes to zero and $\eta^2 = 1 - X = 1 - \omega_p^2/\omega^2$ as before (Equation 2.12). If the propagation of the wave is perpendicular to the magnetic field so that $\theta = \pi/2$, there are three solutions for the refractive index. The 'ordinary' wave, which has an identical refractive index to a plasma without a magnetic field ($\eta^2 = 1-X$), and two 'extraordinary' waves with

$$\eta^2 = 1 - \frac{X(1-X)}{1 - X - Y^2}.$$

Figure 2.2 presents the refractive index results of Equation 2.32 as 'dispersion' diagrams, where the frequency ω of the electromagnetic wave is plotted as a function of the wave number k. The value of ω/k in a dispersion diagram gives the phase velocity of the wave, while the slope $d\omega/dk$ represents the group velocity of the wave. For Figure 2.2, in order to produce plots valid for a range of parameters, the horizontal k axis is plotted as $kc/\omega_p = \eta\omega/\omega_p$, while the vertical frequency axis is shown in units of the plasma frequency ω_p.

The two solutions for propagation parallel to the magnetic field (when $\theta = 0$) have refractive index values

$$\eta^2 = 1 - \frac{X}{1 \pm Y}. \quad (2.33)$$

For $Y \geq 1$ ($\omega \leq \omega_c$) and propagation parallel to the magnetic field, there is an interesting solution known as the 'whistler wave' with the refractive index going to infinity at Y = 1 ($\omega = \omega_c$). The whistler wave is also apparent in Figure 2.2(B) at $\pi/6$ propagation angle to the magnetic field.

Whistler waves propagate at low frequency without a lower frequency limit as is apparent for all other solutions of the Appleton–Hartree formula. The name arises because this type of wave was first detected directly on early telegraph systems at audio frequencies with the telegraph cable acting as an antenna. It is now known that these whistler waves arise largely via electromagnetic wave emission from

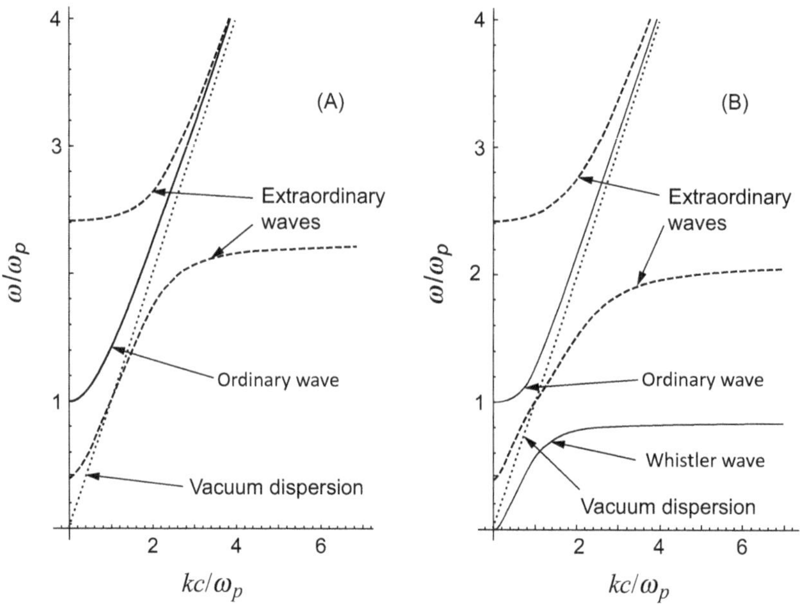

Figure 2.2 Dispersion diagrams showing frequency ω as a function of wavenumber k for electromagnetic radiation in a magnetised plasma. The electron cyclotron frequency ω_c is assumed such that $\omega_c = 2\omega_p$. Propagation (A) perpendicular to the magnetic field lines and (B) at angle $\theta = \pi/6$ to the magnetic field lines is shown. The extraordinary waves are shown as broken curves and the vacuum dispersion is shown as a dotted line. The plots are in multiples of the plasma frequency ω_p (vertical axis) and ω_p/c (horizontal axis).

lightning strikes with the wave travelling along Earth's magnetic field lines to the opposite hemisphere. As they travel, the lower frequencies have a lower group velocity $d\omega/dk$, so arrive last (see the low value of $d\omega/dk$ of the whistler wave on e.g. Figure 2.2(B) at low frequencies). The early telegraph operators heard a whistle of descending pitch. Whistler waves have been used to diagnose plasma conditions in the ionosphere and magnetosphere along magnetic field lines [103]. In laboratory plasmas, whistler waves are sometimes called 'helicons'.

For $Y < 1$ ($\omega > \omega_c$), the two solutions of Equation 2.33 can be shown to have rotating circular polarisation with

$$\frac{E_x}{E_y} = \pm i.$$

Here the $\pm$ depends on whether the $+$ or $-$ solution for the refractive index is applicable. An electromagnetic wave propagating parallel to a magnetic field in a plasma effectively decouples into two waves with 'right' ($E_x/E_y = +i$) and 'left' ($E_x/E_y = -i$) circular polarisation with different refractive indices for the two polarisations. Out of the plasma, the waves can combine back to form e.g. a linear polarisation (if both have the same intensity). The final linear polarisation will rotate with propagation through the plasma, an effect known as Faraday rotation. Faraday rotation of probe electromagnetic waves is used as a diagnostic of magnetic field strength in laboratory plasmas [51].

2.3 Absorption of Light

The treatment of the propagation of electromagnetic waves in a plasma (Section 2.1) does not consider absorption of the radiation. Absorption of light is treated macroscopically by considering changes in the electric field strength **E** of an electromagnetic wave. The electric field strength is related to the irradiance (power per unit area) of the light by Equation 2.15. We consider the macroscopic absorption of light in this section, without detailing the atomic or other microscopic absorption processes. The microscopic treatment of the absorption of light involves studying the individual absorption properties of bound and unbound electrons (and is considered in Sections 5.3 (free electron absorption), 5.5 (photo-ionisation), 10.1.1 (bound electron absorption) and Chapter 3 (light scatter)).

Maxwell's equation governing the propagation of light varies with the electric displacement $\mathbf{D} = \epsilon_m \mathbf{E} = \epsilon_0 \epsilon_r \mathbf{E}$ rather than the electric field strength **E** alone (see Appendix A.2). Here ϵ_m is the permittivity or dielectric constant of the medium, $\epsilon_0 = 8.854 \times 10^{-12}\,\mathrm{Fm}^{-1}$ is the dielectric constant in vacuum and $\epsilon_r (\approx 1)$ is the relative dielectric constant for the medium. In vacuum ϵ_r has a value of unity. The medium in which an electromagnetic wave propagates affects the relationship between the electric displacement **D** and electric field **E** and can consequently change the magnitude of the electric field strength.

There are a number of ways of relating the electric displacement **D** and electric field **E**. We have

$$\mathbf{D} = \epsilon_0 \epsilon_r \mathbf{E} = \epsilon_0 \mathbf{E} + \mathbf{P} = \epsilon_0 (1 + \chi) \mathbf{E} = \epsilon_0 n^2 \mathbf{E}$$

where **P** is known as the polarisation of the medium, χ is known as the susceptibility of the medium and n is the refractive index of the medium. The way of relating **D** and **E** in a given analysis is determined by the problem involved and certain conventions for different areas of study have developed. For example, linear optics uses the refractive index n. The polarisation **P** is particularly important in some solid materials where domains of crystals occur so that a movement of charge is induced across the domain by the electric field from an electromagnetic wave producing a non-zero value of polarisation. The charges induced in the domain produce an electric field opposite in direction to the original electric field, so that the resulting polarisation produces a partly cancelling electric field $-\mathbf{P}/\epsilon_0$.

The relationship between polarisation and susceptibility is useful to employ initially. The atoms, ions and free electrons interacting with an electromagnetic wave lead to the generation of a polarisation **P** as a result of the applied electric field **E** and this is related to the susceptibility χ by

$$\mathbf{P} = \epsilon_0 \chi \mathbf{E}. \tag{2.34}$$

The susceptibility is a function of the frequency ω of the applied field, whose form depends on the details of the energy levels and wavefunctions of the atoms that make up the dielectric. This is where the microscopic details of the individual absorption processes feed into the macroscopic behaviour of a collection of atoms. Propagation of an electromagnetic wave in a dielectric is again described by the wave equation, this time taking into account the polarisation effects,

$$\nabla^2 \mathbf{E} - \mu_0 \epsilon_0 (1 + \chi) \frac{\partial^2 \mathbf{E}}{\partial t^2} = 0. \tag{2.35}$$

This equation reduces to Equation 2.4 for the case of propagation in free space ($\chi = 0$). Solutions of Equation 2.35 are again of the form

$$\mathbf{E} = \mathbf{E_0} \exp[i(kz - \omega t)]. \tag{2.36}$$

Substitution into Equation 2.35 shows us that

$$k^2 = \mu_0 \epsilon_0 \omega^2 (1 + \chi) = \frac{\omega^2}{c^2}(1 + \chi) \tag{2.37}$$

which leads to an expression for the relative dielectric constant $\epsilon_r = (1 + \chi)$ of

$$1 + \chi = \left(\frac{kc}{\omega}\right)^2. \tag{2.38}$$

The relative dielectric constant is only a constant in that it is independent of **E**; its magnitude is still a function of frequency ω, as can be seen from this equation.

The susceptibility is generally a complex quantity and we can write

$$\chi = \chi' + i\chi'', \tag{2.39}$$

where χ' and χ'' are real. It is conventional to write the square root of Equation 2.38 as

$$\sqrt{\epsilon_r} = \sqrt{(1 + \chi)} = \frac{kc}{\omega} = \eta + i\kappa = n \tag{2.40}$$

where η and κ are, respectively, the real and imaginary components of the refractive index n. The imaginary component of the refractive index κ is also referred to as the extinction coefficient. We can compare the real and imaginary parts of Equations 2.39 and 2.40 to find

$$\begin{aligned} \eta^2 - \kappa^2 &= 1 + \chi', \\ 2\eta\kappa &= \chi''. \end{aligned} \tag{2.41}$$

These equations can be used to determine the frequency dependence of η and κ once the frequency-dependent susceptibility is known.

Substituting Equation 2.40 as the expression for k into the solution for the wave equation (Equation 2.36), we find

$$\mathbf{E} = \mathbf{E_0} \exp\left[ik_0\eta z - \frac{\omega\kappa z}{c} - i\omega t\right] = \mathbf{E_0} \exp\left[i\omega\left(\frac{\eta z}{c} - t\right) - \frac{\omega\kappa z}{c}\right] \tag{2.42}$$

where k_0 is the wave vector of the equivalent wave in vacuum. The magnitude of the cycle-averaged irradiance of an electromagnetic wave (see Equation 2.15) is given by

$$I = \frac{1}{2}\epsilon_0 c\eta E_0^2.$$

Substituting Equation 2.42 gives

$$I = \frac{1}{2}\epsilon_0 c\eta E_0^2 \exp\left(-\frac{2\omega\kappa z}{c}\right). \tag{2.43}$$

If we define I_0 as the cycle-averaged irradiance at $z = 0$ we can write

$$I(z) = I_0 \exp(-Kz), \tag{2.44}$$

where

$$K = \frac{2\omega\kappa}{c} = \frac{\omega}{c\eta}\chi''. \tag{2.45}$$

The quantity K defined in this way is called the absorption coefficient. The intensity of an electromagnetic wave falls to $1/e$ of its initial value in a distance $1/K$.

Different atomic processes can contribute to the absorption coefficient K and we can generally simply add absorption coefficients. Absorption in plasma typically has a component due to free electron transitions K_{ff}, bound electrons being photo-ionised K_{bf}, transitions between bound quantum states K_{bb} and scatter K_s. The total absorption coefficient K is an addition of the different absorption contributions:

$$K = K_{ff} + K_{bf} + K_{bb} + K_s. \tag{2.46}$$

These different processes are treated in Section 3 (light scatter), Section 5.3 (free electron absorption), Section 5.5 (photo-ionisation), and Section 10.1.1 (bound electron absorption).

It is perhaps less obvious, but the total refractive index η also comprises additive components due to different processes. Strictly different processes add in the dielectric constant or square of the refractive index, but at small departures from $\eta \approx 1$, refractive index contributions are additive. The most significant contribution for a plasma is usually due to the free electrons (see Section 2.1), but we shall see that bound electrons can contribute to the total refractive index (see Section 3.4).

2.4 Focused Laser Light in Plasmas

Some effects of electromagnetic wave propagation in plasma occur at low irradiance, but start to dominate the energy movement and other processes at the high irradiances present with focused laser light. Focused high-power laser light incident onto a solid surface or into a gas typically produces free electrons in the first few oscillations (that is the first few femtoseconds) of the electric field due to the large electric field associated with even modest focused laser irradiances (see Equation 2.15). In high-power laser-solid target interactions, the bulk of a laser pulse of duration greater than a few femtoseconds interacts with plasma expanding approximately normally to the target surface.

In this section, we explore some of the radiation physics involved with infra-red to ultra-violet light interactions with plasma at high irradiance. High irradiance is typically defined as irradiances $> 10^9\ \mathrm{Wcm^{-2}}$ where significant ablation and ionisation of solid material or the ionisation of gases takes place. With solid targets, plasma material expands normally to the target surface and the light interacts with an almost planar expanding plasma. The light can penetrate the plasma up to the critical density. A similar geometry for radio waves interacting with the approximately planar plasma of the ionosphere occurs with an appropriately much lower critical density.

The ionosphere is an upper layer of Earth's atmosphere that is ionised by solar radiation during the day. It lies 75–1000 km above the surface of the earth with a peak electron density at approximately 160 km. The physics of this section considering the propagation path of radiation in a planar plasma is important in the interaction of radio waves with the ionosphere.

2.4.1 Single Electron Motion in an Electromagnetic Field

Due to the difference in mass of electrons and ions, an electromagnetic wave interacts with the electrons in a plasma and causes only an insignificant direct acceleration of the ions. We first examine the motion of a single free electron in an electromagnetic wave. As before (see Section 2.1), we assume that the electromagnetic wave is linearly polarised and propagates in the z-direction with the electric field directed in the x-direction. The magnetic field of the electromagnetic wave is then directed in the y-direction. Assuming an electric field of amplitude E_0 and angular frequency ω, we can write an expression for the acceleration of an electron of charge $-e$ and mass m_0 such that

$$\frac{d^2x}{dt^2} = -\frac{e}{m_0} E_0 \cos(\omega t).$$

Integrating from time 0 to time t, we obtain an expression for the velocity of the electron in the electromagnetic field

$$\frac{dx}{dt} = -\frac{eE_0}{m_0\omega}\sin(\omega t),$$

and integrating again we find the position of the electron relative to its starting position, namely

$$x = -\frac{eE_0}{m_0\omega^2}(\cos(\omega t) - 1). \tag{2.47}$$

The electron oscillates or 'quivers' in the electric field of the wave. We can calculate an average quiver energy which is often referred to as the 'ponderomotive' energy. At any time the kinetic energy U of the electron is given by

$$U = \frac{1}{2}m_0\left(\frac{dx}{dt}\right)^2 = \frac{e^2E_0^2}{2m_0\omega^2}\sin^2(\omega t).$$

If we average the square of the sine function in time over an oscillation period of the electromagnetic wave, we can obtain a time-averaged quiver energy. Using

$$\frac{\int_0^{2\pi}\sin^2\theta d\theta}{\int_0^{2\pi}d\theta} = \frac{1}{2}$$

gives the time-averaged electron quiver energy $< U >$ in an electromagnetic field as

$$< U > = \frac{e^2E_0^2}{4m_0\omega^2}. \tag{2.48}$$

This average quiver energy of the electrons in an electromagnetic field is known as the ponderomotive potential. Evaluating constants, converting the square of the magnitude of the electric field (E_0^2) to irradiance (Equation 2.15) and the frequency ω to wavelength λ (using $\omega = 2\pi c/\lambda$) gives a numerical expression for the ponderomotive potential

$$< U > \approx \frac{I\lambda_{\mu m}^2}{10^{13}} \quad \text{eV} \tag{2.49}$$

where the wavelength $\lambda_{\mu m}$ is measured in units of microns and the irradiance I is measured in units of Wcm^{-2}. At irradiances greater than 10^{14} $\text{Wcm}^{-2}\mu\text{m}^2$, the ponderomotive potential exceeds 10 eV and starts to become comparable to typical atomic ionisation potentials (e.g. 13.6 eV for hydrogen atoms) and to the temperatures and hence kinetic energy per electron in the plasma.

The motion of the electron in an electromagnetic field, however, is complicated by the presence of the magnetic field. The electron has a velocity $\mathbf{v}$ due to its acceleration in the electric field, which produces a Lorentz force $e\mathbf{v} \times \mathbf{B}$ pointing in the direction of propagation of the electromagnetic wave. We can write for the acceleration in the direction of wave propagation (our z-direction) that

$$\frac{d^2z}{dt^2} = -\frac{e}{m_0}\frac{dx}{dt}B$$

where B is the magnetic field of the electromagnetic wave. The magnetic field B oscillates in phase with the electric field such that

$$B = \frac{E_0}{c}\cos(\omega t).$$

If we assume that any motion of the electron in the z-direction is small compared to the wavelength of the light, the electric field experienced by the electron will be approximately constant with z. We can write for the electron acceleration in the z-direction due to the $e\mathbf{v}$ X $\mathbf{B}$ force that

$$\frac{d^2z}{dt^2} = \frac{e^2E_0^2}{m_0^2\omega c}\sin(\omega t)\cos(\omega t).$$

Integrating from time 0 to time t gives the velocity in the direction of propagation as

$$\frac{dz}{dt} = \frac{e^2E_0^2}{2m_0^2\omega^2 c}\sin^2(\omega t).$$

The position of the electron relative to its starting position is found by integrating again to give

$$z = -\frac{e^2E_0^2}{8m_0^2\omega^3 c}\sin(2\omega t) + \frac{e^2E_0^2}{4m^2\omega^2 c}t. \tag{2.50}$$

We have used here the integration result that

$$\int_0^{\omega t}\sin^2(\theta)d\theta = \frac{1}{2}\left(\omega t - \frac{1}{2}\sin(2\omega t)\right).$$

Equations 2.47 and 2.50 show that an electron in an electromagnetic field executes a figure-eight motion in the xz-plane with a superimposed electron drift in the z-direction (see Figure 2.3). The x-position of the electron is oscillating in position proportionally to $\cos(\omega t)$, while the z-position oscillates proportionally to $\sin(2\omega t)$. These equations for the electron position describe a figure-eight motion relative to a position moving with drift velocity v_d in the direction of electromagnetic propagation (the z-direction). Looking at the second term on the right-hand side of Equation 2.50, we see that the drift velocity v_d is given by

$$v_d = \frac{e^2E_0^2}{4m_0^2\omega^2 c}.$$

From the above expression for the electron drift velocity, we can quickly see that the 'drift momentum' m_0v_d of an electron is given by the average quiver energy divided by the speed of light ($< U >/c$).

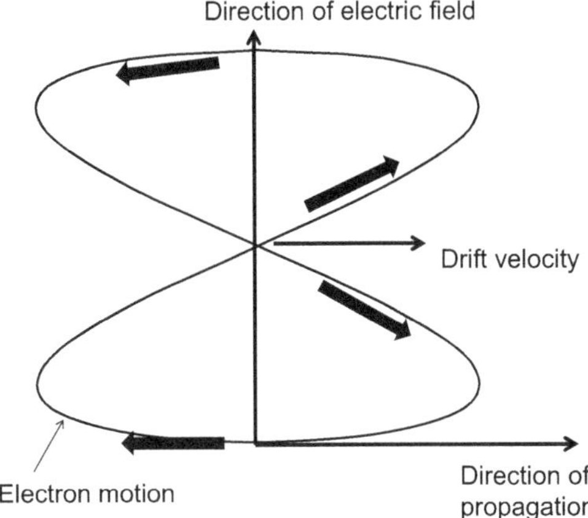

Figure 2.3 The trajectory of an electron in an electromagnetic wave. The electron exhibits a figure-eight motion as well as a superimposed drift in the direction of propagation of the wave.

$\mathbf{J} \times \mathbf{B}$ *acceleration*

The drift velocity of the electrons in a radiation field becomes an important electron-accelerating mechanism at high laser irradiances ($I > 10^{18}$ Wcm^{-2}). Theoretical work on electron acceleration at high irradiance often uses the vector potential **A** description, where for an oscillating electric field **E**, we have $\mathbf{E} = -d\mathbf{A}/dt$. The amplitude of the vector potential A_0 is employed in a reduced form $a_o = eA_0/(m_0c) = eE_0/(m_0c\omega)$ and from our expression for the drift velocity, we see that the electron drift velocity relative to the speed of light $v_d/c = a_o^2/4$. We have, however, neglected any relativistic increase in the electron mass m_0, so this expression is only valid for small a_o. Calculations including relativistic mass increases show that the drift velocity due to electromagnetic radiation is given by

$$\frac{v_d}{c} = \frac{a_o^2}{4 + a_o^2}.$$

The relativistic treatment gives the desired result that the drift velocity cannot exceed the speed of light c. Numerically, we have that the reduced vector potential a_o is related to the laser irradiance I measured in Wcm^{-2} by

$$a_o = \sqrt{\frac{I\lambda_{\mu\mathrm{m}}^2}{1.37 \times 10^{18}}}$$

where $\lambda_{\mu\mathrm{m}}$ is the laser wavelength in microns. At $a_o > 1$, electron acceleration by the drift velocity starts to dominate other electron acceleration mechanisms.

The electron acceleration and consequent heating of laser target material when the electrons collide with a solid target is commonly referred to as $\mathbf{J} \times \mathbf{B}$ 'heating'.

The drift velocity has no dependence on the phase of the electromagnetic field (there are no $\cos(\omega t)$ or $\sin(\omega t)$ terms) and, for a uniform electromagnetic field, no variation with z, so the drift velocity is not dependent on the z-position of the electron as it drifts. As well as considering relativistic mass increases, a more involved treatment can allow for phase changes in the electric and magnetic fields by replacing time t by the variable $t - v_d/c$ in our treatment above, but qualitatively similar motion of the electron is observed provided the reduced vector potential $a_0 < 10$.

2.4.2 The Ponderomotive Force

In a non-uniform electromagnetic field such as a laser focus, the ponderomotive potential or time averaged electron quiver energy $< U >$ given by Equation 2.48 varies spatially. Electrons will tend to move from high $< U >$ to lower values of this potential energy. For any potential field Φ, there exists a force $\mathbf{F}$ given by $\mathbf{F} = -\nabla\Phi$ which serves to homogenise potential energy variations. With a spatially varying ponderomotive potential, there is consequently a time-averaged force on each electron given by

$$\mathbf{F}_p = -\frac{e^2}{4m_0\omega^2}\nabla E_0^2 \tag{2.51}$$

as the value of E_0^2 is the only spatially varying quantity in a non-uniform electromagnetic field (see Equation 2.48) if we assume relativistic effect mass increases are small.

In a laser focus, there is a gradient of intensity and hence a gradient of E_0^2 perpendicular to the direction of laser propagation. On the microscopic level, electrons near the centre of the focus are accelerated outwards away from the focal centre by the electric field of the electromagnetic wave, and then accelerated back to the centre when the oscillating electric field changes direction. However, the value of the electric field is smaller at a distance away from the focal centre, so the electrons do not move back to the centre completely. With many cycles, the net effect is a motion away from the high intensity (high E_0^2) region.

2.4.3 Multi-Photon Absorption

At high laser irradiances, the density of photons is sufficient that several can (i) act together to ionise an electron from a bound state of an atom or (ii) the electric field is sufficient to distort the atomic potential enabling ionisation. These processes produce the first free electrons when a high-power laser interacts with a material such as a solid surface or a neutral gas.

If we estimate the volume of an atom as $8a_0^3$, where a_0 is the Bohr radius, and evaluate the density of photons per unit volume $(I/(c\hbar\omega))$, then for an irradiance I with a laser of 1.06 μm wavelength, the number of photons n_p in the typical volume of an atom is given by

$$n_p = 2 \times 10^{-3} I.$$

A very modest irradiance $I > 10^3$ Wm^{-2} seems sufficient for there to be at least two photons in the atomic volume. (An irradiance level of 10^3 Wm^{-2} is produced by solar radiation on the ground on a hot summer day.) Direct multi-photon absorption scales as the laser intensity squared (I^2) up to irradiances > few 10^{13} Wcm^{-2}, where direct multi-photon absorption can be suppressed. For an ionisation potential of an isolated atom of E_{ion} and n photons absorbed, the requirement for significant multi-photon tunnelling for photons of energy $\hbar\omega$ is given by

$$n\hbar\omega - <U> - E_{ion} > 0.$$

The energy from the absorbed photons needs to be greater than the sum of the ionisation potential and the ponderomotive potential. At irradiances where the ponderomotive potential starts to suppress direct multi-photon ionisation, perturbation of the atomic potential by the electromagnetic wave electric field starts to dominate and electrons can tunnel through a lowered barrier or simply have sufficient energy in their bound state to pass over the lowered potential barrier. The parameter γ_K due to Keldysh [58] defines the transition from direct multi-photon ionisation to this tunnelling ionisation. The Keldysh parameter is given by

$$\gamma_K = \sqrt{\frac{E_{ion}}{2 <U>}} \tag{2.52}$$

with $\gamma_K \gg 1$ implying that direct multiphoton absorption dominates and $\gamma_K \ll 1$ implying that tunneling ionisation dominates.

After the initial ionisation by the multi-photon processes when the initial few free electrons are produced, free electron collisions with ions dominate the production of ionisation for most of the duration of a laser pulse incident on a target.

2.4.4 Laser Penetration into Expanding Plasma

Our treatment of electromagnetic propagation in a plasma has so far implicitly assumed that the plasma is uniform. Plasma produced by laser irradiation of a solid typically expands along a normal to the initial target surface as the aspect ratio of the initial plasma material typically extends over a laser focus in the range 5–100 μm, while the depth of ablated material is typically $\ll 1\,\mu$m. The initial plasma is thin and wide (on a micron scalelength), so the expansion occurs predominantly in the direction of the thinnest dimension, which is the target normal direction. An assumption of a planar target surface over the focal spot dimensions is usually a good approximation even for spherically shaped targets.

In laser-produced plasmas, the Debye length is typically much smaller than the size of plasma, so ions and electrons expand approximately together with fast ions travelling farther. A 'self-similar' plasma expansion can be quickly established as faster ions move farther from the target. The self-similar profile exhibits a linear velocity profile with distance from the target and a density profile which is approximately exponentially decreasing with distance from the target surface. At some distance from the target surface, the electron density drops below the critical density and this is the closest penetration distance of any incoming laser light to the target surface. The interaction of an electromagnetic wave with an increasing electron density was first studied in connection with the propagation of radio waves into the ionosphere before the invention of the laser [82].

To calculate the spatial variation of an electromagnetic field in a non-uniform plasma, we can convert the plasma wave equation (Equation 2.7) into an equation only varying with distance. Performing the differentiations with respect to time on $E = E_0 \exp(-i\omega t)$ and substituting $k = \omega/c$, we re-write Equation 2.7 as

$$\frac{\partial^2 E_0}{\partial z^2} + k^2 E_0 = \frac{\partial^2 E_0}{\partial z^2} + \left(k_0^2 - \frac{\omega_p^2}{c^2}\right) E_0 = 0 \qquad (2.53)$$

where $k_0 = \omega/c = 2\pi/\lambda$ is the vacuum wavenumber. Equation 2.53 is known as the Helmholtz equation for an electromagnetic wave assuming propagation in the z-direction.

Radiation incident at an initial angle θ_0 to the target normal in vacuum is refracted by the decreasing refractive index associated with the increasing electron density closer to the target surface. We consider the z_n-direction along the target normal so that the **k**-vector is initially directed at an initial angle θ_0 to the target normal and has an initial component in the z_n-direction given by

$$k_n = k_0 \cos\theta_0.$$

Equation 2.53 for the z_n-direction becomes

$$\frac{\partial^2 E_0}{\partial z_n^2} + \left(k_0^2 \cos^2\theta_0 - \frac{\omega_p^2}{c^2}\right) E_0 = 0. \qquad (2.54)$$

The electromagnetic wave incident at an initial angle of incidence θ_0 only propagates up to where the wavenumber component in the z_n-direction reduces to zero, that is where

$$\left(k_0^2 \cos^2 \theta_0 - \frac{\omega_p^2}{c^2}\right) = 0,$$

which indicates that propagation in the forward z_n-direction stops when

$$\omega_p^2 = \omega^2 \cos^2 \theta_0.$$

This is equivalent to requiring that the electron density

$$n_e = n_{crit} \cos^2 \theta_0 \tag{2.55}$$

or the refractive index

$$\eta = \sin \theta_0. \tag{2.56}$$

Equations 2.55 and 2.56 represent conditions for a 'turning point' where the radiation incident at angle θ_0 refracts with **k**-vector parallel to the target surface in a laser-produced plasma (and no component in the target normal z_n-direction). The path of the radiation in a laser-produced plasma is schematically illustrated in Figure 2.4. An identical propagation path occurs for radio waves incident at an angle of incidence θ_0 into the ionosphere and is the basis of 'over-the-horizon' radio-wave propagation around the Earth (see Exercise 2.12).

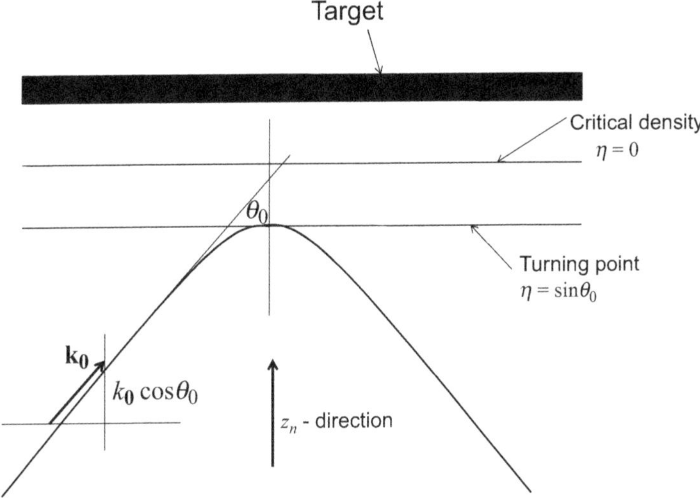

Figure 2.4 A schematic plot of radiation incident at an initial angle θ_0 onto a solid target with a planar plasma expansion in the z_n-direction. The radiation is refracted and penetrates to a turning point of refractive index $\eta = \sin \theta_0$. At the critical density $\eta = 0$. The **k**-vector has an initial component $k_0 \cos \theta_0$ in the target normal z_n-direction.

2.4.5 The Laser Electric Field in the Plasma

We can use the Helmholtz equation (Equation 2.53) to examine the laser electric field variation in a laser plasma with a profile of electron density decreasing away from the target surface. With normal incidence ($\theta_0 = 0$), reflection of light near the critical density produces a backwards-directed electromagnetic field which adds to the total electric field amplitude. We should note here that the electric field will still experience temporal oscillation of form $\exp(-i\omega t)$. We are calculating the spatial variation of the amplitude E_0.

There is a solution of the Helmholtz equation involving a special function (the Airy function) if the electron density has a linear slope. For ease of calculation, we can assume that the critical density is at position $z = L$ and write for the electron density at position z:

$$n_e(z) = n_{crit}\frac{z}{L} \tag{2.57}$$

so that L is a scalelength for the slope of the electron density change away from the critical density. We assume the electron density remains at zero for $z < 0$. The Helmholtz equation (Equation 2.53) for the electric field amplitude $E_0(z)$ as a function of distance z then becomes

$$\frac{\partial^2 E_0(z)}{\partial z^2} + k_0^2\left(1 - \frac{z}{L}\right)E_0(z) = 0.$$

The distance variable z can be replaced with a dimensionless parameter Z_n defined by

$$Z_n = \left(\frac{k_0^2}{L}\right)^{1/3}(z - L). \tag{2.58}$$

The electric field variation is then given by

$$\frac{d^2E_0(Z_n)}{dZ_n^2} - Z_nE_0(Z_n) = 0 \tag{2.59}$$

which is known as the Stokes differential equation. Solutions for the electric field are proportional to the Airy function $A_i(Z_n)$ plotted in Figure 2.5. We see that the amplitude of the electric field has an increase ('swelling') in amplitude peaking at $Z_n = -1$ and that an evanescent decrease of electric field occurs at densities above the critical density (where $Z_n > 0$) penetrating to $Z_n > 2$.

2.4.6 The Laser Electric Field at Oblique Incidence: Resonance Absorption

For electromagnetic waves incident at an angle of incidence θ_0 onto a planar target we can use Equation 2.54 to examine the electric field along the target normal (z_n-direction). We have seen that forward propagation in the z_n-direction ceases when the wavevector component in the z_n direction is equal to zero: the radiation reaches a turning point. We assume a linear density profile (Equation 2.57) and substitute a revised version of Z_θ (see Equation 2.58) for the angle θ such that

$$Z_\theta = \left(\frac{k_0^2}{L}\right)^{1/3} (z_n - L\cos^2\theta_0). \tag{2.60}$$

A Stokes differential equation identical to Equation 2.59 is obtained, namely

$$\frac{d^2E_0(Z_\theta)}{dZ_\theta^2} - Z_\theta E_0(Z_\theta) = 0. \tag{2.61}$$

The turning point for the radiation is now when $Z_\theta = 0$ and the critical density is where $Z_\theta = (k_0L)^{2/3}\sin^2\theta_0$. There is now an evanescent wave from the turning point towards the critical density.

For near spatially uniform conditions (wavelength $\ll$ the spatial variation), the electric field in a medium varies as $E_0/\sqrt{\eta}$, where η is the refractive index. This increase of the electric field with decreasing refractive index η is apparent from Equation 2.15 due to conservation of energy because the intensity I remains constant as the refractive index η reduces below unity. At the critical density in a plasma where the real part of the refractive index η goes to zero, we expect a large increase of the electric field amplitude.

There are two solutions to the Stokes differential equation (Equation 2.61), usually labelled A_i and B_i, which are valid separately or in a linear combination depending on the boundary conditions. The B_i solution increases exponentially at high positive Z_θ so is not valid for normal incidence light (Equation 2.59) as the boundary condition requires the evanescent wave to decay at high positive Z_n. However, the B_i solution contributes to the full solution at oblique incidence where the electric field increases to a large value at the critical density (now at some positive Z_θ). From Figure 2.5, we see that at positive positions of Z_θ, the Airy function solutions of the Stokes equation initially fall off approximately exponentially as $\exp(-Z_\theta)$ before rising, so we can expect the electric field to drop from the turning point to the critical density by a factor $\exp\left(-(k_0L)^{2/3}\sin^2\theta_0\right)$ before increasing due to the resonance at the critical density. The intensity I_{crit} of radiation reaching the critical density is found by allowing for the decrease of the electric field from turning point to critical density and by squaring the electric field reaching the critical density:

$$I_{crit} \propto \exp\left(-2(k_0L)^{2/3}\sin^2\theta_0\right). \tag{2.62}$$

Light incident on a surface is said to be p-polarised if there is a component of the electric field directed into the surface. The ponderomotive force (see Equation 2.51) for p-polarised light arising from the increase of electric field at the critical density (see Figure 2.5) gives rise to electron acceleration away from the critical density.

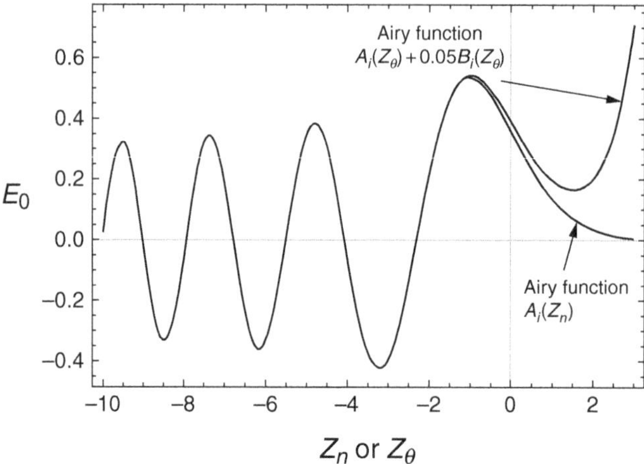

Figure 2.5 The Airy function A_i and a linear combination of the two Airy functions $A_i + 0.05B_i$ are solutions to the Stokes differential equation. The laser electric field E_0 amplitude for light incident from the left onto a linear electron density ramp is proportional to the plotted Airy functions. For normal incidence light, the boundary condition is that the electric field drops to zero at high positive Z_n corresponding to electron densities above the critical density with $Z_n = 0$ in the position of the critical density. A_i represents the solution for normal incidence. For obliquely incident light, the turning point at $Z_\theta = 0$ and $A_i + 0.05B_i$ represents a sample solution with the electric field increasing rapidly at large positive Z_θ (above the turning point density).

With p-polarised light, electrons can oscillate along the direction of the density gradient (away from the target surface) and hence have a net ponderomotive motion due to the decreasing electric field amplitude (as discussed in Section 2.4.2). If the direction of the light polarisation is parallel to the critical density surface (known as s-polarisation), there is no ponderomotive motion of electrons as the electrons oscillate in the electric field in a direction where there is no change of electric field amplitude.

Absorption of the radiation at the critical density can be high for p-polarised light incident at an angle to the target normal due to radiation reaching the critical density by evanescent decay from the turning point with electrons being subsequently accelerated away from the critical density. The amount of absorption varies proportionally to (i) the distance over which absorption can take place which we can take is proportional to the Z_θ distance $(k_0L)^{2/3}\sin^2\theta_0$ between the turning point and the critical density, and (ii) the light intensity reaching the critical density (given by Equation 2.62). We then have for the relative absorption A of the light near the critical density that

$$A \propto (k_0L)^{2/3}\sin^2\theta_0 \exp\left(-2(k_0L)^{2/3}\sin^2\theta_0\right). \tag{2.63}$$

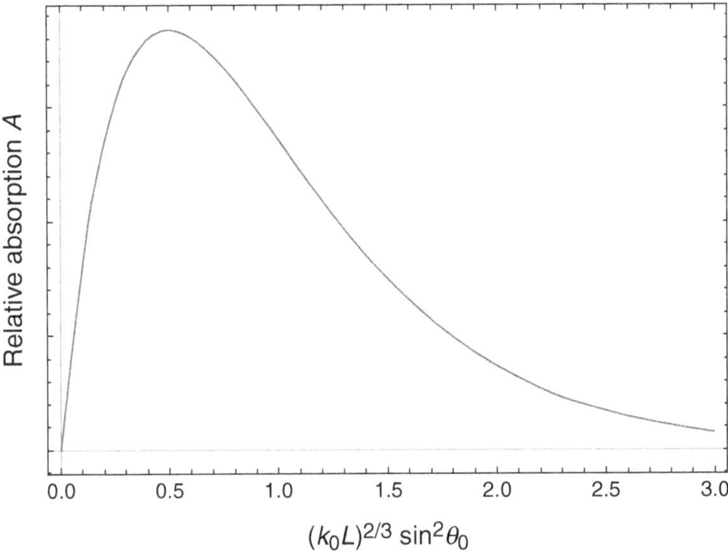

Figure 2.6 Resonance absorption for light of initial angle of incidence θ_0 and vacuum wavenumber k_0 as a function of $(k_0L)^{2/3}\sin^2\theta_0$ for a planar plasma with a linear gradient of electron density of scalelength L (following Equation 2.63).

Forslund et al. [34] showed in simulations – as did others with more detailed analytic analysis (see, for example, Pert [82]) – that the absorption A near the critical density for obliquely incident light changes according to Equation 2.63. The variation of the absorption with the Z_θ distance between the turning point and critical density is shown in Figure 2.6. Due to the increase of electric field magnitude associated with the electric field reaching the critical density (where the plasma frequency equals the laser frequency), the absorption process is known as resonance absorption.

With resonance absorption, electrons are accelerated by an enhanced electric field, so the electrons can be accelerated to high energies and not undergo collisions with ions or other electrons (collision cross-sections drop rapidly for high-velocity particles as particle velocity increases; see Chapter 11). Often superthermal electrons are produced which can be regarded as having a different (much higher) temperature than the thermal temperature associated with the bulk of the plasma. The superthermal electrons can deposit energy at distances from the laser focus or simply cause a loss of energy from the plasma as they escape. In inertial fusion, superthermal electrons are particularly troublesome as they can penetrate large distances and pre-heat the deuterium–tritium fuel, making it harder to compress.

Assuming that angles of incidence θ_0 of the laser light cannot be reduced, Figure 2.6 shows that large scalelengths L in the expanding plasma and high wave number k_0 (i.e. short wavelength) of the laser reduce resonance absorption. Although the electric field propagation and swelling at the critical density is independent of the laser irradiance, resonance absorption only occurs when the electric field is high so that the ponderomotive potential (Equation 2.49) exceeds other energies in the plasma, such as the electron kinetic energies. Equation 2.49 shows that the laser irradiances need to exceed approximately 10^{14} Wcm^{-2} μm^2.

2.5 Radiation and Charge Acceleration

Emission and absorption of electromagnetic radiation occurs when charges are accelerated. A charged body of any mass radiates when accelerated, but the most common emitters and absorbers are electrons, as their low mass results in larger acceleration in plasmas and other media (such as a conducting wire). All electromagnetic radiation emission and absorption occurs due to the acceleration of charge. The acceleration is apparent for unbound particles such as free electrons in a plasma, but is also present when dealing with the time-dependent quantum mechanics of electrons bound in atoms and ions interacting with electromagnetic waves. In this section, we follow a relatively simple treatment by Purcell [87] to determine the radiation produced by an accelerating charge.

The production of electromagnetic radiation when a charge such as an electron accelerates arises due to a dislocation of the radially symmetric electric field from the charge. Immediately before an acceleration impulse, the electric field radially radiates from the initial position in a reference frame moving, say, with the initial charge velocity. Immediately after the acceleration, the electric field radially radiates from the new position in the reference frame. As the electric field lines from the charge are continuous, there is an electric field transverse jump propagating at the speed of light. For the purposes of evaluating an expression for the transverse electric field, we can assume that the velocity jump is small (so that the reference frame is essentially unchanged), but occurs on a short timescale. We also assume that the velocity of the charge is much less than the speed of light, so that relativistic distortion of the electric field lines can be ignored.

We now consider more quantitatively the production of electromagnetic radiation by thinking about the electric field around a charged particle if it suffers an impulse of acceleration to a velocity δv in a time δt at some initial time $t = 0$. At a later time $t > 0$, the electric field lines associated with the initial position extend radially out from the initial position at distances greater than ct, which is the distance that light and the information about the disturbance in the electric field from the charge can travel in the time t. However, at the distance ct, there is a dislocation of the electric field associated with the charge having accelerated to a new position and the consequent need for a transverse movement of the electric field radially emanating

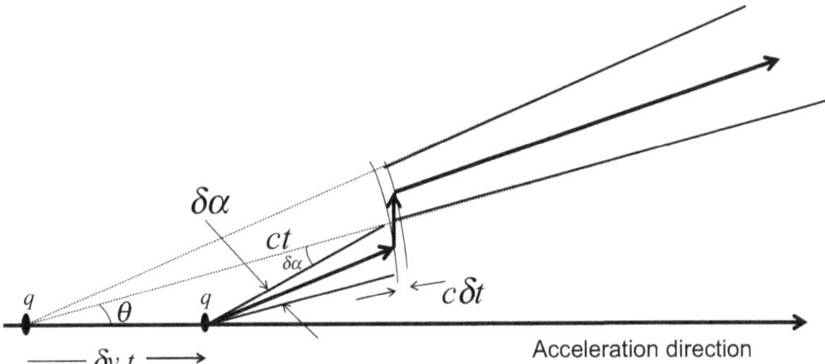

Figure 2.7 The electric field disruption at time t when a charge is accelerated to velocity δv over a time δt at $t = 0$ for an angle θ to the acceleration direction. All electric field lines from the new charge position within the angle $\delta\alpha$ need to move transversely at distance ct to link to the electric field lines established when the charge was in its initial position. The transverse electric field forms a pulse of electromagnetic radiation propagating away from the accelerated charge.

from the new positions to join up with the electric field lines which started before time $t = 0$. The concept is schematically illustrated in Figure 2.7.

The acceleration of a charge q by a velocity δv in a time δt at time $t = 0$ causes a dislocation in the electric field distribution from the charge at a later time t at a distance ct from the charge and produces a transverse electric field. All electric field lines emanating from the later position of the charge move transversely within an annulus of thickness $c\delta t$ to match the field lines emitted before the charge acceleration. Provided that the velocity δv is much less than the speed of light c, the electric field lines from the charge are directed radially from the position of the charge except at the distance ct. We can calculate the enhancement of the electric field transversely over the distance ct to $c(t + \delta t)$ by considering a small angular range $\delta\alpha$ for a particular angle θ to the acceleration direction of the charge. Using Figure 2.7, by simple geometry

$$\frac{\sin\delta\alpha}{\delta v\, t} = \frac{\sin(\pi - \theta - \delta\alpha)}{c\, t}.$$

If $\delta v \ll c$, the angle $\delta\alpha$ is small and we have that

$$\delta\alpha = (\delta v/c)\sin\theta.$$

The enhancement of the electric field in moving transversely through a thickness $c\delta t$ rather than being spread emanating radially from the charge over an angle $r\,\delta\alpha$ is given by

$$\frac{r\,\delta\alpha}{c\delta t} = \frac{r\delta v}{c^2\delta t}\sin\theta.$$

The radial electric field from the charge q at distance r is given by

$$E_r = \frac{q^2}{4\pi\epsilon_0 r^2},$$

so the transverse field at distance r is given by

$$E_t = \frac{q\,\delta v \sin\theta}{4\pi\epsilon_0 c^2(\delta t)r}.$$

The quantity $\delta v/(\delta t)$ is the acceleration of the charge, which we can write as dv/dt. The transverse electric field can then be written as

$$E_t = \frac{q}{4\pi\epsilon_0 c^2 r}\left(\frac{dv}{dt}\right)\sin\theta. \tag{2.64}$$

The acceleration direction of a charge oscillates or rotates in many situations. For example, charged particles orbit around magnetic fields in a plasma in Larmor or gyro-orbits. An electron approaching an ion undergoes a hyperbolic trajectory with a change of direction and hence change of acceleration direction due to the electron–ion Coulomb attraction. In these cases the electric field in a particular direction oscillates sinusoidally (possibly at many frequencies) and consequently creates sinusoidally oscillating magnetic fields (from Ampere's law). Sinusoidally oscillating electric and magnetic fields are simply electromagnetic radiation. The power P_Ω radiated per unit solid angle is related to the intensity $I = \epsilon_0 c E_r^2$ at a distance r by $P_\Omega = I/r^2$, so we have that

$$P_\Omega d\Omega = \frac{q^2}{4\pi\epsilon_0}\frac{1}{4\pi c}\left(\frac{dv/dt}{c}\right)^2 \sin^2\theta d\Omega. \tag{2.65}$$

This expression (Equation 2.65) for the radiation emitted by an accelerating charge is known as the Larmor formula (after Joseph Larmor, 1857–1942, who first developed Equation 2.65). The **k** vector of the radiation is directed radially away from the charge with the electric field directed transversely (in the plane formed by the **k** vector and the acceleration vector). The power P_Ω radiated by an accelerated charge q is proportional to the square of the acceleration and varies with angle θ to the direction of acceleration.

For particles accelerating with velocities approaching the speed of light, relativistic effects become important. We can obtain a qualitative idea of the effects of relativity by considering how the transformation from the frame of reference of the particle to the rest frame of reference affects the behaviour shown in Figure 2.7. When the charge is moving at a constant high velocity approaching the speed of light before the acceleration illustrated in Figure 2.7, the electric field disruption shown relative to the charge is not symmetric for positions in the direction of the initial velocity and in the opposite direction of the initial velocity in the rest frame. Distances for the electric field disruption become 'compressed' in the direction of the velocity and 'elongated' in the opposite direction to the velocity in the rest frame of reference. The net effect is that the power radiated in the rest frame is enhanced in the direction of the velocity and reduced in the opposite direction. At velocities v approaching the speed of light c, there is a 'beaming effect' so that emission is concentrated in the direction of the charge velocity over a range of angles $2/\gamma$, where $\gamma = 1/\sqrt{1 - v^2/c^2}$ (see for example [93]).

Integrating Equation 2.65 for non-relativistic emission over all solid angles gives the total radiated power

$$\begin{aligned} P &= \int P_\Omega d\Omega \\ &= \int_0^\pi P_\Omega 2\pi \sin\theta d\theta \\ &= \frac{q^2}{4\pi\epsilon_0} \frac{1}{4\pi c} \left(\frac{dv/dt}{c} \right)^2 \int_0^\pi 2\pi \sin^3\theta d\theta \\ &= \frac{q^2}{4\pi\epsilon_0} \frac{2}{3c} \left(\frac{dv/dt}{c} \right)^2 . \end{aligned} \tag{2.66}$$

The total power radiated in the frame of reference of the charged particle given by Equation 2.65 is the same as the total power detected in the rest frame, so Equation 2.66 also applies to the total power emitted for charged particles moving at relativistic velocities.

Exercises

2.1 Consider an electromagnetic wave of frequency ω propagating parallel to a magnetic field B_0 in a plasma. Assuming that the refractive indices are close to one, show that the difference in refractive index of two of the solutions from the Appleton–Hartree formula is given by $eB_0/m_0\omega$.

2.2 Carbon dioxide lasers operate at a wavelength of $10.6\,\mu$m. Determine the critical electron density of plasma formed by focusing a high-power carbon dioxide laser onto a target material. [10^{19} cm^{-3}]

2.3 The electron cyclotron frequency ω_c affects electromagnetic wave propagation in a magnetised plasma. Show that an electron of charge e and mass m_0 with any non-zero velocity perpendicular to a magnetic field $\mathbf{B}_0$ rotates around the imposed magnetic field direction with an angular frequency given by

$$\omega_c = \frac{eB_0}{m_0}.$$

2.4 Radio waves directed vertically are found not to reflect back from the Earth's ionosphere if the frequency is greater than 15 MHz. From this information, determine the peak electron density present in the ionosphere. [10^{12} m^{-3}]

2.5 The first observation of a pulsar [45] with radio waves of frequency 81.5 MHz determined a frequency drift in time of −4.9 MHz s^{-1} during each pulse. Estimate the distance of the pulsar from Earth. [2×10^{18} m = 65 parsec]

2.6 Use the Appleton–Hartree formula for the refractive index of a magnetised plasma to show that the refractive index η for a low-frequency whistler wave propagating parallel to a magnetic field line is given by

$$\eta \approx \sqrt{1 + \frac{\omega_p^2}{\omega_c \omega}}$$

where ω_p is the plasma frequency and ω_c is the electron cyclotron frequency.

2.7 Consider a capacitor comprising two large flat plates of area A and plate separation d with a medium of dielectric constant ϵ between the plates. If the area A of the capacitor is large, so that fringing fields around the edges of the plates have negligible effects, consider the work done in charging the capacitor to show that the energy stored in the capacitor is $(1/2)\epsilon E_0^2 Ad$, where E_0 is the electric field between the plates. [Note: This exercise illustrates that the energy stored in an electric field per unit volume is $(1/2)\epsilon E_0^2$.]

2.8 Consider a solenoid with N turns per unit length. Using the integral form of Ampere's law for the magnetic field produced by a current I in the solenoid wire, show that the magnetic field inside the solenoid is given by $B = \mu_0 NI$. The inductance L of the solenoid is defined by the 'back-emf' voltage V produced by a varying current, such that $V = LdI/dt$. Show that the work done when the current in the solenoid is increased from zero to I is $(1/2)LI^2$, and hence that the energy content per unit length of the solenoid is equal to $(1/2)(B^2/\mu_0)A$, where A is the cross-sectional area of the solenoid. [Note: this exercise illustrates that the energy stored in a magnetic field per unit volume is $(1/2)(B^2/\mu_0)$.]

2.9 The classical electron radius r_e is the radius where the electrostatic potential energy of a sphere $e^2/(4\pi\epsilon_0 r_e)$ is equal to the electron rest energy m_0c^2. For an electron accelerating at a rate dv/dt, use Equation 2.66 to show that the total power radiated can be written as

$$P = \frac{2}{3} r_e m_0 c \left(\frac{dv}{dt} / c \right)^2 .$$

2.10 The vector potential **A** considered in the discussion on the acceleration of electrons by a strong electromagnetic wave $\mathbf{E} = \mathbf{E}_0 \exp(i(kz - \omega t))$ is defined by $\mathbf{B} = \nabla \times \mathbf{A}$ and $\mathbf{E} = -\nabla V - d\mathbf{A}/dt$, where V is the scalar potential. Show that these two defining equations for the vector potential imply that

$$\frac{E}{B} = \frac{\omega}{k}$$

with the magnetic field **B** directed perpendicular to **E**.

2.11 Show that the reduced vector potential $a_o = eA_0/(m_0 c)$, where A_0 is the amplitude of the standard electromagnetic vector potential, is dimensionless.

2.12 Consider a radio wave directed horizontally and reflected with a 'turning point' at the peak of the electron density in the ionosphere at a height of $\Delta R = 170\,\mathrm{km}$. Show that the maximum frequency ω_θ that can be reflected is such that

$$\omega_\theta \approx \omega_0 \left(\frac{R}{2\Delta R} \right)^{1/2} = 4.33\,\omega_0$$

where ω_0 is the maximum frequency reflected for normal (vertical) incidence into the ionosphere and $R = 6371\,\mathrm{km}$ is the radius of the Earth. How far can this reflected wave propagate around the Earth's surface with a single ionospheric reflection? [2,900 km]

2.13 Consider an electron in the nth Bohr orbit of the hydrogen atom at radius $a_0 n^2$ from the nucleus, where a_0 is the Bohr radius. Neglecting quantum mechanics, show that the classical rate of radiative power due to the electron acceleration as it orbits the nucleus is given by

$$P = \frac{2\, r_e R_d^2}{3\, c\, m_0\, a_0^2} \frac{1}{n^4}$$

where r_e is the classical electron radius as defined in the previous question and R_d is the Rydberg energy. Allowing quantum mechanics, the $n = 2$ quantum state in hydrogen has a decay rate of $4.699 \times 10^8\,\mathrm{s}^{-1}$. Calculate (i) the quantum mechanical power emission from the $n = 2$ state of a hydrogen atom, and (ii) the classical value neglecting quantum mechanics. [(i) 7.67×10^{-10} W, (ii) 7.25×10^{-10} W]

3

Scattering

Free and bound electrons in a plasma are accelerated by electromagnetic radiation. The interaction with the electrons affects the propagation of the radiation by altering the phase of the oscillating electric and magnetic field of the electromagnetic wave and by absorption of the electromagnetic wave energy as discussed in Chapter 2. As well as affecting a propagating electromagnetic wave, the acceleration of the free and bound electrons in a medium also gives rise to radiation emission: a process referred to as 'scattering'.

As the acceleration of electrons affects the propagation of electromagnetic waves while producing emission of radiation, scattering of light by electrons in a medium can be regarded as determining the optical properties of the medium. Resonances in the responses of free and bound electrons to oscillations from electromagnetic waves tend to have a dominant effect on light propagation. We determined the refractive index arising in plasmas from free electrons and the resonance at the plasma frequency (see Section 2.1). Other resonances associated, for example, with bound electrons also produce refractive index effects.

By determining the refractive index of the medium in which light propagates, scattering processes ultimately govern the reflection and refraction behaviour of light at the junction between materials with different refractive indices. For example, macroscopic particles such as dust in plasmas or water droplets in clouds in the atmosphere reflect light from surfaces (known as Mie scattering). Gradients of refractive index lead to refractive bending of the direction of light propagation.

The fraction of electromagnetic radiation scattered by free electrons is typically a small loss mechanism for radiation of frequency much greater than the plasma frequency, but it is useful for diagnosing conditions in plasmas. For diagnostic measurements of plasmas at optical (ultra-violet to infra-red) frequencies, a laser radiation source is usually employed so that light can be spatially located and the high laser power per unit area ensures that the scattered light is greater than the emission associated with the thermal energy of the plasma. With radio waves and ionospheric scattering, high-power radar systems are employed. In dense plasmas of relevance to inertial fusion, incoherent X-ray sources or free-electron laser sources are used [99].

The scattering cross-section σ_S is determined by the ratio of the average scattered power divided by the average incident power per unit area and can be evaluated for an individual electron, or the cross-section could refer to a grouping of electrons such as from all bound electrons in an atom or ion. The cross-section has dimensions of area and is effectively the area associated with an electron or, for example, an ion where, if incident, an impinging photon is scattered. It is necessary to multiply by the number of scatterers irradiated in order to relate the scattered electromagnetic power P_S to the incident electromagnetic power P_i. We have

$$P_S = N\sigma_S \frac{P_i}{A} \tag{3.1}$$

where N is the number of scattering particles irradiated and A is the cross-sectional area of the beam of radiation. This equation assumes that the scattering 'particles' act independently of each other so that, for example, the electric fields of the scattered light from different scatterers do not add (or cancel). We initially consider an isolated scattering object such as an individual free electron so that we can neglect scattering interactions.

3.1 Scattering by a Free Electron

In the presence of an oscillating electric field $E = E_0 \exp(-i\omega t)$ associated with an electromagnetic wave, an electron experiences an acceleration $-eE/m_0$. Using Equation 2.64, the electric field created by the acceleration of an electron oscillating due to the electromagnetic wave at a distance r from the electron is given by

$$E_\theta = \frac{e}{4\pi\epsilon_0 c^2 r}\left(\frac{eE}{m_0}\right)\sin\theta. \tag{3.2}$$

Using Equation 2.65 the acceleration produces a radiated power P_Ω into an increment of solid angle $d\Omega$ such that

$$P_\Omega d\Omega = \frac{e^2}{4\pi\epsilon_0}\frac{1}{4\pi c}\left(\frac{eE/m_0}{c}\right)^2 \sin^2\theta d\Omega \tag{3.3}$$

where θ is the angle to the incoming electric field direction of the emitted radiation. The classical electron radius r_e (also known as the Compton or Lorentz radius or the Thomson scattering length) is the radius where the electrostatic potential energy of a sphere $e^2/(4\pi\epsilon_0 r_e)$ is equal to the electron rest energy m_0c^2. It is a measure of the 'size' of the electron. We have

$$r_e = \frac{e^2}{4\pi\epsilon_0}\frac{1}{m_0 c^2}. \tag{3.4}$$

Equation 3.3 for the radiated power of an electron accelerated by an electromagnetic wave can be simplified using the classical electron radius. We have

$$P_\Omega d\Omega = r_e^2 c\epsilon_0 E^2 \sin^2\theta d\Omega. \tag{3.5}$$

The instantaneous incident power per unit area (P_i/A) of electromagnetic radiation is given by

$$\frac{P_i}{A} = c\epsilon_0 E^2. \tag{3.6}$$

Consequently, the cross-section $d\sigma_S$ associated with a solid angle $d\Omega$ at angle θ to the electric field of the light is given by

$$d\sigma_S = \frac{P_\Omega d\Omega}{P_i/A} = r_e^2 \sin^2\theta d\Omega$$

and we can write for the differential scattering cross-section of a free electron

$$\frac{d\sigma_S}{d\Omega} = r_e^2 \sin^2\theta. \tag{3.7}$$

The total scattering cross-section of a free electron into any angle is found by integrating over the solid angle. We can write

$$\sigma_S = \int r_e^2 \sin^2\theta d\Omega = 2\pi \int_{\theta=0}^{\pi} r_e^2 \sin^3\theta \, d\theta = \frac{8\pi}{3} r_e^2 \tag{3.8}$$

using $d\Omega = 2\pi \sin\theta d\theta$ for a geometry with a symmetry in θ.

For some later discussion, it is useful to consider the idea of the frequency integrated scattering cross-section. The integration of the total scattering cross-section over all frequencies can be regarded as limited by an upper frequency bound $\omega_{max} \propto c/r_e$ where the wavelength of the light becomes comparable to the classical electron radius. Assuming that the cross-section σ_S is constant up to $\omega_{max} = (3\pi/4)c/r_e$ and then zero at higher frequencies, the total frequency integrated electron scattering cross-section is then given by [93]

$$\int_{\omega=0}^{\omega_{max}} \sigma_S d\omega = 2\pi^2 r_e c. \tag{3.9}$$

Due to quantum electrodynamic effects, the scattering cross-section σ_S actually slowly decreases with increasing energy for energies approaching and greater than the electron rest mass energy $m_0c^2 = 511$ keV (see Figure 3.7), but the integrated value as given in Equation 3.9 is correct.

The classical electron radius r_e is numerically equal to 2.818×10^{-15} m which is much smaller than the dimensions of an atom (for example, the Bohr radius $a_0 = 0.53 \times 10^{-10}$ m) or the wavelength of most electromagnetic radiation (gamma rays of, for example, 1 MeV photon energy have a wavelength of 1.24×10^{-12} m). Using Equation 3.8 and considering the small size of the electron immediately indicates that the scattering of light by isolated electrons is a small effect and that little electromagnetic energy is lost by single electron scatter when light propagates through a plasma. Another feature of single electron scatter is that the scattering cross-sections given by Equations 3.7 and 3.8 do not vary with the frequency of the radiation. The fraction of the power radiated in single electron scatter is independent of the spectral content of the radiation.

Single electron scatter is often referred to as Thomson scatter after the British physicist J. J. Thomson (1856–1940). Thomson scatter can be used as a diagnostic of temperature and density in plasmas (see Section 3.5). At the height of the cold war in 1969, reported high temperatures in a Soviet Union's magnetically confined plasma known as a tokamak were verified by a visiting team of British physicists using Thomson scatter [81]. This Thomson scatter temperature measurement in a tokamak led to the current emphasis on tokamak devices as a way to achieve controlled fusion in the laboratory. The measurement of electron temperatures in plasmas using single electron Thomson scatter is explored further later in this chapter (Section 3.5.2).

3.2 Scattering by Bound Electrons

If an electron is not free, but is bound by the central potential of an ionic or atomic nucleus, the equation of motion of the electron due to the electric field of the light is modified. It is possible to treat the electron motion classically to obtain expressions for the cross-section, though the parameters used in our treatment need to be evaluated using quantum mechanics (see Section 10.3.1 for a quantum mechanical treatment).

A bound electron has an equilibrium distance from the nucleus of an atom or ion which ensures a restoring force on the electron position if perturbed. The bound electron can be regarded as having a resonance behaviour so that if perturbed in position it will oscillate in position at a resonant frequency, say ω_0, with position varying as $x \propto \exp(-i\omega_0 t)$. The acceleration associated with this resonance is obtained from the second derivative of x, that is

$$\frac{d^2x}{dt^2} = -\omega_0^2 x.$$

The energy of an oscillating electron can also be dissipated, resulting in de-acceleration. De-acceleration due to dissipation is similar to friction for macroscopic particles and is proportional to the electron velocity. Dissipative acceleration can be modelled by having an acceleration term $\gamma(dx/dt)$, where γ is a parameter to be determined. The full equation of motion of a bound electron in the electromagnetic field with electric field E is then given by

$$\frac{d^2x}{dt^2} + \gamma\frac{dx}{dt} + \omega_0^2 x = -\frac{e}{m_0}E. \tag{3.10}$$

If the electric field is oscillating in time at frequency ω as $\exp(-i\omega t)$, then x oscillates at the same frequency and we can determine the derivatives of the position x to obtain

$$-\omega^2 x - \gamma i\omega + \omega_0^2 x = -\frac{e}{m_0}E.$$

The position x is given by

$$x = \frac{1}{\omega^2 - \omega_0^2 + i\gamma\omega}\frac{eE}{m_0} \tag{3.11}$$

and the acceleration is given by

$$\frac{d^2x}{dt^2} = \frac{-\omega^2}{\omega^2 + \omega_0^2 + i\gamma\omega}\frac{eE}{m_0}. \tag{3.12}$$

The electric field of the scattered light can be found using Equation 3.2. We have the scattered electric field at distance r and angle θ to the electron oscillation direction

$$E_\theta = \frac{e}{4\pi\epsilon_0 c^2 r}\frac{-\omega^2}{\omega^2 + \omega_0^2 + i\gamma\omega}\left(\frac{eE}{m_0}\right)\sin\theta. \tag{3.13}$$

We can square the modulus of the acceleration to obtain

$$\left|\frac{d^2x}{dt^2}\right|^2 = \frac{\omega^4}{(\omega^2 - \omega_0^2)^2 + (\gamma\omega)^2}\left(\frac{eE}{m_0}\right)^2.$$

Substituting in Equation 3.3 and following our treatment in Section 3.1, the differential cross-section becomes

$$\frac{d\sigma_S}{d\Omega} = r_e^2\frac{\omega^4}{(\omega^2 - \omega_0^2)^2 + (\gamma\omega)^2}\sin^2\theta \tag{3.14}$$

and the total scattering cross-section for a bound electron becomes

$$\sigma_S = \frac{8\pi}{3}r_e^2\frac{\omega^4}{(\omega^2 - \omega_0^2)^2 + (\gamma\omega)^2}. \tag{3.15}$$

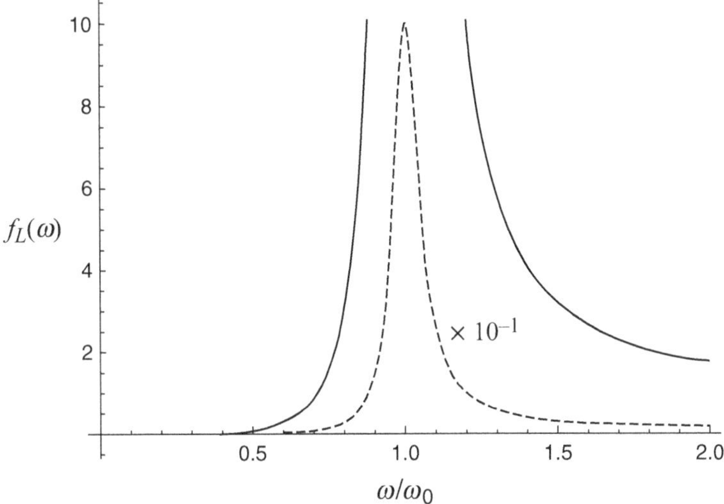

Figure 3.1 The variation of the bound electron scattering cross-section relative to the free electron scattering cross-section. Frequencies ω are in terms of a resonance frequency ω_0 and the damping parameter $\gamma/\omega_0 = 0.1$. The lower curve is the cross-section value $\times 10^{-1}$.

The variation of the bound electron cross-section to the free electron cross-section can be represented by a function $f_L(\omega)$ such that

$$f_L(\omega) = \frac{\omega^4}{(\omega^2 - \omega_0^2)^2 + (\gamma\omega)^2}. \tag{3.16}$$

The frequency dependence of bound scatter represented by $f_L(\omega)$ is shown in Figure 3.1. At high frequencies $\omega \gg \omega_0$, the function $f_L(\omega)$ approaches one and we obtain the Thomson scatter cross-section for a free electron. At low frequencies $\omega \ll \omega_0$, the variation of the bound electron frequency response is such that

$$f_L(\omega) = \left(\frac{\omega}{\omega_0}\right)^4. \tag{3.17}$$

The scattering cross-section has a rapid increase with increasing frequency providing ω is well below a resonance frequency. The British Nobel winner Lord Rayleigh (1842–1919) first used this scattering dependence to explain the blue colour of the sky. Visible light has a photon energy range from approximately $\hbar\omega = 1.65$–$3.26\,\text{eV}$, while the bound electrons of oxygen and nitrogen have resonances at $8.6\,\text{eV}$ and $8.2\,\text{eV}$ respectively. High-frequency (blue) solar radiation around 2.5–$2.75\,\text{eV}$ is preferentially scattered in comparison to the other, lower-frequency components of the solar radiation. The solar radiation at the top of the atmosphere approximates black-body radiation (see Section 4.1) corresponding to the temperature of the Sun's photosphere with a peak emission at approximately 2.2 eV. Higher-frequency violet light (2.75–3.26 eV) from the Sun is scattered even more, but has less flux than blue to red light and, in addition, the visual acuity of the human eye drops rapidly for violet light. We perceive the colour of the sky to be blue. Ultra-violet light is scattered again with an even higher scattering cross-section across the sky, but is outside the human eye response. However, the large level of ultra-violet scattering does mean that it is possible to be over-exposed to ultra-violet light and suffer sunburn even if clouds cover the direct solar flux.

At frequencies close to a resonance $\omega \approx \omega_0$, the frequency dependence of the scattering from bound electrons simplifies to a lineshape known as a Lorentzian lineshape. We can re-write Equation 3.16 as

$$f_L(\omega) = \frac{\omega^2/\gamma^2}{(\omega - \omega_0)^2((1 + \omega_0/\omega)/\gamma)^2 + 1} \approx \frac{\omega^2/\gamma^2}{4(\omega - \omega_0)^2/\gamma^2 + 1}. \tag{3.18}$$

The lineshape has a frequency full width at half maximum of γ.

3.3 Scattering by a Multi-Electron Atom

It is possible to extend the treatment for a single bound electron to take account of scattering of several electrons bound in an ion or atom. We need to add up the electric fields scattered by each electron. If the electric fields of each scattering electron add coherently, the scattering increases rapidly above the rate for the same number of single electron scatterers – increasing at a rate proportional to the square of the number of electrons scattering coherently together. The position $\mathbf{r}'_s$ of each electron relative to the nucleus of the atom is important as we need to keep account of the phase of the scattered light. The total electric field is a variation of Equation 3.13. For an observer at a distant position $\mathbf{r}$ relative to the nucleus of the atom (the scattering centre), the electric field is

$$E(\mathbf{r}, t) = \frac{-e^2}{4\pi\epsilon_0 m_0 c^2} \sum_{s=1}^{Z} \left[\frac{\omega^2 E_i \sin\theta}{\omega^2 - \omega_s^2 + i\gamma_s \omega} \frac{1}{r} \exp\left(-i(\omega t' + \mathbf{k} \cdot \mathbf{r}'_s - \mathbf{k}_i \cdot \mathbf{r}'_s)\right) \right] \tag{3.19}$$

where the incident light is assumed to have wavevector $\mathbf{k}_i$ and electric field strength E_i, and the scattered light is represented by wavevector $\mathbf{k}$. The summation is over all electrons s in the atom, with ω_s representing the resonance frequency and γ_s the dissipative constant (which we saw is equal to the spectral linewidth). The terms in the exponential take account of the phase of light reaching the scattering electron relative to the phase at the scattering centre (the $\mathbf{k}_i \cdot \mathbf{r}'_s$ term) and the phase difference of the scattered light projected to the scattering centre (the $-\mathbf{k} \cdot \mathbf{r}'_s$ term). Time is

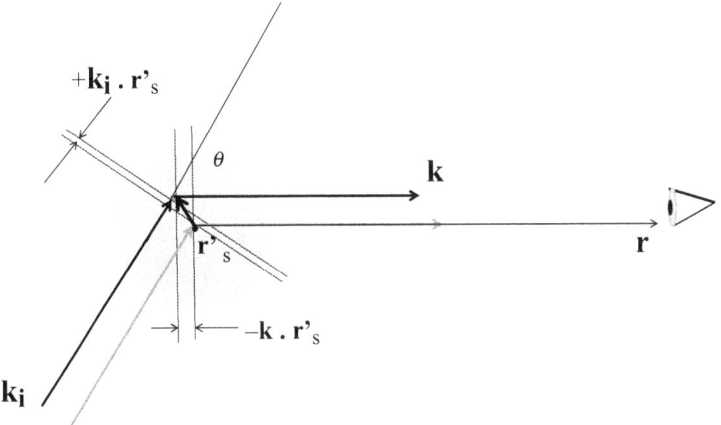

Figure 3.2 A schematic diagram showing the scattering of light from an atom or ion with incident wavevector $\mathbf{k}_i$ and scattered wavevector $\mathbf{k}$. The phase differences in scattering from an electron at position $\mathbf{r}'_s$ relative to notional scattering at the scattering centre (the nucleus) is shown. It is assumed that the observer is a long-distance $\mathbf{r}$ from the atom.

written as retarded time $t' = t - r/c$ to reflect the phase changes between the scattering centre (the nucleus) and the observer (see Figure 3.2).

Using the classical electron radius, the electric field can be written

$$E(\mathbf{r},t) = -r_e \sum_{s=1}^{Z} \left[\frac{\omega^2 E_i \sin\theta}{\omega^2 - \omega_s^2 + i\gamma_s\omega} \frac{1}{r} \exp(-i(\omega(t - r/c) + \Delta\mathbf{k}\cdot\mathbf{r}'_s) \right] \quad (3.20)$$

where $\Delta\mathbf{k} = \mathbf{k} - \mathbf{k}_i$. The scattered electric field is often written in terms of a complex atomic scattering factor $f(\Delta\mathbf{k},\omega)$ with

$$E(\mathbf{r},t) = -\frac{r_e}{r} f(\Delta\mathbf{k},\omega)\, E_i\, \sin\theta\, e^{-i\omega(t-r/c)} \quad (3.21)$$

and

$$f(\Delta\mathbf{k},\omega) = \sum_{s=1}^{Z} \left[\frac{\omega^2 e^{-i\Delta\mathbf{k}\cdot\mathbf{r}'_s}}{\omega^2 - \omega_s^2 + i\gamma_s\omega} \right]. \quad (3.22)$$

The atomic scattering factor represents the scattered electric field amplitude relative to that scattered by a single free electron.

The differential scattering cross-section and the total scattering cross-section are similarly found relative to the free-electron scatter values:

$$\frac{d\sigma_S}{d\Omega} = r_e^2\, |f(\Delta\mathbf{k},\omega)|^2 \sin^2\theta$$

$$\sigma_S = \frac{8\pi}{3}\, |f(\Delta\mathbf{k},\omega)|^2\, r_e^2 \quad (3.23)$$

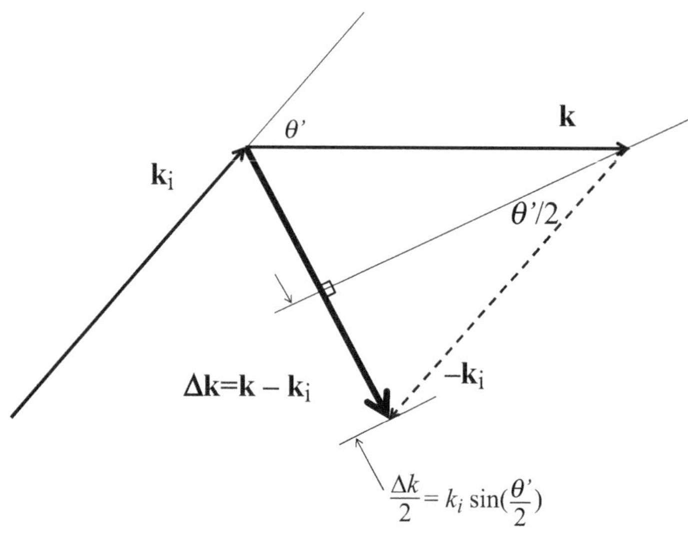

Figure 3.3 The geometry of incident $\mathbf{k}_i$ and scattered $\mathbf{k}$ wavevectors for light scattering at angle θ' to the original direction of light. The geometry illustrates that $\Delta k = |\mathbf{k} - \mathbf{k}_i| = 2k_i \sin(\theta'/2)$.

where θ is the angle of scatter to the electric field direction. As $|\mathbf{k}| = |\mathbf{k}_i|$ the amplitude of the wavevector difference $\Delta k = |\Delta\mathbf{k}| = |\mathbf{k} - \mathbf{k}_i|$ is related to the angle of scattering $\theta' = \pi/2 - \theta$ by simple geometry with

$$\Delta k = 2k_i \sin\left(\frac{\theta'}{2}\right) \tag{3.24}$$

where θ' is the angle of scattering relative to the incident beam (see Figure 3.3).

The solution of Equation 3.22 can be simplified if the phase term $\Delta\mathbf{k} \cdot \mathbf{r}'_s$ is small, so that $e^{i\Delta\mathbf{k}\cdot\mathbf{r}'_s} \approx 1$. This occurs if either the wavevectors are small (the long wavelength limit) or the scattering angle θ' is small (the scattering is forward). The atomic scattering factor then becomes

$$f(\omega) = \sum_{s=1}^{Z} \left[\frac{\omega^2}{\omega^2 - \omega_s^2 + i\gamma_s\omega} \right]. \tag{3.25}$$

If there are several electrons with similar or identical resonance frequencies ω_s, we can group these together and sum up electron groups with different resonance frequencies ω_s. We use the symbol f_s to represent the number of electrons with resonance s and write

$$f(\omega) = \sum_{s=1} \left[\frac{f_s\,\omega^2}{\omega^2 - \omega_s^2 + i\gamma_s\omega} \right]. \tag{3.26}$$

Summing up all the electrons, we must finish with the total number of bound electrons Z in the ion or atom:

$$\sum_s f_s = Z. \tag{3.27}$$

Integrating Equation 3.23 over all frequencies with Equation 3.26 as the atomic scattering factor leads to

$$\int \sigma_S d\omega = 2\pi^2 r_e c f_s \tag{3.28}$$

upon using the integration (Equation 3.9) of the free-electron Thomson scatter cross-section over all frequencies up to a maximum frequency where the wavelength of the light starts to become less than the classical cross-section r_e. Even though the atomic scattering factor has a peak at the resonance (see Figure 3.1) and falls to a lower value below the resonance, the integration over all frequencies is dominated by the higher-frequency constant Thomson scatter value (see Exercises 3.3 and 3.4 at the end of this chapter). Frequency-integrated cross-sections for the absorption of light can be related using Equation 3.28 to the quantity f_s.

The discussion so far has implicitly assumed that each electron has a single resonance frequency; each electron is said to have an 'oscillator strength' f_s of one. Quantum mechanics and the simple Bohr model (discussed in Section 1.5) predict that several resonances or quantum jumps to other quantum states are possible. To allow for this, the concept of an oscillator strength has been carried over into the quantum mechanical calculation of atomic scattering factors (and other quantities such the probability of radiative transitions; see Section 10.1). To allow an electron to finish in different states in the interaction with incoming radiation, quantum mechanical calculations have extended the simple counting of electrons to allow for fractional oscillator strengths representing the relative strength of the particular 'resonance oscillation' (quantum change) that a bound electron can undertake. Interestingly, the highest oscillator strength quantum change for an electron in the ground state of an atom or ion is still commonly known by spectroscopists as the 'resonance transition' (though other resonance transitions with smaller oscillator strengths are possible).

With fractional oscillator strengths, Equation 3.27 still applies: that is the sum of oscillator strengths is equal to the number of electrons. This sum is useful as a check on calculations and is known as the Thomas–Reiche–Kuhn sum rule. General quantum mechanical proofs (see Appendix A.4) and explicit calculations for particular systems confirm the generality of the sum rule, which interestingly was first proposed as outlined here before the modern understanding of quantum mechanics.

3.4 Refractive Index Values

The propagation of electromagnetic waves in a plasma has been treated in Section 2.1, where it was shown that the wave equation involves an electron current density **J** term. We have

$$\nabla^2 \mathbf{E} = \mu_0 \frac{\partial \mathbf{J}}{\partial t} + \epsilon_0 \mu_0 \frac{\partial^2 \mathbf{E}}{\partial t^2}. \tag{3.29}$$

The wave equation is conveniently written in terms of a refractive index n such that

$$\nabla^2 \mathbf{E} = \frac{n^2}{c^2} \frac{\partial^2 \mathbf{E}}{\partial t^2}. \tag{3.30}$$

Equation 2.12 gives the refractive index for a plasma in the absence of a magnetic field and Equation 2.32 is a general formula (the Appleton–Hartree equation) for the more complicated refractive index of a plasma in the presence of an imposed magnetic field. Taking account of bound electrons, we find that the refractive index can vary from the free-electron values due to the presence of ions, atoms or molecules. The following treatment also allows for the calculation of refractive indices due to atoms in media generally.

The current term in Equation 3.29 needs to be evaluated. For a bound electron with a resonance at frequency ω_s and a dissipation term γ_s, the position of the electron varies as given by Equation 3.11. The velocity v of the electron in the direction of the electric field E from an electromagnetic wave is given by

$$v = \frac{e}{m_0} \frac{1}{\omega^2 - \omega_s^2 + i\gamma_s \omega} \frac{\partial E}{\partial t} \tag{3.31}$$

and the current density by

$$J = -e n_i v = -\frac{n_i e^2}{m_0} \sum_s \frac{f_s}{\omega^2 - \omega_s^2 + i\gamma_s \omega} \frac{\partial E}{\partial t} \tag{3.32}$$

upon summing up the contributions from electrons with identical or similar resonances and multiplying by the oscillator strength f_s of each resonance grouping.

Differentiating the current density and substituting into Equation 3.29, we have

$$\nabla^2 \mathbf{E} = \frac{1}{c^2} \left(1 - \frac{n_i e^2}{m_0 \epsilon_0} \sum_s \frac{f_s}{\omega^2 - \omega_s^2 + i\gamma_s \omega} \right)^{1/2} \frac{\partial^2 \mathbf{E}}{\partial t^2}. \tag{3.33}$$

Comparing this equation to Equation 3.30, the refractive index due to the bound electrons is given by

$$n = \left(1 - \frac{n_i e^2}{m_0 \epsilon_0} \sum_s \frac{f_s}{\omega^2 - \omega_s^2 + i\gamma_s \omega} \right)^{1/2} \approx 1 - \frac{1}{2} \frac{e^2 n_i}{m_0 \epsilon_0} \sum_s \frac{f_s}{\omega^2 - \omega_s^2 + i\gamma_s \omega} \tag{3.34}$$

for small deviations of refractive index from unity. For forward scattering, we have an atomic scattering factor given by Equation 3.26, namely

$$f(\omega) = \sum_{s=1} \left[\frac{f_s\, \omega^2}{\omega^2 - \omega_s^2 + i\gamma_s \omega} \right]$$

which can be used in the expression for the refractive index. We can write

$$n = 1 - \frac{1}{2} \frac{n_i e^2}{m_0 \epsilon_0 \omega^2} f(\omega). \tag{3.35}$$

This equation for the refractive index is similar to the expression for the refractive index for free electrons (Equation 2.12). The refractive index due to atoms or ions is equal to the free-electron refractive index after replacing the electron density by the number density of ions n_i and multiplying by the appropriate atomic scattering factor $f(\omega)$ for forward scattering for the atom or ion.

The forward-scattering atomic scattering factor $f(\omega)$ has real and imaginary components consistent with the real component affecting the phase of propagating light and the imaginary components representing absorption of the light (see Section 2.3). Using the forward-scattering atomic-scattering factor is justified as we are dealing with light propagation in the forward direction when considering phase and absorption effects.

For solid unionised material, atomic scattering factors and refractive index values are tabulated [4]. For plasmas, it usually is necessary to calculate these values using Equation 3.26 from tablulated values of the oscillator strength f_s and the damping dissipation term γ_s. If radiative processes dominate the damping of an electron, the value of γ_s is equal to the sum of the radiative transition probabilities from a quantum state (see Section 10.1). As $n_i < n_e$ and $|f(\omega)| < 1$ in the radio to infra-red, visible and ultra-violet spectral ranges, the refractive index of a plasma is dominated by the free electrons with only a small contribution from bound electrons. In the X-ray spectral range, the high value of the frequency in Equation 3.35 ensures that refractive indices n are close to unity for the contributions from both free and bound electrons. The extreme ultra-violet spectral range (photon energy 10–200 eV), however, can exhibit refractive index effects due to bound electrons in plasmas [4].

3.4.1 The Kramers–Kronig Relation

We can expand Equation 3.35 in terms of the real and imaginary components of the scattering factor such that

$$n = 1 - \frac{1}{2} \frac{n_i e^2}{m_0 \epsilon_0 \omega^2} [f_1(\omega) - if_2(\omega)] \tag{3.36}$$

where $f_1(\omega)$ and $f_2(\omega)$ are both real. The real η and imaginary κ components of the refractive index $n = \eta + i\kappa$ were shown in Section 2.3 to affect the phase of propagating light and the irradiance of the light respectively. We have for the absorption coefficient K of an electromagnetic wave that

$$K = \frac{2\omega}{c}\kappa = \frac{n_i e^2}{m_0 c \epsilon_0 \omega^2} f_2(\omega). \tag{3.37}$$

The real and imaginary refractive indices for electromagnetic waves are related by the Kramers–Kronig relations. One of the Kramers–Kronig relations relates the imaginary component of the scattering factor to the real component by

$$f_1(\omega) = Z - \frac{2}{\pi} P \int_0^\infty \frac{u f_2(u)}{u^2 - \omega^2} du \tag{3.38}$$

where Z is the number of electrons per atom. The other Kramers–Kronig relation for the imaginary component in terms of the real component is given by

$$f_2(\omega) = \frac{2\omega}{\pi} P \int_0^\infty \frac{f_1(u) - Z}{u^2 - \omega^2} du. \tag{3.39}$$

In both of these equations the symbol P refers to the principal value of the integral, which means taking appropriate limits at the discontinuity where $u = \omega$. A proof of the Kramers–Kronig relations is given in many texts (for example, see Attwood, p. 91 [4]).

Considering free electrons, we see from Equation 3.36 that n_i becomes the electron density n_e and the real scattering factor $f_1(\omega) = 1$. Integrating Equation 3.39 with $Z = 1$ gives a zero value of $f_2(\omega)$. A more interesting example of the effect of an absorption resonance (the imaginary component of the refractive index) on the real component of the refractive index can be simply assessed by considering a narrow resonance in the electron-scattering cross-section. A moderately broad resonance-scattering cross-section profile $f_L(\omega)$ was plotted in Figure 3.1. If we consider an infinitely narrow scatter profile as represented by the Dirac-delta function $\delta(\omega - \omega_0)$, the integration of Equation 3.38 becomes trivial. A Dirac-delta function $\delta(\omega - \omega_0)$ has a value of zero everywhere except at frequency $\omega = \omega_0$, while integrating over all frequencies gives a value of one. Using Equation 3.37 and noting that the total electron density $n_e = Zn_i$, we can write for the imaginary scattering factor $f_2(\omega)$:

$$f_2(\omega) = K_0 c \frac{Z\omega^2}{\omega_p^2} \delta(\omega - \omega_0)$$

where K_0 is the peak of the absorption coefficient. To simplify this expression, we use the value of the plasma frequency ω_p given by

$$\omega_p^2 = \frac{n_e e^2}{m_0 \epsilon_0}.$$

Substituting the above $f_2(u)$ value in Equation 3.38 gives

$$f_1(\omega) = Z - K_0 c \frac{Z\omega^2}{\omega_p^2} \frac{2}{\pi} \frac{\omega_0}{\omega_0^2 - \omega^2}.$$

The real component of the refractive index is obtained from Equation 3.36. We have

$$\eta = 1 - \frac{1}{2} \frac{\omega_p^2}{\omega^2} \left[1 - K_0 c \frac{\omega^2}{\omega_p^2} \frac{2}{\pi} \frac{\omega_0}{\omega_0^2 - \omega^2} \right].$$

The refractive index is made up of the free-electron refractive index $n = 1 - (1/2)$ ω_p^2/ω^2 (strictly, the refractive index valid for higher frequencies $\omega \gg \omega_p$) and an additional component representing the refractive index values associated with the resonance at frequency ω_0. We can write for the additional component of refractive index $\Delta\eta$ due to the resonance:

$$\Delta\eta = K_0 c \frac{1}{\pi} \frac{\omega_0}{\omega_0^2 - \omega^2} = K_0 \lambda_0 \frac{1}{2\pi^2} \frac{1}{1 - (\omega/\omega_0)^2} \tag{3.40}$$

where λ_0 is the wavelength of the resonance at frequency ω_0. The values of the refractive index associated with a resonance or absorption line with peak absorption coefficient K_0 for an infinitely thin resonance is given here and plotted in Figure 3.4. The discontinuity at $\omega = \omega_0$ disappears if absorption features of finite width are considered. Equation 3.40 is accurate for narrow absorption features where ω is significantly greater than or less than ω_0.

Plots such as Figure 3.4 have relevance for all media in which light propagates. For example, glass absorbs light strongly for photon energies greater than 3.9 eV (wavelengths less than 320 nm), so at lower photon energies in the visible, the refractive index of glass is greater than one (typically $\eta \approx 1.5$). Most plasma densities are below that found in solids and so have corresponding smaller values of absorption K_0 and smaller refractive index $\Delta\eta$ effects, but absorption features do affect the values of refractive indices close to resonances.

3.4.2 Refraction

We have seen in various earlier treatments of the propagation of light (for example, see Equations 2.42 and 2.5) that the electric field of an electromagnetic wave propagating in the z-direction is given by

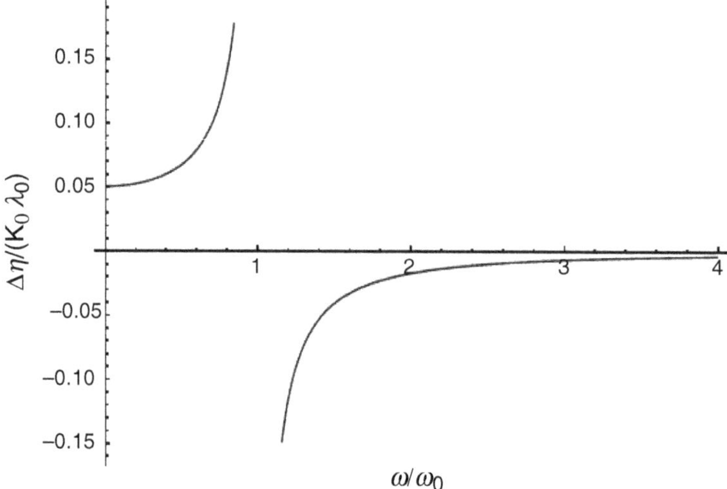

Figure 3.4 The value of the real component of the refractive index $\Delta\eta$ (in units of $K_0\lambda_0$) for a narrow absorption resonance at frequency ω_0 (corresponding to wavelength λ_0) as a function of relative frequency ω/ω_0. The resonance at frequency ω_0 is approximated to be a Dirac-delta function. The absorption coefficient at ω_0 is K_0.

$$\mathbf{E} = \mathbf{E}_0 \exp[i(kz - wt)] = \mathbf{E}_0 \exp[i(k_0\eta z - wt)]$$

where k_0 is the wavevector in vacuum. The phase variation along the propagation path is given by $kz = k_0\eta z$. Light can be regarded as propagating as rays with a phase varying as $k_0\eta z$, provided the wavelength of the radiation is much smaller than the scalelength of any changes in refractive index. The real component of the refractive index η determines the difference in phase from the value of the light propagated in vacuum.

The direction of a ray is perpendicular to surfaces of constant phase, so ray directions can be determined from planes of constant phase. At a sharp junction between two regions of refractive index (η_1 and η_2), two parallel rays incident at an incidence angle θ_1, initially in phase and separated along the junction by a distance δx change direction to an angle of refraction θ_2 (see Figure 3.5(A)). The wavefronts of the two rays stay in phase after the junction if the optical distances $\int k_0\eta dz$ travelled by the rays are equal, which requires that

$$\eta_1\, \delta x\, \sin\theta_1 = \eta_2\, \delta x\, \sin\theta_2.$$

We immediately obtain Snell's law for the angle of refraction θ_2 of a ray in passing from refractive index η_1 to η_2 at angle of incidences θ_1:

$$\eta_1 \sin\theta_1 = \eta_2 \sin\theta_2. \tag{3.41}$$

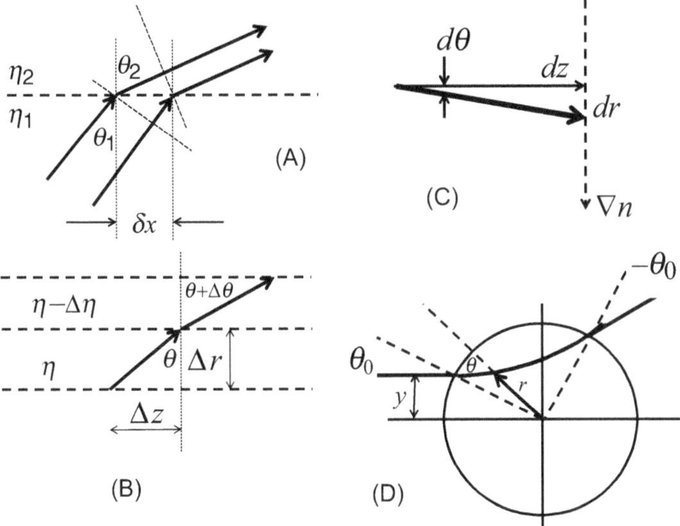

Figure 3.5 Ray paths illustrating refraction effects. (A) The refraction of light at a junction between two refractive indices η_1 and η_2. The angle of refraction θ_2 is determined by the requirement that the optical pathlength of the two rays is identical. (B) The change of angle of incidence $\Delta\theta$ due to a refractive index difference of $\Delta\eta$ between two layers. (C) The change in angle $d\theta$ for a ray incident perpendicular to a density gradient directed in the r direction. (D) Refraction where the refractive index varies with radius r in a cylindrical or spherical geometry. Here θ is the angle of the ray to the radius r for a ray initially incident at distance y from the centre of the cylinder or sphere.

Plasmas have continuously varying values of refractive indices η, which cause rays of light separated spatially to experience different phase velocities. The wavefront of points of constant phase can alter, which means the direction of propagation of a ray (perpendicular to the phase front) also changes – the process known as refraction. We examined in Section 2.4 how light incident obliquely at angle θ_0 into a linear density gradient plasma changes direction, reaches a 'turning point' and ultimately propagates at an angle $-\theta_0$ to the direction of the density gradient (see Figure 2.4). The change of direction of such a light ray is determined by refraction.

An equation for the refraction of light in a continuously varying refractive index can be obtained by considering junctions between a refractive index of η and a slightly smaller refractive index $\eta - \Delta\eta$ (see Figure 3.5(B)). Snell's law for the change in angle of incidence from θ to $\theta + \Delta\theta$ at each junction requires

$$\eta \sin\theta = (\eta - \Delta\eta)\sin(\theta + \Delta\theta). \tag{3.42}$$

Expanding $\sin(\theta + \Delta\theta)$ for small $\Delta\theta$, we have

$$\eta \sin\theta = (\eta - \Delta\eta)(\sin\theta + \Delta\theta\cos\theta).$$

Cancelling $\eta \sin\theta$ from both sides and dividing throughout by $\cos\theta$, we can write

$$\Delta\eta(\tan\theta + \Delta\theta) \approx \Delta\eta \tan\theta = \eta\Delta\theta$$

as $\Delta\theta \ll \tan\theta$. Assuming refractive index layers of thickness Δr, we set

$$\Delta\eta = \frac{\partial\eta}{\partial r}\Delta r$$

where the derivative is the gradient of the refractive index in the r direction perpendicular to the thin layers of changing refractive index. We then have

$$\frac{1}{\eta}\frac{\partial\eta}{\partial r} = \frac{1}{\tan\theta}\frac{\Delta\theta}{\Delta r}.$$

The value of $\tan\theta$ within each of the assumed layers is given by $\tan\theta = \Delta z/\Delta r$ (see Figure 3.5B) so that

$$\frac{1}{\eta}\frac{\partial\eta}{\partial r} = \frac{\Delta\theta}{\Delta z} = \frac{d\theta}{dz} \tag{3.43}$$

taking the limit of infinitely thin layers. Writing $d\theta/dz = d^2r/dz^2$ gives an often-used equation for the propagation of a ray in a medium such as a plasma with a continuously varying refractive index:

$$\frac{1}{\eta}\frac{\partial\eta}{\partial r} = \frac{d^2r}{dz^2}. \tag{3.44}$$

The refractive index of a plasma is usually dominated by the free electron contribution so that

$$\eta = \sqrt{1 - \frac{n_e}{n_{crit}}} \approx 1 - \frac{n_e}{2n_{crit}}$$

where n_e is the electron density and n_{crit} is the critical density. The change in angle θ of the ray due to refraction is then determined by the gradient of the electron density in the direction $\mathbf{r}$. Using Equation 3.43 we can write

$$\frac{d\theta}{dz} \approx -\frac{1}{2n_{crit}}\frac{dn_e}{dr}. \tag{3.45}$$

Fermat's principle states that, compared to nearby paths, light travels between two points along the path that requires the least time. The ray path is determined by the minimum phase difference $\phi_P = k_0 \int \eta dz$ along the paths, where k_0 is the vacuum wavenumber. To deduce the light path requiring minimum time, we find the minimum of a quantity known as the light path integral of the refractive index over ray paths. The light path integral L is an integration of the refractive index η of a medium along a path joining two points, say, z_0 and z. We have

$$L = \int_{z_0}^{z} \eta dz. \tag{3.46}$$

For a continuous refractive index variation as found in a plasma, the Euler equation from the calculus of variations gives the minimum value of L and an equation for the position vector $\mathbf{r}$ of the points on the ray producing the minimum L. For a refractive index of gradient $\nabla\eta$ we have (see [70])

$$\frac{d}{dz}\left(\eta \frac{d\mathbf{r}}{dz}\right) = \nabla\eta. \tag{3.47}$$

This expression extends Equation 3.44 to allow for changing refractive index values along the z-direction. Equation 3.47 can be applied to different geometries. For example, a ray incident perpendicular to a refractive index gradient can have a position vector $\mathbf{r}$ directed parallel to the gradient with a refraction angle θ defined by $\theta = dr/dz$ (see Figure 3.5C). As θ is the angle of refraction, initially $\theta_0 = 0$, so provided $\theta \approx 0$ for the ray path, Equation 3.47 becomes

$$\eta \frac{d\theta}{dz} = |\nabla\eta|$$

in agreement with Equation 3.43.

With a spherical or cylindrical symmetry of the refractive index, so that the refractive index $\eta(r)$ varies with radius r from an origin $r = 0$ to $r = R$ (see Figure 3.5D), the angle θ of the ray relative to r can be evaluated from Equation 3.47. We have with this geometry [70] that

$$\theta = \theta_0 + 2\int_{r_c}^{R} \frac{ydr}{r\sqrt{r^2\eta^2(r) - y^2}} \tag{3.48}$$

where $r_c = y/\eta(r_c)$ is the actual radius of closest approach to the origin and $y = R\sin\theta_0$ is the radius of closest approach of the ray if there were no refraction (with θ_0 representing the initial angle of the ray to the outer radius R).

3.4.3 Interferometry

Interferometry is a family of techniques for determining the phase of an electromagnetic wave by superimposing electromagnetic waves. If the phase of two waves is such that both electric fields are in the same direction, the electric fields add and there is a peak of intensity, while if the electric fields at a position are opposite, the electric fields cancel and there is a trough or dip in light intensity. For measurement purposes, there needs to be a near-constant phase relationship over time and space between the two beams, usually requiring the use of a laser.

Consider two beams with electric field E_1 and $E_2 \exp(i\phi_p)$ in the same plane such that the second beam has a phase difference ϕ_p to the other beam. The electric fields add to produce a total electric field of $E_1 + E_2 e^{i\phi_p}$ and a total intensity using Equation 2.15 given by

$$I = \frac{1}{2}\epsilon_0 c\eta |E_1 + E_2 e^{i\phi}|^2 = \frac{1}{2}\epsilon_0 c\eta \left(|E_1|^2 + |E_2|^2 + 2|E_1|\,|E_2| \cos\phi_p\right). \quad (3.49)$$

If the two beams are equal in magnitude ($|E_1| = |E_2|$), the intensity of light will vary between zero and four times the intensity of each beam separately, varying proportionally to $1 + \cos\phi_p$. The intensity of the interfering beams gives a measure of the phase difference ϕ_P between the beams. A typical interferometer involves measuring the intensity of interfering beams over a time or spatial range, so that changes of phase are measured. Interference intensity peaks (known as fringes) as a function of time or space are measured.

In the interferometry of plasmas, one beam usually passes through the plasma while the other propagates in vacuum. The relative difference in phase ϕ_p between the beams is then given by an integration along the ray path of the beam going through the plasma. We have

$$\phi_p = \int (k_0 - k) dz = k_0 \int (1 - \eta)\, dz \quad (3.50)$$

where k_0 is the vacuum wavevector, k is the wavevector in the plasma and η is the plasma refractive index. For a plasma without an embedded magnetic field, the refractive index is given by Equation 2.12 and in the limit where the electron density n_e is much less than the critical electron density n_{crit}, the phase given by Equation 3.50 reduces to being proportional to an integration of the electron density along the ray path through the plasma:

$$\phi_p \approx \frac{k_0}{2n_{crit}} \int n_e\, dz. \quad (3.51)$$

We see that interferometric measurement of the phase of an electromagnetic wave probing through a plasma can give a direct measure of the electron density.

3.5 Coherent and Incoherent Thomson Scattering by Free Electrons

The discussion of Thomson scatter by free electrons in Section 3.1 considered the electrons to be isolated so that interference of the electric fields associated with the scatter process from separate electrons does not occur. We examined the importance of scatter by electrons in close proximity to each other in the discussion of scatter from bound electrons in an atom or ion (Section 3.2). Overlap of scattered electric fields from free electrons can also occur with higher-density plasmas. In a plasma, if the electrons within a Debye length λ_D move coherently together in response to the electric field of the radiation, then the scattered electric fields from different electrons can have the same approximate phase. The electric field strength increases proportionally to the number N electrons moving in phase so that the intensity of scattered light increases proportional to N^2. In a plasma N can represent the number N_D of electrons in the Debye sphere. We have (see Equation 1.5)

$$N_D = \left(\frac{4}{3}\right)\pi\lambda_D^3 n_e = \left(\frac{4}{3}\right)\pi\left(\frac{\epsilon_0 k_B T}{n_e e^2}\right)^{3/2} n_e.$$

The criterion for coherent or incoherent (single electron) Thomson scatter is determined by the scattering parameter

$$\alpha = \frac{1}{\Delta k\,\lambda_D} = \frac{1}{2k\lambda_D}\frac{1}{\sin(\theta'/2)} = \frac{1}{2k}\left(\frac{\epsilon_0 k_B T}{n_e e^2}\right)^{1/2}\frac{1}{\sin(\theta'/2)} \tag{3.52}$$

where $\Delta\mathbf{k} = \mathbf{k} - \mathbf{k}_i$ for $\mathbf{k}$ the wavenumber of the scattered light and $\mathbf{k}_i$ the wavenumber of the incident light (see Figure 3.3). The angle θ' is the angle of scatter with respect to the direction of the initial light beam $\mathbf{k}_i$. With $\alpha \ll 1$, Thomson scatter is not collective: each electron is separated sufficiently from other electrons that the scattered electric fields do not overlap. The scattering is referred to as being 'incoherent'. For $\alpha \gg 1$, scattering is collective and the phase of scattered light from each electron needs to be considered. With many electrons responding near the same phase point of the incoming oscillating electric field, the scattering with $\alpha \gg 1$ is known as coherent scatter.

3.5.1 Coherent Thomson Scatter in Plasmas

With coherent scatter, the cross-sections can be written in a similar way as considered for bound electrons in an atom or ion (see Equation 3.23) except that we replace the atomic scattering factor $f(\Delta\mathbf{k}, \omega)$ by a spectral density function $S(\Delta\mathbf{k})$. We have for the differential and total cross-section

$$\frac{d\sigma_S}{d\Omega} = r_e^2\,|S(\Delta\mathbf{k})|^2 \sin^2\theta$$

$$\sigma_S = \frac{8\pi}{3}\,|S(\Delta\mathbf{k})|^2\, r_e^2 \tag{3.53}$$

where r_e is the classical electron radius and θ is the angle of scattering relative to the electric field of the incoming radiation. Far from any resonances, the spectral density function $S(\Delta\mathbf{k})$ is obtained from the atomic scattering factor $f(\Delta\mathbf{k}, \omega)$ by dropping all the frequency-dependent terms as the numerator and denominator terms cancel (see Equation 3.22). We can replace the sum of electron positions by an integration over the spatial variation of the density of electrons $\rho(\mathbf{r}'_s)$. When the scattering parameter α is large, the spectral density function can be written as

$$S(\Delta\mathbf{k}) = \int \rho(\mathbf{r}'_s)e^{i\Delta\mathbf{k}\cdot\mathbf{r}'_s}d\mathbf{r}'_s \tag{3.54}$$

where the integration is over the volume of a Debye sphere. Equation 3.54 has the form of a Fourier transform of the electron density variation in space.

When the scattering parameter α is large, light is scattered from electrons associated with longitudinal waves of electrons or ions which can both produce periodic electron density $\rho(\mathbf{r}'_s)$ structures as electrons move with the ions. In fact the change $\Delta\mathbf{k} = \mathbf{k} - \mathbf{k}_i$ in the wavevectors of the incident $\mathbf{k}_i$ and scattered $\mathbf{k}$ light produces a plasma wave with wavevector $-\Delta\mathbf{k}$ due to the need for momentum conservation. An electron density wave of form $\rho(r) \propto \exp(i\Delta kr)$ has a Fourier transform made up of delta functions at wavenumber $\pm\Delta k$, so the spectral density function for a single wavenumber electron plasma wave is of similar form.

Electron plasma waves oscillate at the plasma frequency ω_p (in the zero temperature limit) and frequency $\omega_e = (\omega_p^2 + 3k_e^2\, k_BT_e/m_0)^{1/2}$ (for a 'warm' electron temperature T_e). Here k_e is the wavenumber of the electron plasma wave and ω_e is the frequency of the electron plasma wave. The extra term for a 'warm' plasma arises due to the effect of the electron pressure $(n_ek_BT_e)$ on the restoring force during the wave motion [23].

The oscillating electrons in an electron plasma wave create an oscillating dipole moment $\mathbf{d} = q\mathbf{r}$, where q is the size of the oscillating charge. An oscillating dipole moment creates an oscillating electric field (and associated magnetic field): electromagnetic radiation. The temporal behaviour of the dipole moment determines the frequencies of oscillation and we can define a polarisability α of the plasma medium by the relationship of the dipole moment to the incoming radiation electric field of frequency ω, namely

$$\mathbf{d} = \alpha\mathbf{E} = \alpha\mathbf{E}_0\exp(-i\omega t).$$

In a plasma, the polarisability has an enhanced response at laser frequencies ω near the frequency of electron plasma waves of frequency ω_e. As electron plasma waves have a dispersion relation given by

$$\omega_e^2 = \omega_p^2 + 3k_e^2v_e^2$$

where $v_e = (k_BT_e/m)^{1/2}$ is the electron thermal velocity, we can write that

$$\alpha = \alpha_0 + \alpha_p\cos(\omega_et) = \alpha_0 + (\alpha_e/2)\left(\exp(i\omega_et) + \exp(-i\omega_et)\right).$$

The induced dipole moment then becomes

$$d = [\alpha_0 + (\alpha_{epw}/2)\,(\exp(i\omega_e t) + \exp(-i\omega_e t))]E_0 \exp(-i\omega t)$$

which simplifies to

$$d = \alpha_0 E_0 \exp(-i\omega t) + (\alpha_e/2)E_0 \exp[-i(\omega - \omega_e)t] + (\alpha_e/2)E_0 \exp[-i(\omega + \omega_e)t].$$

The three terms correspond to re-radiation at the incident light frequency ω, and at frequencies $\omega - \omega_e$ and $\omega + \omega_e$.

The main feature of the spectrum of coherent Thomson scatter is consequently two satellites symmetrically separated from the incident light frequency by a frequency separation [49]

$$\omega_e = \pm \left(\omega_p^2 + 3k_B T_e \frac{(\Delta k)^2}{m_0} \right)^{1/2}. \tag{3.55}$$

The width of the satellite features is related to the electron temperatures, while the shift in frequency from the incident light frequency also depends on the plasma frequency $\omega_p^2 = n_e e^2/(m_0 \epsilon_0)$, which varies with the electron density n_e (see Equation 3.55). It is possible to use the scattered spectrum to deduce both the electron density and temperature of the plasma scattering the light. There is also a scattering peak associated with an ion plasma wave (a sound wave, see Equation 1.16), which is spectrally typically two orders of magnitude closer to the original light frequency than the electron plasma wave feature (due to the lower velocity of the ion motion at similar temperature producing waves of larger wave number). With high spectral resolution the ion feature can yield the ion temperature [99].

3.5.2 Incoherent Thomson Scatter in Plasmas

If Thomson scatter is incoherent with each electron separated sufficiently that the scattered electric field from each electon does not overlap, the frequency of scattered photons can be Doppler shifted proportionally to the individual electron velocity. Thomson scattered radiation from an initially moving electron is at a shifted frequency due to the Doppler effect. If the electron is moving at velocity $\mathbf{v}$, we can generally write that the scattered frequency is:

$$\omega_s = \omega_i \frac{1 - \hat{\mathbf{i}}\cdot\mathbf{v}/c}{1 - \hat{\mathbf{s}}\cdot\mathbf{v}/c} \tag{3.56}$$

where the shift in frequency arises from a combination of the shift occurring at the electron due to its motion towards the source (of incident radiation, in direction $\hat{\mathbf{i}}$) and the equivalent shift at the observation point due to electron motion along $\hat{\mathbf{s}}$, relative to the observer. The unit vectors $\hat{\mathbf{i}}$ and $\hat{\mathbf{s}}$ are in the incident light and scattered light directions, respectively.

For $v/c \ll 1$, we can approximate that:

$$\omega_s \approx \omega_i \left(1 - \hat{\mathbf{i}} \cdot \frac{\mathbf{v}}{c}\right)\left(1 + \hat{\mathbf{s}} \cdot \frac{\mathbf{v}}{c}\right)$$
$$\approx \omega_i \left(1 + \frac{(\hat{\mathbf{s}} - \hat{\mathbf{i}}) \cdot \mathbf{v}}{c}\right) \tag{3.57}$$

so that the frequency shift due to the Doppler effect is:

$$\frac{\omega_s - \omega_i}{\omega_i} = \frac{(\hat{\mathbf{s}} - \hat{\mathbf{i}}) \cdot \mathbf{v}}{c} \tag{3.58}$$

There is a shift in detected scattered light frequency which is proportional to the velocity component along the direction $\hat{\mathbf{s}} - \hat{\mathbf{i}}$. The intensity of scattered light as a function of frequency shift $\omega_s - \omega_i$ is thus proportional to the velocity component along $\hat{\mathbf{s}} - \hat{\mathbf{i}}$. The scattered light spectrum gives the velocity distribution function. For a Maxwellian distribution of velocities:

$$f_{\hat{\mathbf{s}}-\hat{\mathbf{i}}} = n_e \left(\frac{m}{2\pi k_B T}\right)^{3/2} \exp\left[-\frac{m_e v_{\Delta k}^2}{2k_B T_e}\right] \tag{3.59}$$

for n_e the total electron density and $v_{\Delta k}$ the velocity component along $\hat{\mathbf{s}} - \hat{\mathbf{i}}$. The scattered spectrum is:

$$\frac{dP}{d\Omega} \propto \exp\left(\frac{c^2}{\omega_i^2}\frac{m_e}{2k_B T_e}(\omega_s - \omega_i)^2\right). \tag{3.60}$$

Measuring the width of Doppler broadened spectrum of scattered light can gives a measurement of the electron temperature T_e as the spectral width of the scattered spectrum is proportional to $\sqrt{T_e}$. The cross-section for Thomson scatter is very small, $\sigma_S = 6.65 \times 10^{-29}$ m^2, which means that very little radiation is scattered. The use of moderately high power lasers as the incident radiation source increases the amount of scattered light and makes incoherent Thomson scattering viable as a plasma diagnostic.

3.6 Scattering of Unpolarised Light and Compton Scattering

In Section 2.5, we saw that the polarisation of light produced by charge acceleration is in the plane formed by the scattered light **k**-vector and the charge acceleration direction. Consequently, we have so far expressed the differential scattering cross-section as a function of the angle θ of the scattered light direction to the electric field of the incoming radiation. Laser light sources used for plasma diagnostics are

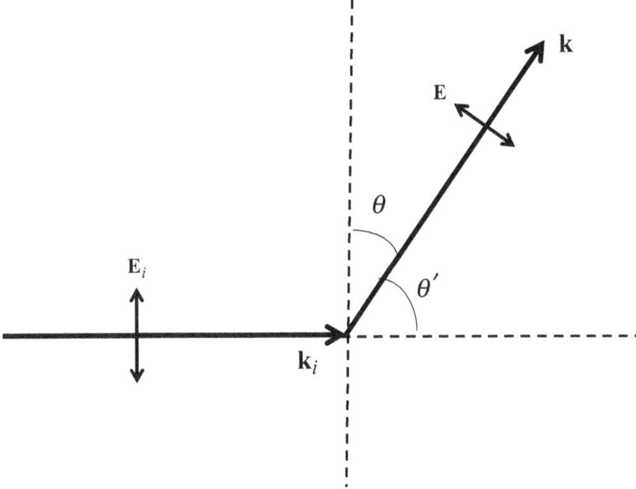

Figure 3.6 Scattering in the plane of the incident $\mathbf{k}_i$ and scattered $\mathbf{k}$ wavevectors showing the angle θ between the scattered wavevector and the electric field $\mathbf{E}_i$ of the incident wave in the plane of the scattering and the angle of scattering θ' (change of wavevector direction). The angle of the scattered wavevector $\mathbf{k}$ to the electric field component in the plane perpendicular to the illustrated plane is $\pi/2$.

often polarised, so such cross-sections are appropriate. However, in dealing with more general scattering such as in astrophysics, the light is usually not polarised.

Unpolarised light can be regarded as the independent supposition of two polarised beams with polarisation direction at angle $\pi/2$ to each other. For scattering with unpolarised light, we can set one of the polarisations to be in the plane formed by the incident $\mathbf{k}_i$-vector and scattered light $\mathbf{k}$-vector (see Figure 3.6), and the other polarisation to be at angle $\pi/2$, 'out of' the plane formed by the incident $\mathbf{k}_i$-vector and scattered light $\mathbf{k}$-vector. For the in-plane polarisation, the angle of scattering θ relative to the electric field is related to the change in angle during scattering θ' (the angle between $\mathbf{k}_i$ and $\mathbf{k}$) by $\theta = \pi/2 - \theta'$ (see Figure 3.6). For the out-of-plane polarisation, the angle of the $\mathbf{k}$-vector of the scatter light to the electric field is $\pi/2$. The total differential cross-section for unpolarised light is the average of the cross-section for the two linear polarisations. We can write

$$\begin{aligned}\left(\frac{d\sigma_S}{d\Omega}\right)_{unpol} &= \frac{1}{2}\left[\left(\frac{d\sigma_S(\theta')}{d\Omega}\right)_{pol} + \left(\frac{d\sigma_S(\pi/2)}{d\Omega}\right)_{pol}\right] \\ &= \frac{1}{2} r_e^2 (\cos^2\theta' + 1). \end{aligned} \tag{3.61}$$

The total scattering cross-section $\sigma_S = (8\pi/3) r_e^2$ for unpolarised light integrated over all scattering angles θ' is the same as for polarised light.

Equation 3.61 shows that the polarised light intensities in the plane and perpendicular to the plane of scattering are in the ratio $\cos^2\theta' : 1$. At a scattering angle of $\theta' = \pi/2$, we have the interesting result that completely unpolarised light scatters to produce fully linearly polarised light. For example, sunlight as well as scattering preferably at short wavelengths in the atmosphere (creating the blue sky; see Section 3.2), is also polarised when scattered as the angle θ' of scatter is large. The polarisation of the sky is less affected than the solar intensity by cloud, mist and fog. Bees and some birds can detect light polarisation and use the polarisation pattern of the sky for navigation [118].

A photon posseses a momentum $\hbar\omega/c$ as well as an energy $\hbar\omega$. The scattering of a photon of light by an electron, consequently, can be regarded as a 'collisional' process with the momentum and energy conserved. The scattering angle θ' of the light now changes the momentum transfer to the electron and the energy of the scattered photon. The requirement for the conservation of energy and momentum results in a change in the scattered photon energy which is best expressed in the wavelength of the incident λ and scattered λ_S light. With some manipulation, we have [93]

$$\lambda_S - \lambda = \lambda_C(1 - \cos\theta') \tag{3.62}$$

where λ_C is the Compton wavelength defined by

$$\lambda_C = \frac{h}{m_0 c} \tag{3.63}$$

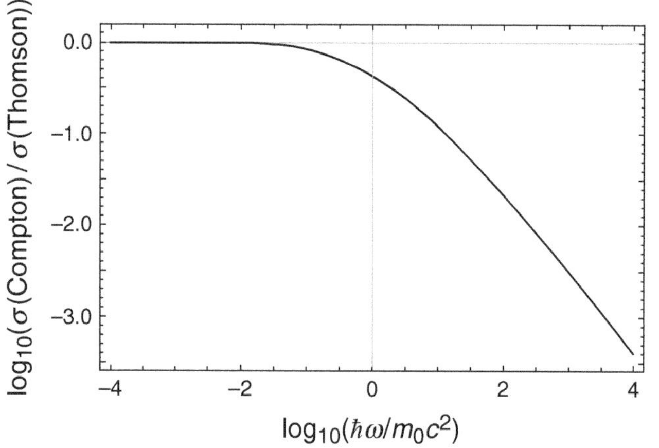

Figure 3.7 The cross-section for Compton scatter relative to the Thomson scatter cross-section as a function of photon energy (in units of the electron rest mass energy $m_0c^2 = 511\,\text{keV}$). See [93] for an expression for the Compton scatter cross-section.

with a numerical value of 2.426×10^{-12} m. For longer wavelengths ($\lambda \gg 10^{-11}$ m), the shift in wavelength by scattering of order of the Compton wavelength is negligible.

In addition to a photon energy shift on scattering, if the photon wavelength is such that the Compton wavelength shift is significant, the Thomson scatter cross-section is reduced below the level of the Thomson value and the scattering is then referred to as Compton scatter. The drop in cross-section starts to become significant when $\hbar\omega > 0.1 m_0 c^2$ (above approximately 50 keV photon energy) and needs to be calculated using quantum electrodynamics. A complete expression for the Compton scatter cross-section is given by, for example, Rybicki (see [93], p. 197, figure 3.7). Integration of the Compton scatter cross-section over all frequency yields Equation 3.9.

Exercises

3.1 A tokamak plasma has a peak electron density of 10^{14} cm^{-3}. A laser beam of 1 cm^2 cross-sectional area is incident into the tokamak plasma. Calculate the fraction of the beam scattered by Thompson scatter per centimetre length of laser propagation at the peak of the electron density. [2×10^{-10}]

3.2 The classical electron radius for Thomson scatter is given by

$$r_e = \frac{e^2}{4\pi\epsilon_0} \frac{1}{m_0 c^2}.$$

Using the expression for the fine structure constant α found in Exercise 1.6, show that

$$r_e = \frac{\hbar\alpha}{m_0 c},$$

and

$$r_e = a_0 \frac{2R_d}{m_0 c^2},$$

where $1R_d$ is the ionisation energy of the ground state of hydrogen and a_0 is the Bohr radius.

3.3 An essential feature of the Thomas–Reiche–Kuhn sum rule derivation in this chapter is that the integration of the scattering cross-section for a bound electron over all frequency is the same as the integration of the free-electron scatter cross-section over all frequency. For an electron with a single resonance at frequency ω_0 and 'dissipation' constant (e.g. radiative transition probability) of γ, show that the ratio of the frequency-integrated bound and free-scattering cross-sections is approximately given by

$$\left(1 - \frac{\omega_0}{\omega_{max}}\right)\left(1 + \left(\frac{\gamma}{\omega_0}\right)^3\right)$$

where $\omega_{max} \approx c/r_e$ is the frequency where the light wavelength approaches the classical electron radius r_e.

3.4 Consider the hydrogen Lyman-α resonance line at 121.567 nm corresponding to a $n = 2$ to $n = 1$ transition. The average transition probability for spontaneous emission of this line is 4.699×10^8 s. Evaluate the ratios ω_0/ω_{max} and $(\gamma/\omega_0)^3$ considered in the previous question. [$\omega_0/\omega_{max} = 1.46 \times 10^{-7}$, $(\gamma/\omega_0)^3 = 6.17 \times 10^{-22}$]

3.5 The frequency integrated Compton cross-section can be accurately evaluated by assuming the Thomson cross-section σ_S is constant up to an upper frequency limit $\omega_{max} = (3\pi/4)c/r_e$ and then zero at higher frequencies. Determine the photon energy at this upper frequency limit. [156 MeV]

3.6 The Lyman-α resonance line of hydrogen has an associated absorption oscillator strength of 0.4162. Determine the addition of the absorption oscillator strengths for all other transitions terminating on the ground state in hydrogen. [0.5838]

3.7 Snell's law (Equation 3.41) implies a constant product of the refractive index η and $\sin\theta$, where θ is the angle of incidence to a refractive index gradient. Differentiate $\eta \sin\theta$ with respect to distance r along the refractive index gradient to show that

$$\frac{1}{\eta}\frac{d\eta}{dr} = \frac{d\theta}{dz}$$

where z is distance perpendicular to the refractive index gradient. [This equation is identical to Equation 3.43.]

3.8 A probe laser beam incident parallel to a target surface is often used to characterise the properties of laser-produced plasmas. Assume that a plasma refractive index variation $\eta(x)$ which increases linearly with distance x from the target surface is present. If a ray in the probe beam is incident at a distance x_0 from the target surface and is at a distance x after travelling a distance z along the target surface, show that

$$x \approx x_0 + \frac{d\eta(x)}{dx} z^2$$

provided the refractive index slope $d\eta(x)/dx$ is small and uniform in the z-direction.

3.9 Capillary discharge lasers are plasma-based lasers with output at wavelength 46.9 nm [91]. A Z-pinch plasma produces lasing in a plasma column typically 19 cm in length with a peak electron density of $10^{19}\,\mathrm{cm}^{-3}$. The peak of the laser output is at a divergence angle of 5 milliradians to the axis of the plasma column. The electron density $n_e(r)$ drops approximately exponentially with distance r from the plasma axis such that

$$n_e(r) = n_e(0)\exp(-r/L)$$

where L is a density scalelength. From the above data for the capillary discharge laser, estimate the value of the scalelength L assuming refraction determines the observed divergence angle. [370 μm]

4
Radiation Emission in Plasmas

In this chapter we initially concentrate on radiation in thermal equilibrium. We derive the Planck radiation law for an equilibrium radiation field by considering the density of photons in a black-body cavity. As well as being applicable to many astrophysical and some laboratory plasmas, the concept of a radiation field in thermal equilibrium is useful in deducing the relationships between inverse processes interacting with a radiation field. The three possible radiative processes (spontaneous emission, photo-excitation and stimulated emission) between two quantum states in an atom or ion in the presence of an equilibrium radiation field are examined. It is shown that a simple thought experiment, where the rates of the three radiative processes are in balance with an equilibrium radiation field, leads to universal relationships between the radiative rates.

4.1 The Planck Radiation Law

If the radiation and electrons in a medium have many interactions through emission and absorption, the radiation field becomes thermalised and can be regarded as having a temperature equal to that of the electrons. In this section, we calculate the form of the thermalised radiation known as the Planck radiation intensity.

An equilibrium radiation field is a collection of photons which follow the rules of statistical mechanics for a collection of bosons. We need to calculate the density of modes for light in a steady-state system. We imagine a cube in space with perfectly conducting (and hence perfectly reflecting) walls and sides of length L. Such a volume with an assumed radiation energy in equilibrium with the walls is known as a black-body cavity (though in inertial fusion work the German word *hohlraum* is often used).

4.1.1 The Density of Modes

The wavelengths λ of light that can exist in a steady state inside a conducting cube of sides L are quantised by the condition that a half-integer number of wavelengths only can exist in directions perpendicular to the walls. Otherwise, the oscillating electric field for the light causes transient interference effects and a steady-state solution for the radiation field does not exist. This condition is best quantified in terms of wavenumber $k = 2\pi/\lambda$, so that k for each direction x, y and z satisfies $k_m = m\pi/L$ and $k_{m+1} - k_m = \pi/L$, where m is an integer. The density of modes per unit volume of k-space is thus $(\pi/L)^3$.

The number $\rho_k dk$ of modes with wavenumber between k and $k + dk$ per unit volume of the cube is given by

$$\rho_k dk = 2\frac{(1/8)4\pi k^2 dk}{(\pi/L)^3 L^3}$$

after allowing for a factor 2 increase in the number of modes due to the two polarisation components of light. The numerator here represents the volume of k-space for modes in the range k to $k + dk$ which is an octant of a sphere of radius k and thickness dk. The octant 1/8 factor arises because only positive k_x, k_y and k_z values of k have meaning as distinct modes. The denominator is the k-space taken up by a single mode ($(\pi/L)^3$) multiplied by the volume of the cube (L^3). We can simplify the above and obtain an equation for the number of modes with wavenumber between k and $k + dk$ per unit volume:

$$\rho_k dk = \frac{k^2}{\pi^2} dk. \tag{4.1}$$

We can convert this expression to the density of modes per unit frequency by noting that angular frequency $\omega = ck$ and hence $d\omega = cdk$, where c is the speed of light. The number of modes $\rho_\omega d\omega$ with frequency between ω and $\omega + d\omega$ per unit volume is given by

$$\rho_\omega d\omega = \frac{\omega^2}{c^3\pi^2} d\omega. \tag{4.2}$$

4.1.2 Quantisation of the Radiation Field

The propagation of light within the cube introduced above is governed by the wave equation which determines the values of electric field $\mathbf{E}$ associated with the radiation. We have for the wave equation

$$\nabla^2\mathbf{E} - \frac{1}{c^2}\frac{\partial^2\mathbf{E}}{\partial t^2} = 0 \tag{4.3}$$

which has solutions of the form $\mathbf{E}(t) = \mathbf{E}_0(t) \exp(i\mathbf{k}.\mathbf{r})$ where $\mathbf{r}$ is a position vector. Substituting this solution into the wave equation (Equation 4.3) gives

$$-k^2\mathbf{E}_0(t) - \frac{1}{c^2}\frac{\partial^2 \mathbf{E}_0(t)}{\partial t^2} = 0$$

which can be re-written as

$$\frac{\partial^2 \mathbf{E}_0(t)}{\partial t^2} = -\omega^2 \mathbf{E}_0(t). \tag{4.4}$$

Solutions of this equation take the form $\mathbf{E}_0(t) = \mathbf{E}_0 \exp(-i\omega t)$ with oscillations as obtained for a harmonic oscillator (see Appendix A.3). To find the correct energy of the radiation field, however, the harmonic oscillation needs to be treated quantum mechanically. Solving the Schrodinger equation in one dimension with the parabolic potential energy variation producing an equation of motion of the form of Equation 4.4 gives quantised harmonic oscillator energies E_{n_p} (see Appendix A.3). We can write

$$E_{n_p} = \left(\frac{n_p + 1}{2}\right)\hbar\omega \tag{4.5}$$

where n_p is an integer such that $n_p = 0, 1, 2, 3, \ldots$. The energies associated with the electric and magnetic fields of the radiation for each ω are quantised in this way with each unit $h\omega$ of quantisation referred to as a photon.

The energy of $(1/2)\hbar\omega$ in Equation 4.5 is known as the vacuum energy or zero-point energy as it exists even when the number of photons n_p is zero. Real effects arise from the vacuum energy. The Casimir effect arises when two unequal black-body cavities are next to each other. There is a measurable force between the cavities as the zero-point energy is different in the two cavities (see [64]). It is thought that dark energy responsible for the acceleration of the expansion of the universe is also associated with the vacuum energy [29].

In thermal equilibrium at temperature T, we can calculate the probability of the radiation field for a particular mode to have n_p photons using the probability distribution (Equation 1.19). Photons are bosons, so the number n_p of photons at each energy $\hbar\omega$ can range from 0, 1, 2, ... upwards and their chemical potential μ is zero as there is no energy cost per boson in adding another boson. The absolute probability P_{n_p} that n_p photons are present is given by

$$P_{n_p} = \frac{\exp(-n_p\hbar\omega/k_BT)}{\sum_{n'} \exp(-n'\hbar\omega/k_BT)}$$

upon cancelling $\exp((1/2)\hbar\omega/k_BT)$ in the numerator and denominator. If we set $U = \exp(-\hbar\omega/k_BT)$ then

$$P_{n_p} = \frac{U_p^n}{\sum_{n'} U^{n'}}.$$

The infinite series $\Sigma_{n'} U^{n'} = 1/(1-U)$ so that

$$P_{n_p} = (1-U)U^{n_p}.$$

The mean number n_{av} of photons

$$n_{av} = \sum_{n_p} n_p P_{n_p}.$$

This equation for n_{av} can be simplified as follows:

$$n_{av} = (1-U)\sum_{n_p} n_p U^{n_p} = (1-U)U\frac{d(\sum_{n_p} U^{n_p})}{dU} = (1-U)U\frac{d(1-U)^{-1}}{dU}.$$

The last equality here simplifies to give

$$n_{av} = \frac{U}{1-U}.$$

The quantity n_{av} is known as the mean number of photons per mode. Substituting in our definition of U gives

$$n_{av} = \frac{1}{\exp(\hbar\omega/k_BT) - 1}. \tag{4.6}$$

The total energy density of radiation can be evaluated by multiplying our expression for the density of modes (Equation 4.2) by the mean number of photons per mode (Equation 4.6) and the energy per photon ($\hbar\omega$). We have for the equilibrium radiation energy $W_p(\omega)d\omega$ per unit volume in the frequency range ω to $\omega + d\omega$:

$$W_p(\omega)d\omega = n_{av}\hbar\omega\rho_\omega d\omega = \frac{\hbar\omega^3}{\pi^2c^3}\frac{d\omega}{\exp(\hbar\omega/k_BT) - 1}. \tag{4.7}$$

Equation 4.7 is the Planck radiation energy density for the thermal distribution of radiation inside a black-body cavity (see Figure 4.1). The maximum in the thermal distribution of the radiative energy can be found by differentiating Equation 4.7. The maximum radiative energy occurs at a photon energy $\hbar\omega_{max}$ given by

$$\hbar\omega_{max} = 2.8k_BT. \tag{4.8}$$

This relationship is known as Wien's displacement law. As well as an increase in the photon energy where the radiative energy is maximum, a rapid increase in the relative values of the radiation energy at the radiative energy maximum with increasing temperature is also apparent from Figure 4.1. We show later that the total thermal radiation energy integrated over all frequencies varies as $(k_BT)^4$.

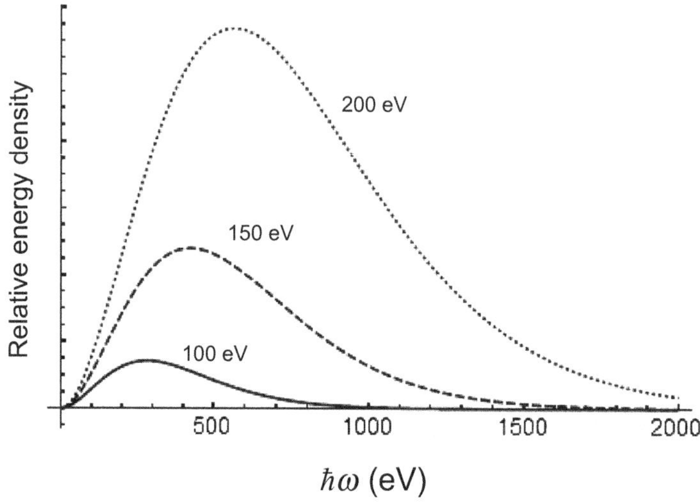

Figure 4.1 The relative energy density per unit photon energy per unit volume for thermal Planckian radiation as a function of photon energy. Three sample radiation temperatures (100 eV, 150 eV and 200 eV) as indicated are plotted.

The Planck radiation energy density is referred to as black-body radiation because of early studies of the absorption and emission of infra-red radiation by Gustav Kirchhoff (1824–1887) and others. Kirchhoff found that a black object is an equally good absorber and emitter of infra-red radiation and coined the term black body. A black object at slightly higher temperature than room temperature (for example, an old fashioned stove) can emit a Planckian spectrum of radiation with a peak of emission in the infra-red and little radiation in the visible. It is a clear sign of over-heating if a black stove starts to emit in the visible (usually red) spectral range. We discuss in Section 4.3 Kirchhoff's law that emission and absorption are proportional.

4.1.3 The Planck Radiation Flux and Intensity

The Planck black-body energy density expression given by Equation 4.7 has radiation uniformly distributed in angle. At the walls of the black-body cavity or if an object is placed inside the cavity, the intensity and flux of the radiation, which are measures of the power per unit area, are important. For angular variations of intensity and flux, we use units of steradian defined as the solid angle subtended by an area of a sphere of unit radius at the centre of the sphere. For example, if the area extends to all the surface of the sphere, that is to all angles, this solid angle is 4π steradian.

The intensity of radiation $< I_p(\omega) >$ represented by the radiation power per unit area integrated over all angles (4π steradian) associated with the Planck radiation energy density is found by multiplying the radiation energy density $W_p(\omega)$ by the speed of light c. We can write that

$$< I_p(\omega) >= cW_p(\omega) = \frac{\hbar\omega^3}{\pi^2 c^2} \frac{1}{\exp(\hbar\omega/k_B T) - 1}. \tag{4.9}$$

We can define the radiation intensity $J_p(\omega)$ per unit steradian by setting $J_p(\omega) =< I_p(\omega) > /(4\pi)$.

If the photon energy $\hbar\omega$ is much less than the radiation temperature $k_B T$, we can expand the exponential in the black-body radiation intensity expressions. For example, with $\hbar\omega \ll k_B T$,

$$< I_p(\omega) >= \frac{\omega^2}{\pi^2 c^2} k_B T. \tag{4.10}$$

This approximation is known as the Rayleigh-Jeans approximation for black-body radiation and is the classical result obtained if Planck's constant $\hbar$ is zero. The intensity at any frequency is proportional to the radiation temperature. At low temperatures where $k_B T \ll \hbar\omega$, the exponential in the Planck radiation intensity becomes large and

$$< I_p(\omega) >= \frac{\hbar\omega^3}{\pi^2 c^2} \exp\left(-\frac{\hbar\omega}{k_B T}\right). \tag{4.11}$$

This expression is known as Wien's formula.

There are several other ways of representing the intensity of black-body radiation. A small aperture in a black-body cavity wall will emit a flux $F_p(\omega)$ of radiation which we can define as the radiated power into 2π steradians per unit area. Consider a black-body cavity with photons impinging on a plane wall. We need to calculate the fraction of photons inside the cavity that impinge on the plane wall in a time Δt to evaluate $F_p(\omega)d\omega$ from the radiation energy density $W_p(\omega)d\omega$. Photons initially at depth $z\Delta t$ along a normal from the wall surface will reach the wall if they are travelling within an angle θ to the wall normal such that $\cos\theta = z/(c\Delta t)$, where c is the speed of light. This means that photons travelling within a cone with half-angle θ will reach the wall in our specified time Δt. The solid angle Ω of such a cone is given by $\Omega = 2\pi(1 - \cos\theta)$. The fraction of photons reaching the wall in time Δt from a depth of z is thus $(1/2)(1 - z/(c\Delta t))$. The radiation energy per unit area impinging on the black-body cavity wall in a time Δt is given by $F_p(\omega)\Delta t$ which is evaluated from the radiation energy density by multiplying by an integration of the fraction of photons reaching the wall with respect to distance z. We have

$$F_p(\omega)\Delta t = W_p(\omega) \int_{z=0}^{c\Delta t} \left(\frac{1}{2}\right)\left(\frac{1-z}{(c\Delta t)}\right) dz = \Delta t \frac{c}{4} W_p(\omega).$$

The flux $F_p(\omega)d\omega$ per unit area into 2π steradian over the frequency range ω to $\omega + d\omega$ is thus given by

$$F_p(\omega)d\omega = \frac{1}{4}cW_p(\omega)d\omega = \frac{\hbar\omega^3}{4\pi^2c^2}\frac{d\omega}{\exp\left(\hbar\omega/k_BT\right) - 1}. \tag{4.12}$$

The radiation intensity per unit steradian $J_p(\omega)$ is related to this flux of radiation impinging on the walls of a black body by

$$J_p(\omega) = \frac{1}{\pi}F_p(\omega). \tag{4.13}$$

We can convert the expression for $F_p(\omega)$ into the radiation intensity $I_{Pedge}(\nu)d\nu$ per unit area per steradian for a frequency range ν to $\nu + d\nu$ with ν measured in Hertz by dividing by 2π (as the flux is into 2π steradian) and noting that $\omega = 2\pi\nu$ and $d\omega = 2\pi d\nu$. The resulting intensity $I_{Pedge}(\nu)$ is such that $I_{Pedge}(\nu)d\nu$ represents the power per unit area per steradian in the spectral range ν to $\nu + d\nu$ impinging on the walls of a black-body cavity. We have

$$I_{Pedge}(\nu)d\nu = \frac{h\nu^3}{c^2}\frac{d\nu}{\exp\left(h\nu/k_BT\right) - 1}. \tag{4.14}$$

For a surface inside a black-body cavity, the intensity incident on the surface is twice this value (with radiation impinging from two directions). The intensity $I_{Pint}(\nu)$ such that $I_{Pint}(\nu)d\nu$ represents the power per unit area per steradian in the spectral range ν to $\nu + d\nu$ impinging from both sides on a surface inside a black-body cavity is given by

$$I_{Pint}(\nu)d\nu = \frac{2h\nu^3}{c^2}\frac{d\nu}{\exp\left(h\nu/k_BT\right) - 1} = J_p(\nu)d\nu. \tag{4.15}$$

The parameter $J_p(\nu)$ is the radiation intensity per unit steradian as defined below Equation 4.9, but over a frequency range of ν to $\nu + d\nu$ (rather than angular frequency).

The total energy density W_{tot} per unit volume integrated over all frequencies can be readily evaluated by integrating Equation 4.7. Setting $x = \hbar\omega/k_BT$ we have

$$W_{tot} = \int_0^\infty W(\omega)d\omega = \frac{(k_BT)^4}{\pi^2c^3\hbar^3}\int_0^\infty \frac{x^3dx}{\exp(x) - 1}.$$

The integral with respect to x here is given in standard tabulations and has a value equal to $\pi^4/15$. The total radiation energy per unit volume for an equilibrium radiation field is then given by

$$W_{tot} = \frac{\pi^2 (k_B T)^4}{15 c^3 \hbar^3}. \tag{4.16}$$

The spectrally integrated radiation flux F_{tot} representing power per unit area impinging on the walls of a black-body cavity and hence the flux escaping through an aperture in a black-body cavity wall is again found using the result that 1/4 of the radiation energy density over the distance $c\Delta t$ from the wall will impinge on the wall in a time Δt. We have

$$F_{tot} = \frac{c}{4} W_{tot} = \frac{\pi^2 (k_B T)^4}{60 c^2 \hbar^3} = \sigma_{SB} T^4 \tag{4.17}$$

where σ_{SB} is known as the Stefan–Boltzmann constant (numerically $\sigma_{SB} = 5.67 \times 10^{-8}\,\mathrm{Wm^{-2}\,K^{-4}}$).

4.2 The Einstein A and B Coefficients

The interaction of radiation with plasma atoms and ions involves a lower and upper quantum state. Our knowledge of the Bohr model for hydrogen and hydrogen-like ions is sufficient here to think about a generic energy level model for electrons bound in a potential.

Einstein originally introduced the concept of stimulated emission, where a photon interacting with an excited atom causes the de-excitation of the atom and the emission of an identical photon to the stimulating photon. As we see in this section, it was not possible to get agreement between the Planck black-body radiation energy density and the form of the equilibrium radiation density arising from a rate equation balance of photo-absorption and spontaneous emission alone. For many years after Einstein proposed the existence of stimulated emission, it was a curiosity necessarily used in models of radiative processes. This changed with the invention of the laser in 1960 and the later development of applications using the extremely bright, collimated and coherent radiation possible with lasers [98]. The word laser is an acronym for light amplification by stimulated emission of radiation.

The rates of the three radiative optical process – spontaneous emission, absorption and stimulated emission – are given by rate equations involving the population densities of the upper N_2 and lower N_1 quantum states and the radiation energy density $W(\omega)$. Energy conservation requires that the relevant radiation energy density is at photon energy $\hbar\omega = E_2 - E_1$, where E_2 and E_1 are the upper and lower quantum state energies respectively. We do not need to consider the detail of atomic energy levels and just need to consider the upper (labelled 2) and lower (labelled 1) states involved in a set of related radiative processes (see Figure 4.2).

The rate of change of the population densities is linearly dependent on the population density and for absorption and stimulated emission is also linearly

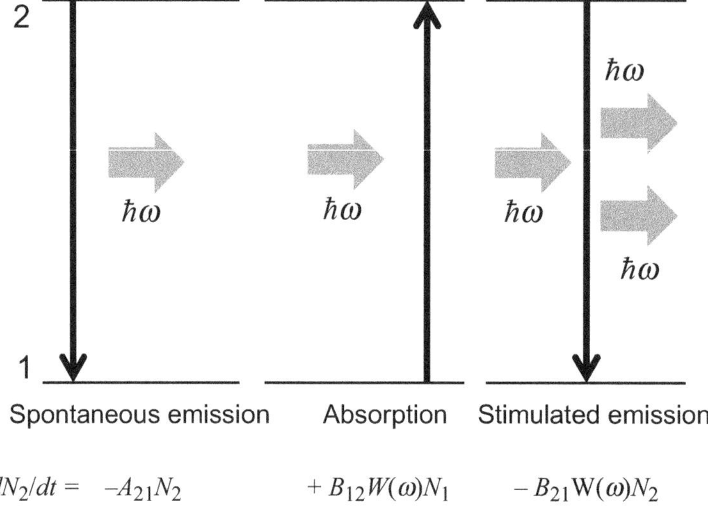

Figure 4.2 A schematic of the three radiative processes between two generic energy levels. The rate of change of the population of the upper quantum state is shown expressed in terms of the Einstein A and B coefficients.

dependent on the radiation energy density. The rate equations for spontaneous emission, absorption and stimulated emission can be added to give the total rate of change of, say, the upper quantum state due to radiative processes:

$$\frac{dN_2}{dt} = -A_{21}N_2 + B_{12}W(\omega)N_1 - B_{21}W(\omega)N_2. \tag{4.18}$$

The proportionality constant for the rate of spontaneous emission (A_{21}) is known as the transition probability or Einstein A-coefficient, while similar constants (B_{12} and B_{21}) for absorption and stimulated emission are simply referred to as the Einstein B-coefficients. As implied by the names of these constants, Einstein originally proposed the details of this rate equation approach.

In the absence of a radiation field $W(\omega)$, we have

$$\frac{dN_2}{dt} = -A_{21}N_2$$

which has a solution

$$N_2(t) = N_2(0)\exp(-A_{21}t). \tag{4.19}$$

Any population initially in the upper quantum state decays exponentially in time with a time constant of $1/A_{21}$.

It is useful to follow Einstein's original treatment and consider what happens for atoms in an equilibrium state. Equilibrium now means that the populations

N_2 and N_1 are in equilibrium with the population ratio given by the Boltzmann ratio and that the radiation field is the equilibrium black-body Planck distribution. Equilibrium also means that the populations are in steady state with $dN_2/dt = 0$. Re-arranging Equation 4.18, we can write an expression for the radiation energy density

$$W(\omega) = \frac{A_{21}}{(N_1/N_2)\, B_{12} - B_{21}}.$$

The Boltzmann population ratio depends on the temperature T of the system such that

$$\frac{N_2}{N_1} = \frac{g_2 \exp(-E_2/k_B T)}{g_1 \exp(-E_1/k_B T)} = \frac{g_2}{g_1} \exp\left(-\frac{\hbar\omega}{k_B T}\right)$$

where g_1 and g_2 are the degeneracies of the energy levels 1 and 2 respectively. It is convenient to group quantum states with similar (or identical energies), so the degeneracies represent the number of quantum states at energy E_1 and E_2, respectively. The above equation for $W(\omega)$ can be re-written as

$$W(\omega) = \frac{A_{21}}{(g_1/g_2) \exp(\hbar\omega/k_B T) B_{12} - B_{21}}. \tag{4.20}$$

Assuming equilibrium means that the radiation energy density is given by the Planck black-body energy density (see Equation 4.7) with radiation energy density between frequency ω and $\omega + d\omega$ given by

$$W_p(\omega) d\omega = \frac{\hbar\omega^3}{\pi^2 c^3} \frac{d\omega}{\exp(\hbar\omega/k_B T) - 1}.$$

Equating $W_p(\omega)$ to $W(\omega)$ (from Equation 4.20) can give us information regarding the relationships between the Einstein coefficients A_{21}, B_{12} and B_{21}. The Planck radiation energy density and Equation 4.20 can both be correct only if

$$B_{12} \frac{g_1}{g_2} = B_{21} \tag{4.21}$$

and

$$A_{21} = \frac{\hbar\omega^3}{\pi^2 c^3} B_{21}. \tag{4.22}$$

If stimulated emission did not exist, Equation 4.20 would have $B_{21} = 0$ in the denominator and it would be impossible to equate Equation 4.20 with the Planck black-body radiation energy density. It was this potential inconsistency that led Einstein to propose the existence of the stimulated emission process.

The relationships between the Einstein coefficients (Equations 4.21 and 4.22) have been obtained by considering a system of radiation and matter in thermal equilibrium. However, the exact rates of A_{21}, B_{12} and B_{21} are determined by quantum mechanics with parameters determined by the central potential arising from the nucleus of an atom and the electric field of the radiation. Consequently, the relationships (Equations 4.21 and 4.22) are independent of the state in which atoms are embedded. It does not matter whether the population distributions of photons and quantum state populations are in equilibrium or not. The relationships between the Einstein coefficients apply for materials not in equilibrium, including the highly non-equilibrium systems used for laser production.

Equations 4.21 and 4.22 are particularly useful as they mean that it is only necessary to calculate one of the Einstein coefficients to obtain the others. Calculating the absorption rate B_{12} is more straightforward and can be undertaken semi-classically (with a quantum mechanical treatment of the atom and a classical treatment of the radiation field; see Section 10.1). As only one Einstein coefficient is needed, published tabulations for particular atomic and molecular transitions almost always only list the transition probability A_{21}. For example, if a transition does not occur (often referred to as being 'strictly forbidden'), then $A_{21} = B_{12} = B_{21} = 0$.

4.3 Emission and Absorption

The intensity or irradiance I of electromagnetic radiation can be defined as the power of radiation per unit area normal to the direction of propagation (see Equation 2.15). It is often necessary to delineate the radiation frequency range of interest by defining the intensity at a particular frequency, so that the units of intensity become power per unit area per unit frequency. Unless an observer is inside a large isotropic plasma or a black-body cavity (see Section 4.2), radiation intensity also varies with angle, so that units of power per unit area per unit frequency per solid angle are used. We use the solid angle unit of steradian, which is obtained by considering the angle subtended by an area on a sphere of unit radius at the centre of the sphere.

If radiation absorption is not important, the intensity I of radiation emission from a plasma can be obtained by integrating a quantity known as the emission coefficient along a line of sight. The emission coefficient ϵ is the radiated power per unit volume in, say, a frequency range ω to $\omega + d\omega$. In the absence of absorption, the intensity in a short length dz will change by dI such that

$$dI = \epsilon dz.$$

The emission coefficient ϵ of a spectral line is related to the upper quantum state population density N_2 and the transition probability A_{21}. If the intensity of the radiation is measured in power per unit area per 4π steradian per unit frequency range, then the emission coefficient ϵ of a spectral line is determined by the emission in the frequency range ω to $\omega + d\omega$ given by

$$\epsilon\, d\omega = N_2 A_{21} \hbar\omega_{21} f(\omega) d\omega. \tag{4.23}$$

The function $f(\omega)$ is known as the lineshape function. The parameter $f(\omega)d\omega$ is the probability of emission of radiation in the frequency range ω to $\omega + d\omega$, assuming that the radiative transition does occur at some frequency. Assuming that the spectral line is emitted at some frequency means that

$$\int_0^\infty f(\omega) d\omega = 1.$$

We examine the lineshape function in some detail in Section 10.3.

If a medium is only absorbing radiation and not spontaneously emitting, the rate of change dI of intensity after passing along a distance dz is proportional to the original intensity I and is given by

$$dI = -KIdz$$

where K is known as the absorption coefficient. We can obtain an expression for the absorption coefficient if we evaluate dI/dz for the absorption process. As stimulated emission acts in a very similar way to absorption – but creates photons instead of destroying them – the processes of absorption and stimulated emission are best considered together, rather than trying to consider stimulated emission alongside spontaneous emission. To evaluate K, we can consider the absorption of a beam of radiation directed along an axis z. We use the population of the upper (N_2) and lower (N_1) quantum states to work out the absorption of the beam.

For a spectral line, the change in intensity ΔI in unit time within a volume of unit cross-section and small thickness Δz is related to the population rate of change with time dN_2/dt due to the absorption (and stimulated emission) in the volume Δz of the incident beam of intensity I. As dN_2/dt is measured in units of number density per unit volume per unit time and ΔI is measured in units of energy per unit volume per unit time, we need to multiply the number of transitions in the volume Δz by the photon energy $\hbar\omega$ per transition. We have $\Delta I = -\hbar\omega\,(dN_2/dt)\Delta z$ so that in the limit where ΔI and Δz become infinitessimally small

$$\frac{dI}{dz} = -\hbar\omega_{21}\frac{dN_2}{dt}. \tag{4.24}$$

The rate of change with time of the upper quantum state population dN_2/dt is found using the Einstein B-coefficients. We write for the rate of change of the upper population density due to absorption minus the rate due to stimulated emission

$$\frac{dN_2}{dt} = (B_{12}N_1 - B_{21}N_2) \int_0^\infty W(\omega) f(\omega) d\omega. \tag{4.25}$$

The quantities dN_2/dt and $W(\omega)$ here are those relating to the intensity change dI/dz of interest along a direction z.

We need to explicitly consider a radiation energy density $W(\omega)$ which may not be spectrally broad (like spectral line emission; see Equation 4.23). To consider spectrally narrow incident radiation, we have used the lineshape parameter $f(\omega)$ in Equation 4.25 within an integration of the lineshape multiplied by the radiation energy density. If the radiation energy density is spectrally much broader than the lineshape function, this equation reduces to Equation 4.18 (neglecting spontaneous emission) as we can move $W(\omega)$ from within the integral and the lineshape function alone integrates to unity. If the radiation energy density is spectrally much narrower than the lineshape function then the integral in Equation 4.25 reduces to $f(\omega) \int_0^\infty W(\omega) d\omega$. The radiation energy density $W(\omega)$ affecting the intensity along a line of sight is related to the intensity $I(\omega)$ along the line of sight by $I(\omega) = cW(\omega)$. Using Equations 4.21 and 4.22, we can re-write Equation 4.25 as

$$\frac{dN_2}{dt} = \frac{\pi^2 c^3}{\hbar \omega_{21}^3} A_{21} \left(\frac{g_2}{g_1} N_1 - N_2 \right) \int_0^\infty \left(\frac{I(\omega)}{c} \right) f(\omega) d\omega. \tag{4.26}$$

Assuming that we are interested in the change of intensity over a narrow spectral range (much narrower than the lineshape width), we can set

$$\int_0^\infty \left(\frac{I(\omega)}{c} \right) f(\omega) d\omega = f(\omega) \int_0^\infty \frac{I(\omega)}{c d\omega} = \frac{f(\omega) I}{c}$$

by defining $I = \int_0^\infty I(\omega) d\omega \approx I(\omega) d\omega$ for a narrow spectral range $d\omega$. Using the definition of the absorption coefficient K (i.e. $dI = -KIdz$) and Equation 4.24, we obtain an expression for the absorption coefficient

$$K = \frac{\pi^2 c^2}{\omega_{21}^2} A_{21} \left(\frac{g_2}{g_1} N_1 - N_2 \right) f(\omega). \tag{4.27}$$

The frequency integrated cross-section for absorption is found by dividing the component of Equation 4.27 associated with absorption by the number density N_1 of atoms (dropping the stimulated emission contribution). We have

$$\int \sigma_{abs} d\omega = \frac{\pi^2 c^2}{\omega_{21}^2} A_{21} \frac{g_2}{g_1}. \tag{4.28}$$

In Chapter 3, we found another expression for the frequency integrated cross-section for light absorption (Equation 3.28), namely

$$\int \sigma_S d\omega = 2\pi^2 r_e c f_s$$

where f_s originally proposed as the number of electrons associated with atomic absorption is known as the oscillator strength. The parameter r_e is the classical electron radius determined by the radius where the electrostatic potential of the electron is equal to the electron rest mass energy $m_0 c^2$ (see Equation 3.4). Equating the free-electron frequency integrated absorption cross-section involving the oscillator strength to Equation 4.28, the radiative transition probability A_{21} can be written in terms of the oscillator strength. We have

$$A_{21} = \frac{g_1}{g_2} \frac{2\omega_{21}^2}{c} r_e f_{12} \tag{4.29}$$

where f_{12} is now the oscillator strength for absorption from quantum state one to two.

The oscillator strength representation of the rate of radiative emission (and absorption) is useful as the Thomas–Reiche–Kuhn sum rule obtained for classical absorption applies: each electron has a total oscillator strength of one and the total oscillator strength for an atom or ion is equal to the number of bound electrons (see Equation 3.27). There may be many possible absorption 'resonances' for an electron from a quantum state to other states, but the sum of the oscillator strengths for all transitions is equal to one. Quantum mechanical calculations are required (see Section 10.1.1) to determine the individual oscillator strengths, but the Thomas–Reiche–Kuhn sum rule is a useful check sum and requires an upper bound of one for an oscillator strength. The quantum mechanics of the Thomas–Reiche–Kuhn sum rule is also considered in Appendix A.4.

4.4 Introducing the Equation of Radiative Transfer

It is possible to evaluate changes of intensity when both absorption and spontaneous emission occur along a line of sight by combining the expressions used above for emission and absorption. The key equation is known as the equation of radiative transfer and is written (in differential form) as

$$\frac{dI}{dz} = \epsilon - KI. \tag{4.30}$$

We can define a quantity known as the optical depth τ by considering the optical depth change $d\tau = Kdz$ over an incremental length (dz) along the line of sight. We then have

$$\frac{dI}{d\tau} = \frac{\epsilon}{K} - I = S - I \tag{4.31}$$

where $S = \epsilon/K$ is known as the source function. In a large uniform plasma with $dI/dz = 0$, the radiation intensity $I = \epsilon/K = S$. For a system in thermal equilibrium, the radiation intensity is given by the Planck radiation law and the source function is then also equal to the Planck radiation intensity. This relationship between the emission and absorption coefficients is known as Kirchhoff's law of thermal radiation.

For a plasma with populations in equilibrium, but not necessarily an equilibrium radiation field, the relation $I = \epsilon/K$ can be used to deduce the value of the absorption coefficient K if, say, the emission coefficient ϵ is known. The intensity I is set to the Planck radiation intensity for radiation within a black-body cavity at the electron temperature of the plasma. Similarly, the emissivity ϵ could be deduced from the absorption coefficient K. The condition where populations of bound and free electrons are in equilibrium, but the radiation field is not in equilibrium is known as local thermodynamic equilbrium (often abbreviated LTE).

By considering a thought experiment (a *gedanken* experiment in German) for a plasma with a complete thermal equilibirum between quantum state populations and a radiation field, Einstein was able to relate the coefficients for spontaneous emission, stimulated emission and photon absorption between bound states (see Section 4.2). This is an example where we can set the source function to be equal to the Planck black-body radiation field to give the relationship of the emissivity ϵ to the absorption coefficient K. Such relationships are more generally applicable (e.g. they apply to a system not in complete thermal equilibrium) and can enable atomic coefficients to be determined which are applicable for any equilbrium or non-equilbrium situation. For a particular temperature T and angular frequency ω, the source function is equal to the Planck black-body radiation intensity:

$$S = \frac{\epsilon}{K} = < I_p(\omega) > \tag{4.32}$$

where $< I_p(\omega) >$ is the Planck black-body radiation intensity as given by, for example, Equation 4.9. We return to use this relationship between emission and absorption later to relate expressions between bound and continuum (free) electron states and between free-electron emission and absorption. Equation 4.32 is an example of a detailed balance relationship between inverse processes.

The processes of absorption and stimulated emission are implicitly grouped together in Equation 4.32 in the absorption coefficient K, while the emission coefficient ϵ refers only to spontaneous emission. If we write the absorption coefficient as being equal to the coefficient due to photon absorption K_0 minus the coefficient due to stimulated emission K_S (as stimulated emission creates photons rather than destroying them), re-arranging Equation 4.32 and substituting for the Planck radiation intensity (Equation 4.9) gives

$$K = K_0 - K_S = \epsilon \frac{\pi^2 c^2}{\hbar\omega^3}\left(\exp\left(\frac{\hbar\omega}{k_B T}\right) - 1\right). \tag{4.33}$$

The absorption coefficient associated with photon absorption is given by

$$K_0 = \epsilon \frac{\pi^2 c^2}{\hbar\omega^3} \exp\left(\frac{\hbar\omega}{k_B T}\right),$$

and the coefficient associated with stimulated emission is given by

$$K_S = \epsilon \frac{\pi^2 c^2}{\hbar\omega^3}.$$

Exercises

4.1 A laser cavity with two spatially separated flat parallel mirrors can be regarded as a one-dimensional black-body cavity. In a steady-state laser cavity there are an integral number of half wavelengths of the electric field oscillation between the two mirrors just as required for the three-dimensional cavity used in the derivation of the Planck radiation energy distribution. Show that the modes present in the one-dimensional laser cavity are separated in frequency by $c/2L$, where L is the spatial separation of the two mirrors of the cavity.

4.2 The National Ignition Facility (NIF) is a laser system in California capable of producing 1.5 MJ of laser power. An estimated efficiency for the conversion of the laser power to soft X-rays of 70% is obtained by irradiating the walls of a black-body cavity (known as a hohlraum). If a cylindrical hohlraum the size of a can of Coca-Cola is used, calculate the temperature of the black-body radiation. [Assuming a can of 375 ml, 120 eV.]

4.3 Inertial fusion research at the NIF requires radiation temperatures of 300 eV. Determine an estimate of the more typical volume of hohlraum that is used for inertial fusion studies. [9 ml]

4.4 The Sun emits as a black body with a surface temperature of ≈ 5800 K. The solar radius is 6.957×10^8 m, while the Earth's orbit around the Sun is at radius 1.496×10^{11} m. Neglecting atmospheric absorption, calculate the power of the solar radiation striking a unit area positioned normal to the Sun's rays at the Earth. [1390 Wm^{-2}]

4.5 (a) On a summer's day at a mid-latitude location, the solar radiation power striking a tarmac road is 500 Wm^{-2}. Calculate the equilibrium temperature of the road in degrees Celsius. (b) In winter, many mid-latitude locations have approximately 200 Wm^{-2} of incident solar radiation. Determine the equilibrium temperature in winter of the road if it is only heated by the Sun. [(a) 33 °C, (b) –29 °C]

4.6 The cosmic microwave background (CMB) has a spectrum of black-body radiation with a radiation temperature of 2.73 K. Calculate (i) the frequency where the CMB has maximum energy density in an energy density as a function of frequency plot, and (ii) the frequency integrated irradiance of the CMB. [(i) 8.5×10^{10} Hz, (ii) 3.15×10^{-6} $\mathrm{Wm^{-2}}$]

4.7 If black-body radiation was actually emitted according to Wien's formula given by

$$W(\omega) = \frac{\hbar\omega^3}{\pi^2 c^2} \exp\left(-\frac{\hbar\omega}{k_B T}\right),$$

show that stimulated emission could not exist. Show that the transition probability A_{21} for spontaneous emission, however, is still related to the Einstein coefficient B_{12} by

$$A_{21} = \frac{\hbar\omega^3}{\pi^2 c^2} \frac{g_1}{g_2} B_{12}.$$

4.8 Most laser media involve the bound states of excited atoms or ions. Gain in a laser occurs when the stimulated emission process occurs at a greater rate than absorption. The gain coefficient G is related to the absorption coefficient K obtained for Equation 4.27 by $G = -K$. Show that the gain coefficient for a laser medium is given by

$$G = \frac{\lambda^2}{8\pi} A_{21} \left(N_2 - \frac{g_2}{g_1} N_1\right) f(\nu)$$

where λ is the lasing wavelength and the line profile $f(\nu)$ is written in terms of frequency ν (rather than angular frequency ω).

4.9 Use Equation 4.29 to show for a spectral line of wavelength λ_{21} (measured in metres) that the transition probability A_{21} is related to the absorption oscillator strength f_{12} by the numerical relationship

$$A_{21} = 6.67 \times 10^{-5} \frac{g_1}{g_2} \frac{1}{\lambda_{21}^2} f_{12}$$

where g_1 and g_2 are the degeneracies of the lower and upper quantum states.

5

Radiation Emission Involving Free Electrons

The emission of radiation is one of the most important tools for diagnosing conditions in plasmas and can play a significant role in moving energy. For plasmas at all but the highest densities, we can assume that electrons are either in bound quantum states with energies dominated by the central potential of ionic nuclei, or they are unbound, occupying a continuum of free-electron states. We have seen that free electrons do not really occupy a true energy continuum (see Section 1.3), but a free electron means that the density per unit energy of quantum states is high, so that we can often consider the free-electron energies as continuous.

Radiation arising from free electrons dominates in low atomic number plasmas. Radiation transition probabilities scale rapidly with increasing atomic number: for example, as Z^4 for hydrogen-like ions, while radiation for free-electron transitions scales at a lesser rate proportional to Z^2 (see Section 5.2). In complete thermal equilibrium with the radiation field in equilibrium with particle temperatures, the emission of radiation is given by the Planck black-body formulas derived in Chapter 4. However, complete thermal equilibrium is rare in laboratory plasmas and the more tenuous astrophysical plasmas as radiation absorption within the dimensions of the plasmas is small.

We consider the radiation processes involving free electrons in this chapter. A full quantum mechanics understanding is generally not required to model emission from free electrons, so we discuss the emission of radiation in this chapter using largely classical non-quantum treatments. The Bohr model for bound energy states is utilised in considering emission from an electron making a transition from a free to bound state.

5.1 Cyclotron Radiation

Astrophysical plasmas such as those near a neutron star have strong embedded magnetic fields and even in interstellar space there is a weak magnetic field

($\approx 10^{-10}$ Tesla). The plasmas studied in magnetic fusion research have magnetic fields in the range 0.1–10 Tesla confining the plasma. Confinement works as charged particles orbit with helical-shaped trajectories around an imposed magnetic field $\mathbf{B_0}$. The velocity component $v_\perp$ perpendicular to the magnetic field causes a $\mathbf{v} \times \mathbf{B_0}$ acceleration perpendicular to $v_\perp$ due to the Lorentz force

$$\mathbf{F} = m_0 \mathbf{v} \times \mathbf{B_0}.$$

For electrons with a velocity component, $v_\perp$, perpendicular to the magnetic field and, $v_\parallel$, parallel to the magnetic field, the acceleration due to the Lorentz force is given by

$$\frac{dv}{dt} = \frac{e\, v_\perp B_0}{m_0}. \tag{5.1}$$

We can balance this acceleration produced by the $\mathbf{v} \times \mathbf{B_0}$ force with the centrifugal acceleration $v_\perp^2/r$, where r is the radius of the circular orbit around the magnetic field line (and is known as the Larmor or gyroradius). Cancelling a $v_\perp$ on both sides of the balance gives the angular frequency of the electron orbit

$$\omega_c = \frac{v}{r} = \frac{eB_0}{m_0}$$

which is known as the cyclotron frequency, Larmor frequency or gyrofrequency. The electrons undergo circular cyclotron motion of frequency $\omega_c = eB_0/m_0$ in the frame of reference traveling $v_\parallel$, parallel to the magnetic field, so in the laboratory frame, trace out a helical orbit.

If the electron velocity is at an angle, α, to the magnetic field, we can write that $v_\perp = v \sin\alpha$ and

$$\frac{dv}{dt} = \frac{e\, B_0\, v \sin\alpha}{m_0} \tag{5.2}$$

where dv/dt is directed perpendicular to the magnetic field. Using Larmor's formula (Equation 2.66), the radiated electromagnetic power associated with the acceleration is given by

$$\begin{aligned} P &= \frac{e^2}{4\pi\varepsilon_0} \frac{2}{3c} \frac{e^2 B_0^2}{m_0^2} \frac{v^2 \sin^2\alpha}{c^2} \\ &= \frac{e^2}{4\pi\varepsilon_0} \frac{2}{3c} \omega_c^2 \left(\frac{v}{c}\right)^2 \sin^2\alpha \end{aligned} \tag{5.3}$$

and is known as 'cyclotron radiation'. As an electron executes its orbit around the magnetic field, the electric field of the radiation detected at a position far from the orbiting electron varies in time. For non-relativistic electron motion with an observation angle approximately perpendicular to the magnetic field lines, the radiated instantaneous power peaks when the acceleration is directed perpendicular to the line of sight ($\theta = \pi/2$) and then oscillates as $\sin^2\theta$ (see Equation 2.65) with 'lobes' of equal power emitted perpendicular to the acceleration. The angular frequency of this radiation is consequently equal to the orbiting frequency $\omega_c = eB_0/m_0$.

If cyclotron emission is observed at an angle varying from $\pi/2$ to the magnetic field line, harmonics will be observed. Viewing at an angle to the magnetic field normal, the observer will see peaks of radiated electromagnetic power which alternate high and low emission as the θ angular dependence of the power shown in Figure 2.65 sweeps around with the changing acceleration direction perpendicular to the magnetic field. The Fourier transform of the high/low pattern comprises harmonics of frequency $m\omega_c$, where m is an integer.

Relativistic effects cause the lobes of instantaneous emitted power to become non-symmetric so that the power in the direction of the velocity vector is enhanced. An observer far from the orbiting electron will see peaks of electric field which occur with a frequency of the cyclotron frequency, but with an enhanced instantaneous power when the velocity is towards the observer. The Fourier transform of this non-sinusoidal pattern is no longer a single frequency ω_c, but rather can also comprise ω_c and a range of harmonics at frequencies $m\omega_c$, where m is an integer.

If the electron velocity distribution is isotropic, we can average the expression for $\sin\alpha$ over the surface of a sphere and obtain for the radiated electromagnetic power

$$P_{av} = \int_0^{\pi} P\, 2\pi \sin\alpha d\alpha \tag{5.4}$$

$$= \frac{e^2}{4\pi\varepsilon_0}\frac{2}{3c}\omega_c^2\left(\frac{v}{c}\right)^2\int_0^{\pi} 2\pi\sin^3\alpha\, d\alpha$$

$$= \frac{e^2}{4\pi\varepsilon_0}\frac{2}{3c}\omega_c^2\left(\frac{v}{c}\right)^2\frac{8\pi}{3}. \tag{5.5}$$

Cyclotron radiation is an energy-loss mechanism in magnetised plasmas. From the form of Equation 5.5, we have that the power loss is proportional to the square of the electron speeds v^2 and hence the loss is proportional to the electron kinetic energy. The kinetic energy E_{kin} of the electrons will decay exponentially from a plasma such that $E_{kin}(t) = E_{kin}(0)\exp(t/t_0)$ with time constant

$$t_0 = 1\Bigg/\left[\frac{e^2}{4\pi\varepsilon_0}\frac{2}{3c}\omega_c^2\left(\frac{1}{c}\right)^2\frac{8\pi}{3}\frac{2}{m_0}\right] = \frac{9}{8}\epsilon_0\frac{c^3m_0^3}{e^4B_0^2} \approx \frac{0.3}{B_0^2}\text{ seconds.}$$

In practise, we find that cyclotron radiation is readily re-absorbed when emitted unless the plasma density is low, so that the time t_0 of kinetic energy decay is a lower bound.

We can estimate the absorption coefficient K_C for cyclotron radiation at electron density n_e using detailed balance between emission ϵ and absorption. We have $\epsilon/K_C = n_e P_{av}/K_C = < I_P(\omega) > \Delta\omega$, where $\Delta\omega$ is a measure of the spectral width of the cyclotron radiation. We assume that each individual electron emits independently and not coherently with other electrons. The frequencies of cyclotron emission are in the infra-red so $\hbar\omega \ll k_B T$ and hence the black-body radiation $< I_P(\omega) >$ in the low frequency Rayleigh-Jeans limit (see Equation 4.10) can be used. We obtain for the peak absorption coefficient for cyclotron radiation

$$\begin{aligned} K_C &= n_e \frac{e^2}{4\pi\epsilon_0}\left(\frac{16\pi^3}{9}\right)\frac{v}{k_B T}\frac{1}{\Delta\omega} \\ &\approx n_e \frac{e^2}{4\pi\epsilon_0}\left(\frac{16\pi^3}{9}\right)\frac{c}{k_B T\,\omega_c} \\ &\approx 4\times 10^{-18}\frac{n_e}{k_B T\,\omega_c} \end{aligned} \tag{5.6}$$

assuming that the spectral width associated with the cyclotron radiation is Doppler broadened so that $\Delta\omega \approx \omega_c(v/c)$ and then evaluating in SI units. Due to the large absorption of the cyclotron radiation at electron densities $n_e > 10^{19}\,\mathrm{m}^{-3}$ with say $\omega_c = 1.8\times 10^{11}$ radians s^{-1} Tesla^{-1} and $k_B T \approx 1\,\mathrm{keV} = 1.6\times 10^{-16}$ joules, the net emission is given by the black-body expression (Equation 4.10). In the Rayleigh-Jeans limit, the emitted intensity is proportional to the plasma electron temperature, T.

5.1.1 Electron Cyclotron Emission as a Diagnostic of Temperature in Tokamaks

A tokamak is a device that uses a magnetic field to confine plasma in the shape of a torus. Current-carrying coils around the torus combined with an induced electron current flow around the torus produce helical magnetic field lines directed around the torus. Charged particles orbit around the magnetic field lines (as discussed in Section 5.1), so are confined to some extent within the magnetic field topology. There is a good potential for energy production from fusion of the deuterium and tritium isotopes of hydrogen in the confined plasma [35]. To produce the electron current around the torus, an electrical coil in the central 'hole' of the torus acts as the primary coil in a transformer configuration, with the toroidally shaped plasma acting as the secondary of the transformer. The current carrying coils around the torus produce the dominant magnetic field and are all placed on a radius from the centre of the central 'hole' in order to produce a magnetic field perpendicular to the radius. These coils cause the magnetic field to vary as $1/R$ where R is the distance from the centre of the toroidal central 'hole' (R is known as the major radius).

We have seen that the net emission from cyclotron radiation at the densities found in laboratory plasmas is given by the Planck black-body radiation formula in the low-frequency limit, so the emitted intensity is proportional to the plasma electron temperature, T. As the toroidal magnetic field in a tokamak varies as $1/R$ where R is the major radius, the frequencies of the cyclotron radiation $\omega_c = (eB_0/m_0)$ also vary spatially. The spectral width of harmonic emission is usually significantly smaller than the variation in harmonic frequency associated with the spatial variation of the magnetic field. Frequency variations in optically thick emission in tokamaks can thus be associated with different spatial regions of the plasma and it is possible to associate a spatial region of magnetic field strength B_0 with a temperature determined by the absolute intensity $I(\omega_c)$ of cyclotron emission, i.e:

$$\begin{aligned} k_B T &= 8\pi^3 c^2 \left(\frac{1}{\omega_c}\right)^2 I(\omega_c) \\ &= 8\pi^3 c^2 \left(\frac{m_0}{eB_0}\right)^2 I(\omega_c). \end{aligned} \tag{5.7}$$

In a tokamak, given that $B \propto 1/R$, we have that:

$$k_B T = I\, 8\pi^3 c^2 \left(\frac{R}{\omega_0 R_0}\right)^2 \tag{5.8}$$

where ω_0 is the cyclotron frequency at major radius R_0. An instrument detecting the emitted cyclotron emission will have a spectral resolution $\Delta\omega_i$. We can write that:

$$\frac{\Delta\omega_i}{\omega_0} = \frac{\Delta R}{R_0} \tag{5.9}$$

where ΔR corresponds to the spatial resolution associated with the spectral resolution.

5.2 Bremsstrahlung

The acceleration of an electron in the field of an ion of charge $+Z_i e$ depends on the distance of closest approach p of the electron to the ion, which for small angle collisions is the same as the 'impact factor' (the closest distance of the electron to the ion if undeviated by the Coulomb attraction). We have

$$\frac{dv}{dt} = \frac{1}{4\pi\epsilon_0} \frac{Z_i e^2}{m_0 p^2}.$$

For small angle collisions, the time over which a collision occurs can be regarded as $2p/v$. The bremsstrahlung pulse lasts for a time t_a and so when Fourier analysed produces a dominant contribution to that part of the frequency spectrum with

$$\omega = \frac{1}{t_a} = \frac{v}{2p}.$$

Using Equation 2.66 for the power P radiated by an accelerated charge, the energy radiated in a single collision is given by Pt_a where

$$Pt_a = \left(\frac{e^2}{4\pi\epsilon_0}\right)^3 \frac{4}{3c^3}\frac{Z_i^2}{m_0^2 p^3 v}.$$

The number of collisions per unit time per electron with impact parameter between p and $p + dp$ is given by $n_{Z_i} v 2\pi p dp$, where n_{Z_i} is the ion density. As $p = v/(2\omega)$, we have that $dp = -(v/2\omega^2)d\omega$, so the power radiated in the frequency range ω to $\omega + d\omega$ due to a single electron is given by

$$\frac{dE(\omega)}{dt}d\omega = Pt_a n_{Z_i+1} v 2\pi p dp = \left(\frac{e^2}{4\pi\epsilon_0}\right)^3 \frac{4}{3c^3}\frac{Z_i^2}{m_0^2 v} 2\pi n_{Z_i} d\omega. \tag{5.10}$$

Calculating the power radiated due to all electrons requires that we integrate over the velocity distribution of all electrons which have sufficient energy to produce photons of energy $\hbar\omega$. In equilibrium the electron velocity distribution is given by a Maxwellian distribution. The number of electrons per unit volume with speeds between v and $v + dv$ for a Maxwellian distribution is given by

$$f_v(v)dv = n_e \left(\frac{m}{2\pi k_B T}\right)^{3/2} 4\pi v^2 \exp\left(-\frac{m_0 v^2}{2k_B T}\right) dv \tag{5.11}$$

where n_e is the electron density. To obtain the total power $\epsilon_{tot}(\omega)d\omega$ emitted over the frequency range ω to $\omega + d\omega$, we need to undertake the integration

$$\epsilon_{ff}(\omega) = \int_{(2\hbar\omega/m_0)^{1/2}}^{\infty} \frac{dE(\omega)}{dt} f_v(v)dv.$$

The total power emitted between frequency ω and $\omega + d\omega$ per unit time per unit volume can then be written as

$$\epsilon_{ff}(\omega)d\omega = 16\sqrt{\pi}\left(\frac{e^2}{4\pi\epsilon_0}\right)^3 \frac{4}{3c^3}\frac{Z_i^2}{m^2} n_e n_{Z_i} \left(\frac{m_0}{2k_B T}\right)^{1/2} I_{ff}\, d\omega \tag{5.12}$$

where we set

$$I_{ff} = \int_{\sqrt{\hbar\omega/k_B T}}^{\infty} x \exp(-x^2)dx.$$

Evaluating the integral gives

$$\epsilon_{ff}(\omega)d\omega = 8\sqrt{\pi}\left(\frac{e^2}{4\pi\epsilon_0}\right)^3 \frac{4}{3c^3}\frac{Z_i^2}{m_0^2} n_e n_{Z_i} \left(\frac{m_0}{2k_B T}\right)^{1/2} \exp\left(-\frac{\hbar\omega}{k_B T}\right) d\omega. \tag{5.13}$$

This expression shows the dependence of bremsstrahlung radiation from a plasma. It is possible to measure the electron temperature T by recording the variation of power emitted as a function of frequency as there is a simple exponential decrease of power with frequency.

In high-temperature plasmas of low atomic number, the dominant radiation energy loss arises from bremsstrahlung. The total bremsstrahlung emission can be used for radiation-loss calculations [35]. Integrating Equation 5.13 over all frequencies gives

$$\epsilon_{fftot} = \int_{\omega=0}^{\infty} \epsilon_{ff}(\omega) d\omega = 4\sqrt{2\pi} \left(\frac{e^2}{4\pi\epsilon_0} \right)^3 \frac{4}{3\hbar c^3} \frac{Z_i^2}{m_0^{1.5}} n_e n_{Z_i} (k_B T)^{1/2}. \tag{5.14}$$

The dependencies here on ion charge Z_i, temperatures T and densities (plus the physical constants such as c, $\hbar$) agree with more exact quantum mechanical evaluations. To have the correct numerical multiplier, the expression needs to be multiplied by $< G_{ff} >$ and by $1/\sqrt{3}$, where $< G_{ff} >\approx 1$ is the frequency-averaged Gaunt factor [57, 73]. Gaunt factors are evaluated taking into account the quantum processes of the electron–ion collision. A factor $1/\sqrt{3}$ arises if a precise classical evaluation of the radiation emission, rather than, for example, our assumptions leading to Equation 5.10, is employed. Kramers [62] was the first to obtain the exact classical bremsstrahlung expression for an electron in a Coulombic field and so the classical expressions for Equations 5.13 and 5.14 with a $1/\sqrt{3}$ multiplier are often referred to as the Kramers values. Some authors suggest a slightly different multiplier. For example, Freidberg [35] indicates that for Coulombic collisions of free electrons with ionised hydrogen (i.e. a proton), an accurate expression for the total radiation emission including quantum mechanical effects is obtained if Equation 5.14 is multiplied by $2/\pi = 0.64$ (which is close to $1/\sqrt{3} = 0.58$).

5.3 Inverse Bremsstrahlung Absorption

In bremsstrahlung, an electron radiates electromagnetic energy as it is accelerated in the field of an ion. The process of inverse bremsstrahlung occurs when the electron in the field of an ion absorbs energy from an electromagnetic wave. When the radiation field is in equilibrium, the rates of bremsstrahlung and inverse bremsstrahlung are in detailed balance. We can consequently obtain an expression for the absorption coefficient for inverse bremsstrahlung from the power emitted by bremsstrahlung. The expression for the absorption coefficient K_{ib} for inverse bremsstrahlung is obtained using

$$K_{ff} < I_p(\omega) >= \epsilon_{ff}(\omega)$$

where $< I_p(\omega) > d\omega$ is the black-body radiation intensity between frequency ω and $\omega + d\omega$. We have that

$$< I_p(\omega) >= \frac{\hbar\omega^3}{\pi^2 c^2} \frac{1}{\exp\left(\frac{\hbar\omega}{k_B T}\right) - 1}.$$

The absorption coefficient for inverse bremsstrahlung can now be readily calculated. We have

$$K_{ff} = 8\sqrt{\pi}\left(\frac{e^2}{4\pi\epsilon_0}\right)^3 \frac{4}{3c^3}\frac{Z_i^2}{m_0^2} n_e n_{Z_i} \left(\frac{m_0}{2k_B T}\right)^{1/2} \frac{\pi^2 c^2}{\hbar\omega^3}\left(1 - \exp\left(-\frac{\hbar\omega}{k_B T}\right)\right). \tag{5.15}$$

The absorption of radiation due to free-free inverse bremsstrahlung varies as $n_e n_{Z_i} \omega^{-3} T^{-1/2}$. The exponential term in the bracket represents the negative effect on absorption (i.e. the production of photons) associated with stimulated emission. Inverse bremsstrahlung is the dominant laser absorption process for laser heating of plasma in inertial fusion research. A simplified form of Equation 5.15 for the absorption coefficient for inverse bremsstrahlung is often used in laser-produced plasma research (see Exercise 5.8).

5.4 Radiative Recombination

An electron can be radiatively captured by an ion with the electron finishing bound in a quantum state of the ion. All the electron's kinetic energy plus the energy difference of the final quantum state and the ionisation energy of the ion is released as photon energy. The emission is known as recombination radiation or free-bound radiation because it results from electrons recombining from free, unbound states to bound, discrete quantum states.

To obtain an expression for the power radiated by the process of radiative recombination, we can use the result obtained for the bremsstrahlung power and invoke the correspondence principle to take into account the essential quantum mechanical nature of the process.[1] When an electron recombines from the continuum of free states to a discrete quantum state within an ion, we assume that classical theory (in particular Equation 5.12) predicts the radiated power, but with (i) a simple adjustment of the photon energy emitted to take account of the extra ionisation energy released in the recombination process, and (ii) the use of an estimate of the appropriate free-electron velocity range for a given photon energy.

The initial free electron velocity range dv (which will contribute to photon emission of a particular energy E_p) when recombining to a quantum state with energy E_{ion} below the ionisation limit of the ion occurs is given by:

[1] The correspondence principle states that a more general theory (such as quantum mechanics) must reduce to the less general (for example, classical theory) when applied to circumstances where it is known that the less general theory is valid. Classical theory must be valid when the density of quantum states is large (i.e. there is a continuum of states).

$$dv = \frac{1}{m_0 v} dE_p \tag{5.16}$$

where the energy range dE_p has to be such that the correspondence principle is satisfied. We assume initially that the quantum states into which an electron can recombine are similar to those of a hydrogen-like ion with energies scaling proportionally to the square of the initial charge of the ion Z_i^2. For recombination into high-lying quantum states (large principal quantum number n), this is a reasonable approximation as these quantum states have potential energy defined by a central potential of nuclear charge minus the shielding effect of some un-ionised (low-lying) electrons. It is also an accurate assumption for recombination to form a hydrogen-like ion in any quantum state n. We can write:

$$E_{ion} = R_d \frac{Z_i^2}{n^2} \tag{5.17}$$

where n is the principal quantum number to which recombination has occurred and R_d is the Rydberg constant. The energy range dE_p is then:

$$\begin{aligned} dE_p &\approx R_d Z_i^2 \left[\frac{1}{(n-\frac{1}{2})^2} - \frac{1}{(n+\frac{1}{2})^2} \right] \\ &= R_d Z_i^2 \frac{2n}{(n^2 - \frac{1}{4})^2} \\ &\approx \frac{2 R_d Z_i^2}{n^3}. \end{aligned} \tag{5.18}$$

Determining the energy range of recombining electrons using Equation 5.18 is not accurate for low-lying levels (small n), but is a reasonable approximation for recombination into high-lying quantum states (large n) where the quantum state energies are close together. To partly correct the inaccuracy for low-lying levels, we later substitute exact ionisation energies for the factor $R_d Z_i^2/n^2$ for a final expression for recombination radiation (Equation 5.29).

Using Equation 5.12 as a template, the total power radiated when an electron recombines to a quantum state n is given by multiplying the expression (Equation 5.10) for the power radiated by bremsstrahlung by the Maxwellian distribution for the number of electrons at the appropriate speed (Equation 5.11). We have

$$\epsilon_{fb}(\hbar\omega) = \frac{dE(\omega - \frac{E_{ion}}{\hbar})}{dt} f_v(v) dv.$$

The electron speed range dv can be written in terms of the photon energy range dE_p (Equation 5.16) and the notional range of bound quantum state energies (Equation 5.18). Substituting the Maxwellian distribution of speeds $f_v(v)$ (Equation 5.11) and a modified expression $dE(\omega - E_{ion}/\hbar)/dt$ for the power radiated in an electron–ion collision (following Equation 5.10) gives

$$\epsilon_{fb}(\hbar\omega) = 8\sqrt{\pi}\left(\frac{e^2}{4\pi\epsilon_0}\right)^3 \frac{4}{3c^3}\frac{Z_i^2}{m_0^2} n_e n_{Z_i+1}\left(\frac{m_0}{2k_BT}\right)^{1/2}\left[\frac{2R_dZ_i^2}{n^3}\frac{e^{\frac{E_{ion}}{k_BT}}}{k_BT}\right] e^{-\frac{\hbar\omega}{k_BT}}. \tag{5.19}$$

This expression for the power radiated by recombination radiation is the same as our expression for the power radiated by bremsstrahlung (Equation 5.12) multiplied by the quantities in the square brackets. Our simple treatment gives an answer close to that for a full quantum mechanical treatment. The dependences on photon energy $\left[\propto \exp\left(-\frac{E_p}{k_BT}\right)\right]$ electron temperature $\left[\propto \exp\left(-\frac{E_p}{k_BT}\right)/(k_BT)^{\frac{3}{2}}\right]$, ion charge $\left[\propto Z_i^4\right]$ and electron and ion density $\left[\propto n_e n_{Z_i+1}\right]$ are correct. The dependence on photon energy is the same as for bremsstrahlung, which is useful in the deduction of electron temperature from the inverse slope of the bremsstrahlung and recombination radiation as it is not necessary to know what process (free-free or free-bound) is producing the radiation. However, recombination radiation only contributes emission for photon energies greater than the ionisation energy E_{ion} of the quantum state involved and so produces a characteristic step in the continuum emission at the ionisation energy.

Our discussion on photo-ionisation, the inverse process to recombination, in Section 5.5 illustrates that to take account of quantum mechanical effects we need to multiply the expression (Equation 5.19) by $G_{fb}/\sqrt{3}$, where $G_{fb} \approx 1$ is known as the Gaunt factor. Equation 5.19 becomes

$$\epsilon_{fb}(\hbar\omega) = \frac{4\sqrt{2\pi}}{\sqrt{3}}\left(\frac{e^2}{4\pi\epsilon_0}\right)^3 \frac{4}{3c^3}\frac{1}{m_0^{3/2}}\frac{2R_dZ_i^4}{n^3} n_e n_{Z_i+1}\Theta(T) \tag{5.20}$$

where $\Theta(T)$ represents the temperature-dependent parameters for free-bound emission. We have

$$\Theta(T) = G_{fb}\left(\frac{1}{k_BT}\right)^{3/2} \exp\left(\frac{E_{ion} - \hbar\omega}{k_BT}\right). \tag{5.21}$$

As well as producing continuum radiation emission, free-bound recombination reduces the number of free electrons and populates the bound states of ions. We need to consider the rate of this process in evaluations of quantum state populations (see Chapter 12). Using the concept of a rate coefficient A_{Cp} for free-bound recombination, the rate of radiative recombination to a bound quantum state p in a Z_i-charged ion can be written as $dN_p/dt = n_e n_{Z_i+1} A_{Cp}$. An appropriate expression for the transition probability for free-bound recombination is then found using Equation 5.20. We note that $\epsilon_{fb} = (dN_p/dt)\hbar\omega$ and integrate the emission of the free-bound radiation over the range of possible photon energies. The photon energy integration leads to an integral of form

$$\int_{E_{ion}}^{\infty} \frac{1}{\hbar\omega} \exp\left(-\frac{\hbar\omega}{k_B T}\right) d(\hbar\omega) = E_i\left(\frac{E_{ion}}{k_B T}\right)$$

where $E_i(E_{ion}/k_B T)$ is known as the exponential integral given by

$$E_i(y) = \int_{y}^{\infty} \frac{e^{-x}}{x} dx.$$

The rate coefficient for radiative recombination to a bound state is then given by

$$A_{Cp} = \frac{4\sqrt{2\pi}}{\sqrt{3}} \left(\frac{e^2}{4\pi\epsilon_0}\right)^3 \frac{4}{3c^3} \frac{1}{m_0^{3/2}} \frac{2R_d Z_i^4}{\hbar n^3} \Theta_I(T) \tag{5.22}$$

where the temperature-dependent parameters for free-bound recombination after integration over all photon energies are given by

$$\Theta_I(T) = \tilde{G}_{fb} \left(\frac{1}{k_B T}\right)^{3/2} \exp\left(\frac{E_{ion}}{k_B T}\right) E_i\left(\frac{E_{ion}}{k_B T}\right) \tag{5.23}$$

where $\tilde{G}_{fb} \approx 1$ is a frequency-averaged Gaunt factor. The exponential and exponential integral component $(\exp(E_{ion}/k_B T)E_i(E_{ion}/k_B T))$ in the dependence of $\Theta_I(T)$ increases slowly with temperature $k_B T$ at a rate approximately proportional to $0.5 k_B T/E_{ion}$ (see Figure 5.1). The overall temperature dependence of the rate

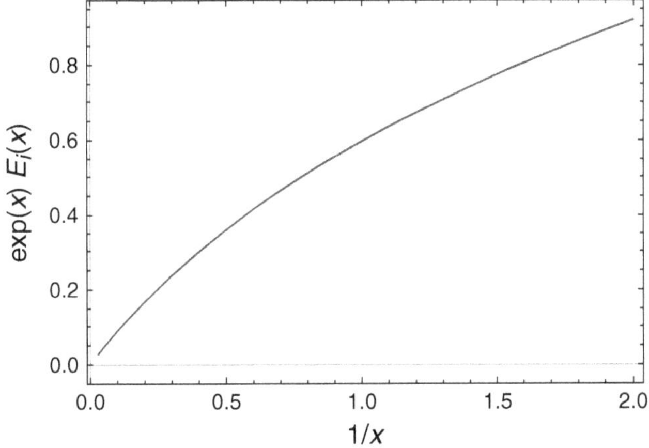

Figure 5.1 The exponential multiplied by the exponential integral component $(\exp(x)E_i(x))$ in the dependence of $\Theta_I(T)$ (Equation 5.23) for the frequency-integrated free-bound recombination rate coefficient as a function of the temperatures/ionisation energy $1/x = (k_B T/E_{ion})$.

coefficient for free-bound recombination to single quantum state is consequently approximately $\propto T^{-1/2}$.

5.5 Photo-Ionisation

We can obtain an expression for the absorption coefficient for photo-ionisation from the power emitted by recombination radiation. As for bremsstrahlung and inverse bremsstrahlung, in an equilibrium plasma the absorption coefficient K_{pi} for photo-ionisation is related by detailed balance to the power $\epsilon_{rec}(\omega)$ emitted in radiative recombination. We have

$$K_{bf} < I_p(\omega) >= \epsilon_{fb}(\omega)$$

where $< I_p(\omega) > d\omega$ is the black-body radiation intensity between frequency ω and $\omega + d\omega$. Using Equation 5.19, we obtain for the photo-ionisation absorption coefficient

$$K_{bf} = 8\sqrt{\pi}\left(\frac{e^2}{4\pi\epsilon_0}\right)^3 \frac{4}{3c^3}\frac{Z_i^2}{m_0^2} n_e n_{Z_i+1}\left(\frac{m_0}{2k_B T}\right)^{1/2} \frac{\pi^2 c^2}{\hbar\omega^3}\left(1 - e^{-\frac{\hbar\omega}{k_B T}}\right) C_{bf} \quad (5.24)$$

where the C_{bf} term is given by

$$C_{bf} = \left[\frac{2R_d Z_i^2}{n^3}\frac{e^{\frac{E_{ion}}{k_B T}}}{k_B T}\right].$$

We need to note that $n_{Z_i+1} = n_{Z_i} n_{Z_i+1}/n_{Z_i}$ and use the Saha-Boltzmann ratio for n_{Z_i+1}/n_{Z_i} to obtain the absorption coefficient in terms of the population n_{Z_i} of the ion absorbing the radiation. We obtain

$$K_{bf} = 32\pi^2 \frac{g_{Z_i+1}}{g_{Z_i}}\left(\frac{e^2}{4\pi\epsilon_0}\right)^3 \frac{4}{3c^3}\left(\frac{1}{h}\right)^3 \frac{\pi^2 c^2}{\hbar\omega^3}\left(1 - \exp\left(-\frac{\hbar\omega}{k_B T}\right)\right)\left[\frac{2R_d Z_i^4}{n^3}\right] n_{Z_i} \quad (5.25)$$

where g_{Z_i+1} and g_{Z_i} are the degeneracies of the final and initial bound quantum states respectively.

The contribution from the $\exp(-\hbar\omega/k_B T)$ term allows for stimulated emission. It is convenient to allow for the process of stimulated emission as a negative effect on the photo-ionisation absorption coefficient (see the discussion in Section 4.4). The cross-section σ_{bf} for photo-ionisation neglecting stimulated emission for a single ion can be readily seen to be independent of temperature with an expression

$$\sigma_{bf} = \frac{K_{bf}}{n_{Z_i}}$$

so that

$$\sigma_{bf} = 32\pi^2 \frac{g_{Z_i+1}}{g_{Z_i}} \left(\frac{e^2}{4\pi\epsilon_0}\right)^3 \frac{4}{3c^3} \left(\frac{1}{h}\right)^3 \frac{\pi^2 c^2}{\hbar\omega^3} \left[\frac{2R_d Z_i^4}{n^3}\right]$$
$$= \frac{16}{3\pi} \frac{g_{Z_i+1}}{g_{Z_i}} \alpha^3 \frac{\pi^2 c^2}{\hbar\omega^3} \left[\frac{2R_d Z_i^4}{n^3}\right] \tag{5.26}$$

upon substituting for the parameters of the fine structure constant (see Exercise 1.6):

$$\alpha = \frac{e^2}{4\pi\epsilon_0 \hbar c} \approx \frac{1}{137}.$$

As the photo-ionisation cross-section is an atomic parameter, it should be independent of the plasma environment (e.g. the density and temperature) as we have found. If the photon energy $\hbar\omega$ is less than the ionisation energy, the cross-section is zero, so the above equation only applies for photon energies greater than the ionisation energy.

Although approximate due to the simplifications made in determining the expressions for bremsstrahlung and recombination radiation, Equation 5.26 does predict the correct scaling of the photo-ionisation cross-section for hydrogen-like ions with frequency ($\propto 1/\omega^3$) and atomic number ($\propto Z_i^4$). For the ionisation of hydrogen-like ions, the ratio of degeneracies $g_{Z_i+1}/g_{Z_i} = 1/2n^2$, so that the correct $1/n^5$ scaling with the principal quantum number n of the initial energy level is obtained. Karzas and Latter [57] undertook exact quantum mechanical calculations of the photo-ionisation cross-section for hydrogen-like ions and found in units of $\text{m}^2(\text{rad s}^{-1})^{-1}$ here that

$$\sigma_{bf} = \frac{16}{3\pi} \alpha^3 \frac{\pi^2 c^2}{\hbar\omega^3} \left[\frac{R_d Z_i^4}{n^5}\right] \frac{G(\omega, n, l, Z_i)}{\sqrt{3}} \tag{5.27}$$

where $G(\omega, n, l, Z_i) \approx 1$ is a quantum mechanical correction known as the Gaunt factor. The exact expression for the photo-ionisation cross-section of hydrogen-like ions (Equation 5.27) differs from Equation 5.26 by the factor $G(\omega, n, l, Z_i)/\sqrt{3}$. The Gaunt factor $G(\omega, n, l, Z_i)$ depends on the principal n and angular quantum numbers l and on the atomic number Z_i and radiation frequency ω. Values are typically close to unity, but more accurate values can be found in published tabulations or from computer codes [56, 57].

5.6 Generalised Expressions for Radiative Processes Involving Free Electrons

Writing the ionisation energy E_{ion} as $E_{ion} = R_d Z_i^2/n^2$ and substituting for the wavelength λ of the radiation, the expression for the photo-ionisation cross-section can be reduced to a particularly simple form. We have if $\hbar\omega \geq E_{ion}$ that

$$\sigma_{bf} = \frac{8}{3\pi}\frac{g_{Z_i+1}}{g_{Z_i}}\alpha^3\lambda^2\left[\frac{E_{ion}Z_i^2}{\hbar\omega\, n}\right]\frac{G_{bf}}{\sqrt{3}}. \tag{5.28}$$

where G_{bf} is the appropriate Gaunt factor for photo-ionisation.

We can combine and generalise the expressions for free-free emission (Equation 5.13) and free-bound recombination emission (Equation 5.19) found for hydrogen-like ions taking account of the quantum mechanical correction and utilising the expression for the fine structure constant α. The total continuum emission associated with an arbitrary ion of number density n_{Z_i+1} associated with bremsstrahlung and recombination to ion charge Z_i is then given by

$$\epsilon(\omega) = \epsilon_{ff}(\omega) + \epsilon_{fb}(\omega)$$

so that

$$\epsilon(\omega) = \frac{32\sqrt{\pi}}{3\sqrt{3}}\alpha^3\hbar^3\left(\frac{m_0}{2k_BT}\right)^{1/2}(Z_i+1)^2 n_e n_{Z_i+1}\left(e^{-\frac{\hbar\omega}{k_BT}}\right)C_{ff/fb} \tag{5.29}$$

where

$$C_{ff/fb} = \left[G_{ff} + \sum_n G_{fb}\frac{2E_n}{n}\left(\frac{u_{Z_i}}{g_{Z_i}}\right)\frac{e^{\frac{E_n}{k_BT}}}{k_BT}\right].$$

We have replaced the hydrogen-like ionisation energies $R_dZ_i^2/n^2$ with an energy E_n to represent the actual ionisation energy for an energy level of principal quantum number n in the $C_{ff/fb}$ expression. The degeneracy g_{Z_i} of the lower bound level and the number of holes u_{Z_i} in this level before the recombination are now explicitly included in Equation 5.29. The rate of recombination for ions which are not hydrogen-like is reduced by u_{Z_i}/g_{Z_i} as some of the quantum states are already occupied for non-hydrogen-like ions. (For hydrogen-like ions $u_{Z_i}/g_{Z_i} = 1$.) The Gaunt factors G_{ff} and G_{fb} representing the appropriate quantum mechanical corrections for free-free and free-bound processes respectively plus the $1/\sqrt{3}$ multiplier are also included. Detailed balance between photo-ionisation and radiative recombination ensures that the two Gaunt factors are equal: $G_{bf} = G_{fb}$. The summation is over all levels n with non-zero holes u_{Z_i} per level.

For free-free inverse bremsstrahlung absorption, Equation 5.15 can be multiplied by $G_{ff}/\sqrt{3}$, where G_{ff} is the free-free Gaunt factor. Evaluating the numerical constants and allowing for the effect of a non-unity refractive index η gives K_{ib} in units of cm^{-1} at frequency $\nu = \omega/2\pi$ such that

$$K_{ff} = 1.75\times 10^6\frac{n_e n_{Z_i} Z_i^2}{\eta\nu^3(k_BT)^{1/2}}G_{ff}\left(1-\exp\left(-\frac{h\nu}{k_BT}\right)\right). \tag{5.30}$$

We need to input the electron and ion densities (n_e and n_{Z_i+1}) in units of cm^{-3}, temperatures k_BT in electron volts and frequencies ν in Hz.

Exercises

5.1 The interstellar region of space has a magnetic field of approximately 10^{-10} Tesla. Over the lifetime of the universe (13.8×10^9 years) and assuming that the interstellar magnetic field has remained constant, determine the fraction of kinetic energy of interstellar plasma still remaining if the only loss mechanism has been cyclotron radiation. [0.986]

5.2 In 1977 the neutron star Hercules X-1 was found to be emitting an emission line at photon energy of 55 keV [112]. If this emission arises from electron cyclotron radiation, determine the magnetic field at the surface of the neutron star. [4.7×10^8 Tesla assuming no relativistic mass increase of the electrons]

5.3 The ITER tokamak is designed to have a peak magnetic field of 11.8 Tesla. What is the minimum wavelength of the fundamental cyclotron emission? [0.9 mm]

5.4 Consider radiative recombination of a fully stripped ion to form the ground state of a hydrogen-like ion of atomic number Z. From the expressions given in this chapter, show that the ratio $\epsilon_{fb}/\epsilon_{ff}$ of this emission to that of bremsstrahlung (free-free emission) is given by

$$\frac{\epsilon_{fb}}{\epsilon_{ff}} = \frac{G_{fb}}{G_{ff}} \frac{R_d Z^2}{k_B T} \exp\left(\frac{R_d Z^2}{k_B T}\right).$$

5.5 Evaluate the rate coefficient for free-bound recombination (Equation 5.22) to show that

$$A_{Cp} = 5.21 \times 10^{-20} \frac{Z_i^4}{n^3} \tilde{G}_{fb} \left(\frac{R_d}{k_B T}\right)^{3/2} \exp\left(\frac{E_{ion}}{k_B T}\right) E_i(\frac{E_{ion}}{k_B T}) \ \ \mathrm{m^3 s^{-1}}.$$

5.6 Hydrogen gas surrounding a star is ionised to produce fully stripped hydrogen ions where there are no bound electrons (denoted by HII). Assume that the star emits S^* photons per second with sufficient energy (>13.6 eV) to photo-ionise hydrogen and that radiative free-bound recombination balances this photo-ionisation. Show that the radius of gas which is ionised by the radiation from the star is given by

$$R_s = \left(\frac{3S^*}{4\pi n_H^2 A_{Cp}}\right)^{1/3}$$

where n_H is the hydrogen ion number density and A_{Cp} is the rate coefficient for radiative recombination. An ionised gaseous nebula surrounding a star as described here is known as a Strömgren sphere and sometimes as planetary nebulae: early astronomers with small telescopes thought the hydrogen plasma surrounding some stars looked like planets.

5.7 Use Figure 5.1 and the answers to Exercises 5.5 and 5.6 to estimate the radius of a Strömgren sphere of ionised hydrogen of density $10^8\,\mathrm{m}^{-3}$ and temperature 1 eV surrounding a star with output $S^* = 10^{49}$ photons per second. [2.9×10^{17} m = 9.2 parsec]

5.8 In laser-produced plasmas, inverse bremsstrahlung is often the major absorption mechanism for the laser light. Given that electron temperatures k_BT in such plasmas are typically much greater than the laser photon energy $\hbar\omega$, use Equation 5.15 to show that the absorption of laser light by inverse bremsstrahlung varies as

$$K_{ff} = 2.7 \times 10^{-50} \frac{\lambda^2 n_e n_{Z_i} Z_i^2}{(k_BT)^{3/2}} \quad \mathrm{m}^{-1}$$

where λ is the laser wavelength in microns and the temperature k_BT is measured in eV. The electron density n_e and ion density n_{Z_i} are in units of m^{-3}.

6

Opacity

The opacity of a medium is its impenetrability to radiation. For electromagnetic radiation in plasmas, the opacity arises due to absorption by free and bound electrons or due to scattering (where some radiation is re-emitted). We discussed scattering in Chapter 3, absorption of radiation in Chapter 4 and the particular physics of absorption by free electrons in Chapter 5.

Radiation opacity in plasma is important at higher densities or in larger plasmas. However, even low-density plasma may have some frequencies of radiation, say in the microwave or radiofrequency spectral range where radiation absorption is significant. Plasma opacity can limit the ability to diagnose conditions within the interior of a plasma and plasma opacity can slow the outflow of energy within a plasma and lead to non-local radiation heating.

Radiation transport calculations use a measure of dimensionless distance known as the optical depth. The optical depth of a thickness of material is the natural logarithm of the ratio of the incident to transmitted radiation power through the material. In the absence of any source of radiation, the radiation power falls exponentially with optical depth. When the optical depth approaches zero, a medium is said to be optically thin for the specified radiation and the opacity can be neglected. Large values of optical depth $\gg 1$ for a plasma or other medium suggest that opacity and radiation transport is significant: a condition known as 'optically thick'.

We explore in this chapter the issue of opacity and the movement of radiation energy in plasmas. After some general treatment of opacity, we consider plasmas which are in, or close to, thermal equilibrium, so that the radiation field can be described by the Planck black-body distribution. Opacity modeling and radiation transport in plasmas close to thermal equilibrium is important in solar and stellar radiation modelling and in laser-produced plasmas.

6.1 The Equation of Radiative Transfer

The radiation intensity within a plasma exchanges energy with the bound and free electrons by emission and absorption processes. Electron transitions between bound discrete quantum states associated with the energy states of an ion or atom produce spectral-line emission, while transitions from free electrons to other free quantum states or to bound quantum states produce continuum emission. The emission processes can be represented by an emission coefficient ϵ, which is the radiated power per unit volume in a frequency range ω to $\omega + d\omega$. Radiation can be absorbed by electrons bound within ions or atoms or by free electrons with resulting conversion of the photon energy to kinetic energy (for free electrons) or a transition of an electron to a higher quantum state (for bound electrons). The rate of radiation absorption can be represented by an absorption coefficient K with dimensions of inverse length.

The equation of radiative transfer was introduced in Section 4.4 by starting with an expression for the change in intensity dI within a small length dz along a line of sight through a medium:

$$dI = \epsilon dz - KIdz. \tag{6.1}$$

The quantity τ is the optical depth and can be defined by considering the optical depth change $d\tau = Kdz$ over an incremental length (dz) along the line of sight. We then have

$$\frac{dI}{d\tau} = \frac{\epsilon}{K} - I = S - I \tag{6.2}$$

where $S = \epsilon/K$ is known as the source function.

For the radiation discussion in this chapter we assume that unless stated otherwise, the source function S and intensity I are measured per unit frequency per unit solid angle, so that the source function is typically in units of $\mathrm{Wm^{-2}\,(Hz)^{-1}\,sr^{-1}}$ and intensity in units of $\mathrm{Wm^{-2}\,(Hz)^{-1}\,sr^{-1}}$. In other chapters (for example, Chapters 4 and 12), we have evaluated the source function, emission coefficient and intensity per 4π solid angle as it is then easier to evaluate expressions for rate coefficients affecting the populations of quantum states (as we do not need to multiply our expressions by 4π).

An important issue in radiative transfer is to understand that instead of thinking of real space z to evaluate radiative-transfer equations, we think of how the source function S varies in optical depth τ space. For a uniform plasma, optical depth τ and distance z are proportional. However, we do not need to restrict the discussion to a uniform medium, though some assumptions regarding the geometry in optical-depth space, such as the assumption of a planar structure, can be useful.

The radiative-transfer equation can be integrated to the form

$$I = \int_0^{\tau} S(\tau') \exp(-\tau') d\tau' + I(0) \exp(-\tau) \tag{6.3}$$

where τ is the optical depth measured through the plasma along a line of sight and $I(0)$ is the intensity incident onto the plasma. For many laboratory plasmas, there is no external source of radiation so that $I(0) = 0$. If the source function is constant with optical depth then $S(\tau') = S_0$ everywhere and

$$I = S_0(1 - \exp(-\tau)). \tag{6.4}$$

For a large plasma with large τ, we see that $I = S_0$ as discussed in Section 4.4.

If a source function varies linearly with optical depth from a surface, we can write for the source function

$$S(\tau) = S_0 + f\tau.$$

The equation of radiative transfer then becomes

$$I = S_0 \int_0^{\tau} \exp(-\tau') d\tau' + f \int_0^{\tau} \tau' \exp(-\tau') d\tau' = S_0 + f + \exp(-\tau)(f(\tau - 1) - S_0).$$

If τ is large

$$I \approx S_0 + f.$$

This expression tells us that the intensity of radiation with a large linearly increasing source function as optical depth increases from the edge of the medium is equal to the source function at an optical depth $\tau = 1$. This is known as the Eddington-Barbier relation. Allied to this result is the concept of the mean optical depth given by

$$\tau_{av} = \int_0^{\infty} \tau' \exp(-\tau') d\tau' = 1.$$

The 'average' or mean optical depth can be regarded as unity.

6.2 Intensities in an Optically Thick Planar Geometry

Many radiative-transfer issues can be approximated to a planar geometry. If the optical depth is large, the depth that needs to be considered in detail is only of the order of unity (as discussed above), so even a non-uniform geometry can approximate to a planar geometry on the important optical depth distances.

We consider a planar geometry where radiation is emitted from uniform planes within a plasma running parallel to the surface. For radiative-transfer calculations, the spatial dimensions are best considered in units of optical depth τ typically measured from the edge of the plasma closest to the observer (where $\tau = 0$). The radiation propagating at an angle of θ to the normal to the plane subtends a solid angle given by $2\pi\tau \sin\theta d\theta$, where τ is the optical depth of the planar element along the angle θ. The intensity of radiation emitted from the plane of thickness $d\tau$ at an optical depth of τ in the angular range θ to $\theta + d\theta$ is $S(\tau)\exp(-\tau)2\pi\tau \sin\theta d\theta d\tau$.

We can evaluate an average intensity $< I >$ of radiation by integrating over optical depth distance and over angle. We obtain

$$< I >= \int_{\theta=0}^{\theta=\pi} \int_{\tau=0}^{\infty} S(\tau)\exp(-\tau)2\pi\tau \sin\theta d\theta d\tau.$$

The integration over angle includes radiation propagating in all directions in the plasma. Integrals as above are more readily solved by letting $u = \cos\theta$, so that $\sin\theta d\theta = -du$. We have

$$< I >= 2\pi \int_{u=-1}^{u=1} \int_{\tau=0}^{\infty} S(\tau)\tau \exp(-\tau)d\tau du. \tag{6.5}$$

For a constant source function S_0, $< I >= 4\pi S_0$ as

$$\int_0^{\infty} \tau \exp(-\tau)d\tau = 1$$

and

$$\int_{u=-1}^{u=1} du = 2.$$

We see that the angle-integrated intensity for a constant source function is equal to the source function value times 4π steradian.

The total flux F escaping per unit area from the plasma is calculated by integrating the function $S(\tau)\cos\theta \exp(-\tau)2\pi\tau \sin\theta d\theta d\tau$ over optical-depth distance and over angle. This is a similar integration to that for Equation 6.5 except with an additional $\cos\theta$ term so as to evaluate the component of the radiation intensity in a plane parallel to the assumed planar geometry. The $\cos\theta$ term adjusts for the radiation being distributed over an area on a plane which increases as $1/\cos\theta$ with increasing angle θ to the plane normal. We obtain

$$F = 2\pi \int_{u=-1}^{u=1} \int_{\tau=0}^{\infty} S(\tau) u\tau \exp(-\tau) d\tau du. \tag{6.6}$$

If we let $I(u) = \int_{\tau=0}^{\infty} S(\tau)\tau \exp(-\tau)d\tau$, then

$$< I >= 2\pi \int_{u=-1}^{u=1} I(u)du = 4\pi J$$

and

$$F = 2\pi \int_{u=-1}^{u=1} I(u)udu = 4\pi H$$

upon defining integrals

$$J = \frac{1}{2} \int_{u=-1}^{u=1} I(u)du \tag{6.7}$$

and

$$H = \frac{1}{2} \int_{u=-1}^{u=1} I(u)udu. \tag{6.8}$$

6.3 Radiation Pressure in a Planar Geometry

For any particle with kinetic energy much greater than its rest mass energy (m_0c^2), the kinetic energy is given by pc where the particle momentum is p. Considering photons as such particles means that each photon has a momentum equal to the photon energy divided by c, so the rate of energy arrival (i.e. intensity) divided by c gives the momentum change upon absorption which is equivalent to pressure. Consequently, radiation of intensity I exerts a pressure of $I\cos^2\theta/c$ when absorbed onto a surface at angle θ to the radiation propagation direction. The $\cos^2\theta$ variation arises because (i) the radiation is distributed over an area on the plane which increases as $1/\cos\theta$ with increasing angle θ to the plane normal (and hence the intensity per unit area on the surface drops as $\cos\theta$) and (ii) the component of the radiation momentum normal to the plane varies as $\cos\theta$. The average radiation pressure for radiation emanating from a planar surface is thus

$$P_{rad} = \frac{2\pi}{c} \int_{u=-1}^{u=1} I(u)u^2du = \frac{4\pi}{c}K$$

where

$$K = \frac{1}{2} \int_{u=-1}^{u=1} I(u) u^2 du. \tag{6.9}$$

If we assume that the radiation intensity $I(u)$ is a power series in u with terms only up to linear, we can write that

$$I(u) = a + bu.$$

Evaluating the J, H and K integrals introduced (Equations 6.7–6.9) gives

$$J = \frac{1}{2} \int_{-1}^{1} (a + bu) du = a,$$

$$H = \frac{1}{2} \int_{-1}^{1} (au + bu^2) du = \frac{1}{3} b$$

and

$$K = \frac{1}{2} \int_{-1}^{1} (au^2 + bu^3) du = \frac{1}{3} a.$$

We see that $K = (1/3)J$ which is known as the Eddington approximation. We have that the average radiation pressure ($P_{rad} = (4\pi/c)K$) is related to the average intensity ($< I >= 4\pi J$) such that

$$\frac{P_{rad}}{< I >} = \frac{1}{3c}.$$

6.4 Radiation Diffusion in a Planar Geometry

It is possible to re-write the equation of radiative transfer ($dI/d\tau = S - I$) for a planar geometry using angle-integrated parameters. We need to consider the optical depth (say τ_n) in a direction normal to the planes of symmetry. For any radiation propagation angle to the normal of the planes of symmetry, we then have an optical depth $\tau = \tau_n/u$ and the equation of radiative transfer becomes

$$\frac{d(Iu)}{d\tau_n} = S - I.$$

Integrating over all angles, we have

$$\int_{-1}^{1} \frac{d(Iu)}{d\tau_n} du = \int_{-1}^{1} S du - \int_{-1}^{1} I du.$$

Multiplying throughout by 1/2 and substituting our definitions of J (Equation 6.7) and H (Equation 6.8), gives that

$$\frac{dH}{d\tau_n} = S - J \tag{6.10}$$

assuming that the source function S is isotropic (independent of angle).

In a similar manner, if we multiply the equation of radiative transfer by u and integrate over all angles, we obtain for the 'first moment' of the equation of radiative transfer

$$\int_{-1}^{1} \frac{d(Iu)}{d\tau_n} u du = \int_{-1}^{1} Su du - \int_{-1}^{1} Iu du,$$

which simplifies to

$$\frac{dK}{d\tau_n} = -H \tag{6.11}$$

upon substituting for K (Equation 6.9) and H (Equation 6.8). The integration of Su over all angles for an isotropic source function S is zero.

If we use the Eddington approximation ($K = (1/3)J$) and Equation 6.11, we can obtain

$$\frac{1}{3}\frac{d^2 J}{d\tau_n^2} = -\frac{dH}{d\tau_n} = J - S \tag{6.12}$$

and

$$H = -\frac{1}{3}\frac{dJ}{d\tau_n} \tag{6.13}$$

upon substituting Equation 6.10 for $dH/d\tau_n$. Equation 6.12 has the form of a steady-state (time-independent) diffusion equation for angle-averaged intensity J as a function of optical depth τ_n along the planar normal.

Radiation heat waves are known as Marshak waves and can play an important role in the transport of energy in astrophysical and inertial confinement fusion plasmas [44]. The frequency and angularly integrated radiation intensity J for radiation approaching a black-body spectrum varies with temperature T as T^4 (see Equation 4.17) and opacity K drops rapidly with increasing temperature T. A high-incident radiation intensity consequently propagates into a cooler optically thick plasma with a steep radiation temperature gradient referred to as a 'heat front'. The radiation heat-front shape is determined using $J \propto T^4$ in Equation 6.12 so that

$$\frac{d^2 T^4}{d\tau_n^2} \approx 0 \tag{6.14}$$

as close to steady-state $J \approx S$ for an optically thick medium. The intensity J and source function S also both tend to be small compared to the derivative of the intensity gradient with respect to optical depth τ_n for a propagating heat front. Equation 6.14 has solutions for the temperature T as a function of optical depth τ_n from, say, a hot surface with temperature T_s given by

$$T = T_s \left(1 - \frac{\tau_n}{\tau_F}\right)^{1/4} \tag{6.15}$$

where τ_F is the optical depth from the hot surface to the heat front. Converting optical depth τ_n to distance z, the position z_F of a heat front using Equation 6.15 can be shown [44] to increase in time t proportionally to $t^{1/2}$ if it is assumed that the opacity of the material after the heat front has passed is constant.

6.5 The Rosseland Mean Opacity

We now specifically investigate the flux of radiation over all frequencies by considering the frequency and angularly integrated flux F_{tot} of radiation impinging on a surface. Considering the definition of H (Equation 6.8), we can write for the frequency and angle-integrated flux of radiation

$$F_{tot} = 4\pi \int_{\omega=0}^{\infty} H(\omega) d\omega$$

where the integration is over all frequencies ω. The optical-depth change $d\tau_n$ is related to distances z which are normal to our assumed planes of symmetry by $d\tau_n = K dz$, where K is the absorption coefficient. Using Equation 6.13, we then have for the spectrally and angularly integrated flux

$$F_{tot} = -\frac{4\pi}{3} \int_{\omega=0}^{\infty} \frac{1}{K} \frac{dJ}{dz} d\omega = -\frac{4\pi}{3} \int_{\omega=0}^{\infty} \frac{1}{K} \frac{dJ}{dT} \frac{dT}{dz} d\omega.$$

The plasma temperature T variation with distance z does not depend on frequency ω, so can be taken outside the integration to give

$$F_{tot} = -\frac{4\pi}{3} \left[\int_{\omega=0}^{\infty} \frac{1}{K} \frac{dJ}{dT} d\omega \right] \frac{dT}{dz}.$$

The spectrally and angularly integrated flux of radiation can thus be expressed as

$$F_{tot} = -\frac{4\pi}{3K_{Ros}} \left[\int_{\omega=0}^{\infty} \frac{dJ}{dT} d\omega \right] \frac{dT}{dz} \tag{6.16}$$

where we define a mean absorption coefficient (known as the Rosseland mean absorption coefficient) by

$$\frac{1}{K_{Ros}} = \left[\int_{\omega=0}^{\infty} \frac{1}{K} \frac{dJ}{dT} d\omega \right] / \left[\int_{\omega=0}^{\infty} \frac{dJ}{dT} d\omega \right]. \tag{6.17}$$

This evaluation for the Rosseland mean absorption coefficient gives a value of $1/K_{Ros}$ which is a spectral average of $1/K$. For the denominator in the definition of the Rosseland mean absorption coefficient, we can undertake the integration of the angle integrated radiation flux J with frequency ω before we differentiate with respect to temperature T.

For plasmas with high opacity, the radiation flux will approach a Planckian distribution and then $J = F_p/\pi$ (see Equation 4.13) and following Equation 4.17 the integration over all frequency gives

$$\int_{\omega=0}^{\infty} F_p d\omega = \sigma_{SB} T^4$$

where σ_{SB} is the Stefan–Boltzmann constant. Differentiating with respect to temperature gives that

$$\frac{d\left(\int_{\omega=0}^{\infty} F_p d\omega\right)}{dT} = 4\sigma_{SB} T^3.$$

We can now write an expression for the spectrally and angularly integrated flux of radiation for a plasma with radiation field close to Planckian. We have

$$F_{tot} = -\frac{16\sigma_{SB}}{3K_{ros}} \frac{dT}{dz} \tag{6.18}$$

with the Rosseland mean absorption coefficient K_{Ros} given by

$$\frac{1}{K_{Ros}} = \frac{1}{4\sigma_{SB} T^3} \int_{\omega=0}^{\infty} \frac{1}{K} \frac{dF_p}{dT} d\omega. \tag{6.19}$$

We have the flux of radiation determined by the gradient of temperature with distance (dT/dz).

The Planck radiation flux can be analytically differentiated for use in Equation 6.19. Setting $x = \hbar\omega/k_B T$, we can write that

$$\frac{dF_p}{dT} = \frac{k}{4\pi^2 c^2} \left(\frac{k_B T}{\hbar}\right)^2 \frac{x^4 e^x}{(e^x - 1)^2}. \tag{6.20}$$

The Rosseland mean opacity becomes

$$\frac{1}{K_{Ros}} = \frac{1}{4\sigma_{SB}} \frac{k}{4\pi^2 c^2} \left(\frac{k}{\hbar}\right)^3 \int_0^\infty \frac{1}{K} \frac{x^4 e^x}{(e^x - 1)^2} dx$$
$$= \frac{\int_0^\infty \frac{1}{K} \frac{x^4 e^x}{(e^x-1)^2} dx}{\int_0^\infty \frac{x^4 e^x}{(e^x-1)^2} dx}$$
$$= \int_0^\infty \frac{1}{K} W_{Ros}(x) dx. \tag{6.21}$$

For the inverse of the Rosseland mean opacity evaluation, the inverse of the absorption coefficient K is multiplied by a weighting function

$$W_{Ros}(x) = \frac{15}{4\pi^4} \frac{x^4 e^x}{(e^x - 1)^2}$$

and integrated over all positive x. The integration of the weighting function on its own from zero to ∞ is equal to one and is plotted in Figure 6.1.

Free-free inverse bremsstrahlung absorption varies with frequency ω and temperature T as $\omega^{-3} T^{-1/2}$. Converting frequency to our integration parameter x shows that this dependence results in the Rosseland mean opacity K_{Ros} for free-free absorption having a variation with temperature proportional to $T^{-3.5}$ (a result known as the Kramers opacity dependence). Bound free photo-ionisation varies as ω^{-3} without an explicit temperature dependence, so the Rosseland mean opacity then varies proportionally to T^{-3}, though changes in ionisation with temperature have a signficant effect on the bound-free opacity.

When the radiation distribution is not given by a Planck distribution, it is sometimes assumed that a Planck distribution over a limited frequency range is appropriate, so that a frequency-averaged Rosseland opacity over groups of opacity ranges, say ω_i to ω_{i+1}, is employed [46]. The flux of radiation within each group is calculated based on the radiation temperature gradient and the average opacity in the frequency range. The different frequency groups can have different radiation temperatures to allow for departures of the radiation field from the Planck distribution.

6.6 Intensities Absorbed in a Thin Layer

Radiation can impinge onto the surface of much denser material. In the denser material, the absorption coefficient is much higher and the radiation is absorbed in a thin layer at the surface. We wish to have a measure of the fraction f_A of the radiation flux absorbed in a layer of thickness Δz with a frequency-varying absorption coefficient K. If the radiation intensity is given by the flux $F_P(\omega)$ of radiation incident per unit area on the surface angle (see Equation 4.12), we have

$$f_A = \frac{\int_{\theta=0}^{\pi/2}\int_{\omega=0}^{\infty} F_p(\omega)\cos\theta\,(1-\exp(-K\,\Delta z/\cos\theta))\,d\omega d\theta}{\int_{\theta=0}^{\pi/2}\int_{\omega=0}^{\infty} F_p(\omega) d\omega d\theta} \tag{6.22}$$

where the integrals average the radiation flux absorbed over all frequencies ω and angles θ to the surface normal. The radiation flux for rays at angle θ on the surface reduces proportionally as $\cos\theta$ and the distance for the ray to pass through the thin layer increases as $\Delta z/\cos\theta$. For small thickness Δz, we can expand the exponential to obtain

$$f_A \approx \frac{\int_{\omega=0}^{\infty} F_p(\omega) K d\omega}{\int_{\omega=0}^{\infty} F_p(\omega) d\omega}\Delta z = K_P \Delta z \tag{6.23}$$

where

$$K_P = \frac{\int_{\omega=0}^{\infty} F_p(\omega) K d\omega}{\int_{\omega=0}^{\infty} F_p(\omega) d\omega} = \frac{\int_{\omega=0}^{\infty} F_p(\omega) K d\omega}{\sigma_{SB} T^4} \tag{6.24}$$

is known as the Planck mean opacity. As before, σ_{SB} is the Stefan-Boltzmann constant. Changing the variable of integration from ω to $x = \hbar\omega/k_B T$ and substituting for the Planck radiation flux (Equation 4.12) enables Equation 6.24 to be written in a similar fashion to Equation 6.21. We have

$$K_P = \frac{1}{\sigma_{SB}}\frac{k}{4\pi^2 c^2}\left(\frac{k}{\hbar}\right)^3 \int_0^{\infty} K\frac{x^3}{e^x - 1}dx = \frac{\int_0^{\infty} K\frac{x^3}{e^x-1}dx}{\int_0^{\infty}\frac{x^3}{e^x-1}dx} = \int_0^{\infty} K W_P(x) dx. \tag{6.25}$$

For the evaluation of the Planck mean opacity, the absorption coefficient K is multiplied by a weighting function

$$W_P(x) = \frac{15}{\pi^4}\frac{x^3}{e^x - 1}$$

which is equivalent to a 'reduced' Planck black-body radiation variation. The integration of the weighting function for the Planck opacity on its own from zero to ∞ is equal to one. Comparing Equations 6.21 and 6.25, we see that the weighting functions for the Rosseland mean opacity and the Planck mean opacity are similar, peaking at 3.8 and 2.8 respectively (see Figure 6.1). Substituting the frequency and temperature variations for free-free opacity ($\omega^{-3}T^{-1/2}$), it is also clear that the Planck mean opacity for free-free absorption varies with temperature as $T^{-3.5}$, which is the same dependence as the Rosseland mean opacity.

6.7 Relationships between the Frequency-Averaged Opacities

It is possible to determine relationships between the frequency-averaged opacities which are useful for checking more detailed calculations. The Schwartz inequality states for a and b arbitrary functions of x that

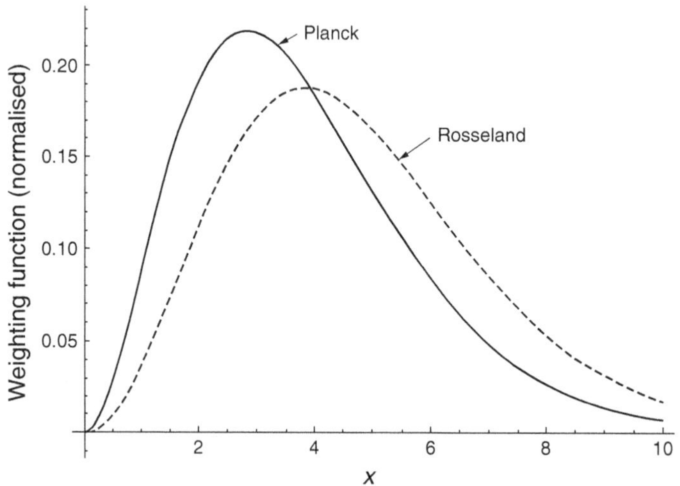

Figure 6.1 The weighting functions for the evaluation of the Rosseland and Planck mean opacities. The weighting functions are plotted in a normalised form so that the integral of the weighting function over all $x = \hbar\omega/k_B T$ is one.

$$\left(\int ab dx\right)^2 \leq \left(\int a^2 dx\right)\left(\int b^2 dx\right). \tag{6.26}$$

If we write

$$\frac{K_P}{K_{Ros}} = \left(\int_0^\infty K W_P(x) dx\right)\left(\int_0^\infty \frac{1}{K} W_{Ros}(x) dx\right)$$

then

$$\frac{K_P}{K_{Ros}} \geq \int_0^\infty \left(W_{Ros}(x) W_P(x)\right)^{1/2} dx = 0.974 \tag{6.27}$$

upon numerically evaluating the integral. We have that the Planck mean opacity is always greater than the Rosseland mean opacity.

We can use the Schwarz inequality to write that

$$\left(\int_0^\infty \frac{1}{K} W_{Ros}(x) dx\right)\left(\int_0^\infty K dx\right) \geq \left(\int_0^\infty \left(W_{Ros}(x)\right)^{1/2} dx\right)^2 = 10.903$$

upon evaluating the right-hand integral. As the first integral gives the Rosseland mean opacity, we have that

$$\int_0^\infty K dx \geq 10.903 K_{Ros}. \tag{6.28}$$

This result shows that simple integration of the absorption coefficient K over all frequencies gives an integrated opacity at least ten times greater than the Rosseland mean opacity.

The integration of an absorption cross-section over all frequencies has been considered previously in our treatment of the scattering of light by electrons (see Equation 3.27). If the density of ions is n_i, the integration of the absorption coefficient is related to the integration of the absorption cross-section, which has a value determined by the Thomas-Reiche-Kuhn sum rule:

$$\int_0^\infty K dx = \frac{\hbar}{k_B T} \int_0^\infty K d\omega = n_i \frac{\hbar}{k_B T} \int_0^\infty \sigma_s d\omega = n_i \frac{\hbar}{k_B T} 2\pi^2 r_e c Z \tag{6.29}$$

where Z is the number of free and bound electrons per ion and r_e is the classical electron radius (see Section 3). A simple evaluation of the right-hand expression given in Equation 6.29 produces a value of the spectrally integrated absorption coefficient and an upper bound (using Equation 6.28) to the Rosseland mean opacity.

The relationships between frequency-averaged opacities were set out in 2003 by the polymath physicist Freeman Dyson (1923) [9] (also see [5]). Dyson indicated [9] that he developed the relationships between frequency-averaged opacities while working, from 1957–1961, on the Orion project, which sought to investigate the possibility of space flight using nuclear explosions for thrust. The Thomas-Reiche-Kuhn sum rule is widely employed to verify opacity calculations. For example, the sum rule with a treatment similar to Equation 6.29 has been used [53] to check experimental results [6] showing high opacities for iron at the densities ($\approx 0.1\,\mathrm{g\,cm^{-3}}$) and temperatures ($\approx 200\,\mathrm{eV}$) found in the Sun at a radius of ≈ 0.75 of the photosphere radius (at the boundary of the radiative and convection zones).

Exercises

6.1 A thin foil with a mirrored surface on one side and a black surface on the other side is exposed to sunlight incident normally onto the mirrored side in a vacuum vessel with a pressure of 1/100 of an atmosphere. A net pressure is exerted on the foil thought to be due to radiation pressure. A competing process could be gas pressure due to a temperature difference on either side of the foil. Determine the temperature difference in the background gas either side of the foil which is required for the gas pressure difference to equal the radiation pressure in bright sunlight of irradiance 1,000 $\mathrm{W\,m^{-2}}$. [$\approx 5 \times 10^{-6}\,^\circ\mathrm{C}$]

6.2 In laser-produced plasmas, the profile of density with distance from the target surface often exhibits a steepening at the critical density. For a laser-produced plasma at irradiance 3×10^{16} W cm^{-3} with a profile steepened to electron density 10^{22} cm^{-3} at a temperature of 100 eV, calculate (i) the gas pressure due to the electrons, and (ii) the radiation pressure at the critical density assuming normal incidence and full reflection of the laser light. [Electron gas pressure 8×10^{11} N m^{-2}, radiation pressure 2×10^{12} N m^{-2}.]

6.3 The radiation escaping from a black-body cavity through a small aperture will have a frequency-integrated flux into 2π steradians given by

$$F_{tot} = \sigma_{SB} T^4$$

where σ_{SB} is the Stefan-Boltzmann constant. Assuming perfect absorption of radiation at the cavity walls, show that the radiation pressure on the walls of the black-body cavity is given by

$$P_{rad} = \frac{\sigma_{SB} T^4}{3c}.$$

6.4 The shape of a planar radiation heat front of radiation temperature T is determined by Equation 6.14, namely

$$\frac{d^2 T^4}{d\tau_n^2} = 0$$

where τ_n is the optical depth normal to the symmetry plane. Verify by differentiation that this equation has solutions of form (Equation 6.15):

$$T = T_s \left(1 - \frac{\tau_n}{\tau_F}\right)^{1/4}.$$

6.5 If ρ is the plasma-material density, using Equation 6.29 show that the maximum value of the Rosseland mean opacity per unit of density K_{Ros}/ρ is given in units of cm^2 g^{-1} by

$$\frac{K_{Ros}}{\rho} = 6 \times 10^6 \frac{Z}{A} \frac{1}{k_B T}$$

where the temperature $k_B T$ is measured in electron volts, Z is the atomic number and A the atomic mass of the ions in the plasma.

7

Discrete Bound Quantum States: Hydrogen and Hydrogen-Like Ions

An important type of emission from plasmas consists of spectral lines originating from transitions between bound quantum states. Quantum mechanics gives information on the energies of the quantum states and can give information on the intensity of emission. The absorption of radiation resulting in transitions between bound quantum states can dominate the calculation of absorption coefficients, particularly for higher atomic number ions where the spectral density of absorption and emission lines is large and the radiative-transition probabilities are high (for example, hydrogen-like ion transition probabilities scale proportionally to Z^4).

The Bohr model of the atom where electrons are said to orbit the nucleus like planets orbiting the Sun was introduced in Section 1.5. The Bohr model gives a good approximation of the energies of hydrogen and hydrogen-like ions. It is also a reasonable model for the energies of excited states of higher-atomic-number atoms and ions. Excited electrons of multi-electron atoms and ions have orbits some distance from the nucleus and other electrons, so their kinetic and potential energy is determined by the near point-like net charge near the nucleus as occurs with hydrogen-like ions. The Bohr model, however, fails to predict the correct angular momentum of orbiting electrons and fails to predict the fine structure of the energy levels. Fine structure causes, for example, the energy states to split and produce two or more closely spaced spectral lines, rather than a single line. The Bohr model also does not enable a satisfactory method of evaluating the rates of radiative absorption and emission (which can be done with time-dependent quantum mechanics; see Section 10.1).

We consider the quantum mechanics of hydrogen and hydrogen-like ions in this chapter. Hydrogen and hydrogen-like ions only differ in the charge Z of the nucleus, so it is possible to treat both almost identically. We commence with solutions of the Schrodinger equation, treat the effects leading to the fine structure and mention hyperfine energy splitting arising from electron interaction with the nuclear spin.

There are many books dealing with the details of the quantum mechanics of hydrogen. We refer the reader to Haken and Wolf [43] for further reading.

7.1 A Quantum Mechanical Treatment of Atoms and Ions

A quantum mechanical treatment of atoms and ions can start with the time-dependent Schrodinger equation. Here we write that

$$\hat{H}\Psi = i\hbar\frac{d\Psi}{dt} \tag{7.1}$$

where $\hat{H}$ is the quantum mechanical Hamiltonian (or energy operator) and Ψ is the time-dependent wavefunction. An isolated atom has a Hamiltonian or energy operator given by the addition of the kinetic energy operator and potential energy operator, such that

$$\hat{H}_0 = -\frac{\hbar^2}{2m_0}\nabla^2 + V \tag{7.2}$$

where m_0 is the electron mass and V is the potential energy of the electron in the field of the nucleus.

For an isolated atom, we can write that the time-dependent wavefunction Ψ is a product of the two functions, one of which is a function of time and the other a function of spatial co-ordinates. We write

$$\Psi = \psi(\mathbf{r})\phi(t).$$

If we substitute the last two equations here into the time-dependent Schrodinger equation, we obtain that

$$\frac{1}{\psi(\mathbf{r})}\left[-\frac{\hbar^2}{2m_0}\nabla^2 + V(\mathbf{r})\right]\psi(\mathbf{r}) = -\frac{\hbar}{i\phi(t)}\frac{d\phi(t)}{dt}$$

upon dividing throughout by Ψ. The left-hand side of this equation is a function of spatial co-ordinates only and the right-hand side is a function of time only, so the two sides must be separately equal to a constant, which we denote by E. We can obtain two equations, such that

$$\left[-\frac{\hbar^2}{2m_0}\nabla^2 + V(\mathbf{r})\right]\psi(\mathbf{r}) = E\psi(\mathbf{r}) \tag{7.3}$$

which is known as the time-independent Schrodinger equation and

$$i\hbar\frac{d\phi(t)}{dt} = E\phi(t).$$

In quantum mechanics, measurable quantities are represented by operators. A measurement of a particular quantity yields an eigenvalue of the corresponding operator. For example, if we are interested in the value of a quantity a, then the relevant eigenvalue equation is

$$\hat{A}\psi = a\psi$$

where ψ is the wavefunction describing the system on which the measurement takes place, $\hat{A}$ is the relevant operator and a is the eigenvalue–the value of the measurement. The operators can be found by replacing the more familiar classical expressions for the measurement with quantum mechanical analogues. These quantum mechanical analogues can be derived using the following: for position a vector $\mathbf{r}$ and for momentum the operation $-i\hbar\nabla$. For example, the kinetic energy operator is derived from the momentum p of a particle as energy is equal to $p^2/2m_0$. The kinetic energy operator becomes $-\hbar^2/2m_0\nabla^2$.

Returning to the two equations above for the spatial and temporal variation of a wavefunction associated with an atom, we can readily solve the equation for the temporal variation $\phi(t)$. This integrates to give

$$\phi(t) = \exp\left(-i\frac{Et}{\hbar}\right)$$

assuming the normalisation that the modulus of $\phi(t)$ is unity.

The other time-independent Schrodinger equation involving the spatial variation of the wavefunction $\psi(\mathbf{r})$ is an example of an eigenvalue equation with the operator being the Hamiltonian or energy operator. This immediately gives us that the constant E introduced has a physical meaning–namely that it is the energy value of the system. The wavefunction value also has a direct physical interpretation. The quantity

$$\psi^*(x, y, z)\psi(x, y, z)dxdydz$$

where ψ^* represents the complex conjugate of the wavefunction ψ is equal to the probability that the particle represented by the wavefunction will be found in the volume between (x, y, z) and $(x + dx, y + dy, z + dz)$. This is only valid if we assume the normalisation for $\phi(t)$ that its modulus is unity.

7.2 The Hydrogen Atom

The time-independent Schrodinger equation for the hydrogen atom can be solved analytically if we consider only the Coulomb potential of the electron in the field of the nucleus. This potential energy $V(\mathbf{r})$ for hydrogen or the hydrogen-like ion of atomic number Z is given by

$$V(\mathbf{r}) = -\frac{Ze^2}{4\pi\epsilon_0 r}. \tag{7.4}$$

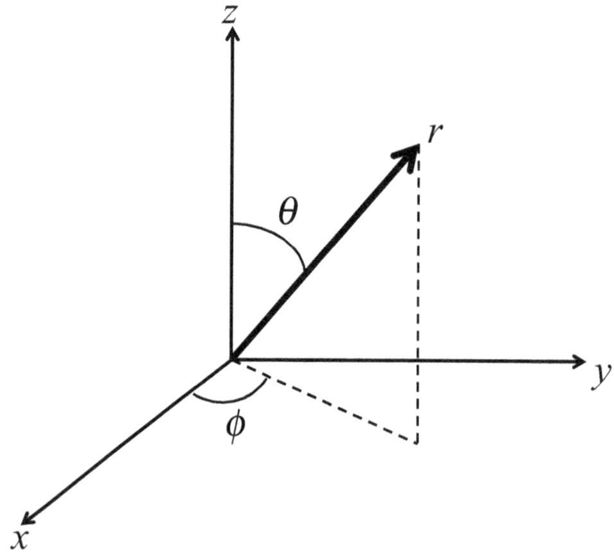

Figure 7.1 The axes and symbols used for spherical polar co-ordinates.

The Hamiltonian is then

$$\hat{H}_0 = -\frac{\hbar^2}{2m_0}\nabla^2 - \frac{Ze^2}{4\pi\epsilon_0 r}. \tag{7.5}$$

As the potential $V(\mathbf{r})$ is spherically symmetric, the solution of the time-independent Schrodinger equation for the electron in hydrogen and hydrogen-like ions needs to be solved in spherical polar co-ordinates around the nucleus, which is fixed as the co-ordinate system origin. Positions in spherical polar co-ordinates from the nucleus are determined by the distance r from the nucleus (the origin) and two angles, θ, the angle to the z-axis and ϕ, the angle to the x-axis of a vector $\mathbf{r}$ from the origin to a position (see Figure 7.1). The wavefunction ψ can be expanded as two functions varying with the radius r (the magnitude of the vector $\mathbf{r}$) and the two angles. We can write

$$\psi(\mathbf{r}) = R(r)Y(\theta, \phi).$$

Substituting into the time-independent Schrodinger equation enables the Hamiltonian to be evaluated. The Laplace operator in the Schrodinger equation $\nabla^2 = \partial^2/\partial x^2 + \partial^2/\partial y^2 + \partial^2/\partial z^2$ in spherical polar co-ordinates (see Figure 7.1 and Appendix A.1) can be written as

$$\nabla^2 = \frac{1}{r^2}\left[\frac{\partial}{\partial r}\left(r^2\frac{\partial}{\partial r}\right) + \frac{1}{\sin\theta}\frac{\partial}{\partial\theta}\left(\sin\theta\frac{\partial}{\partial\theta}\right) + \frac{1}{\sin^2\theta}\frac{\partial^2}{\partial\phi^2}\right].$$

Using the spherical polar co-ordinate Laplace operator, the electron kinetic energy operator can be written as a sum of the rotational energy $L^2/(2m_0r^2)$, where L is the angular momentum of an orbiting electron, and an energy which can be associated with electron motion in the radial direction. We have

$$-\frac{\hbar^2}{2m_0}\nabla^2 = \frac{1}{2m_0r^2}\hat{L}^2 - \frac{\hbar^2}{2m_0}\frac{1}{r^2}\frac{\partial}{\partial r}\left(r^2\frac{\partial}{\partial r}\right)$$

with the operator for the square of angular momentum given by

$$\hat{L}^2 = -\hbar^2\left[\frac{1}{\sin\theta}\frac{\partial}{\partial\theta}\left(\sin\theta\frac{\partial}{\partial\theta}\right) + \frac{1}{\sin^2\theta}\frac{\partial^2}{\partial\phi^2}\right].$$

The time-independent Schrodinger equation ($\hat{H}\psi = E\psi$) can now be re-written using

$$\hat{H}\psi = Y(\theta,\phi)\left[-\frac{\hbar^2}{2m_0}\frac{1}{r^2}\frac{\partial}{\partial r}\left(r^2\frac{\partial}{\partial r}\right) + V(r)\right]R(r) + \frac{R(r)}{2m_0r^2}\hat{L}^2Y(\theta,\phi).$$

We know that the operator for the square of the electron orbital angular momentum will have eigenvalues, so we write the eigenvalues as $l(l+1)\hbar^2$ where l is to be determined. We shall see when we examine the angular momentum that the eigenvalues are quantised so that $l = 0, 1, 2, \ldots,$ but at this stage we do not need to be more specific than to just require an eigenvalue for the square of the angular momentum.

Substituting the eigenvalue for the square of the electron orbital angular momentum, the Schrodinger equation becomes a function of the radial wavefunction only as we substitute $\hat{L}^2Y(\theta,\phi)$ by $l(l+1)\hbar^2Y(\theta,\phi)$ and divide throughout by $Y(\theta,\phi)$. We have

$$\left[-\frac{\hbar^2}{2m_0}\frac{1}{r^2}\frac{\partial}{\partial r}\left(r^2\frac{\partial}{\partial r}\right) + V(r) + \frac{l(l+1)\hbar^2}{2m_0r^2}\right]R(r) = ER(r). \tag{7.6}$$

This equation is considerably easier to solve as it only varies with the radial distance r.

The total angular momentum of the electron in the hydrogen atom was not calculated correctly in the simple Bohr model. We can see from our treatment that the total angular momentum is actually $\sqrt{(l(l+1)}\hbar$. Solving the eigenvalue equation for the square of the total angular momentum

$$\hat{L}^2Y(\theta,\phi) = l(l+1)\hbar^2Y(\theta,\phi) \tag{7.7}$$

gives solutions for the eigenvalues. Valid eigenfunctions $Y(\theta,\phi)$ for Equation 7.7 require that l is an integer with values $l = 0, 1, 2, \ldots.$ The values of l are referred to as the orbital angular-momentum quantum numbers.

The angular-momentum component in a particular direction (say the z-direction) is also important and can be found using an additional eigenvalue equation. Classically, the angular momentum L_z of an orbiting particle around the z-axis in spherical polar co-ordinates is given by

$$L_z = p_{xy} r \sin\theta$$

where p_{xy} is the component of the momentum of the particle parallel to the xy plane. Converting to quantum mechanics, we use the momentum operator ($\hat{p} = -i\hbar\nabla$) so that the operator for momentum in the xy plane becomes $\hat{p}_{xy} = -i\hbar/(r\sin\theta)\frac{\partial}{\partial\phi}$ upon expanding ∇. The eigenvalue equation for the component of angular momentum around the z-axis then becomes

$$\hat{L}_z Y(\theta,\phi) = -i\hbar\frac{\partial}{\partial\phi}Y(\theta,\phi) = m\hbar Y(\theta,\phi) \tag{7.8}$$

where $m\hbar$ is written to represent the eigenvalue. Again valid eigenfunctions require that m is an integer, but in the range $-l \leq m \leq l$. In atomic measurements, a z-axis will be defined by the imposition of, say, a magnetic field. There are small energy differences depending on the component of angular momentum along the magnetic field, so m is commonly referred to as the magnetic quantum number (and should not be confused with the electron mass, which we are writing as m_0).

The wavefunction variation $Y(\theta,\phi)$ can be found by solving the two eigenvalue equations for the square of total angular momentum and the angular momentum along the z-axis. They take a form known as spherical harmonic functions, which can be written

$$Y(\theta,\phi) = \exp(im\phi)P_l^m(\cos\theta) \tag{7.9}$$

where $P_l^m(\cos\theta)$ is a special function known as a Legendre polynomial (for $m = 0$) or an associated Legendre function (for $m \neq 0$). It is possible to verify by substitution that Equation 7.9 is a solution of Equation 7.8. The form of $P_l^m(\cos\theta)$ changes with the value of m, but does not have m as an explicit parameter. The spherical harmonic wavefunctions Y_{lm} vary as follows

$$\begin{aligned}
Y_{00} &= \frac{1}{2\sqrt{\pi}} \\
Y_{10} &= \frac{1}{2}\sqrt{\frac{3}{\pi}}\cos\theta \\
Y_{1\pm1} &= \mp\frac{1}{2}\sqrt{\frac{3}{2\pi}}\sin\theta\exp(\pm i\phi) \\
Y_{20} &= \frac{1}{4}\sqrt{5\pi}(3\cos^2\theta - 1)
\end{aligned}$$

$$Y_{2\mp1} = \pm\sqrt{\frac{15}{2\pi}}\sin\theta\cos\theta\exp(\pm i\phi)$$

$$Y_{2\pm2} = \frac{1}{4}\sqrt{\frac{15}{2\pi}}\sin^2\theta\exp(\pm i2\phi). \tag{7.10}$$

The $Y^*_{lm}Y_{lm}$ values of the spherical harmonics give the angular dependence for the probability density of the electron – the shape of the electron 'distribution' around the nucleus. We can immediately see that the ground state of hydrogen where Y_{00} is the only available spherical harmonic wavefunction is spherically symmetric; there is no dependence on θ or ϕ. Indeed the values of $Y^*_{lm}Y_{lm}$ are all seen to be independent of ϕ and vary only with θ the angle to the z-axis (with the direction of the z-axis determined, for example, by the direction of an imposed magnetic field). Representations of the spherical harmonic wavefunctions are shown in Figure 7.2.

In the absence of an additional directional field superimposed on the central radial Coulomb field from the nucleus, the energies of different m quantum numbers are the same. Indeed, the solutions of the radial wavefunctions of hydrogen

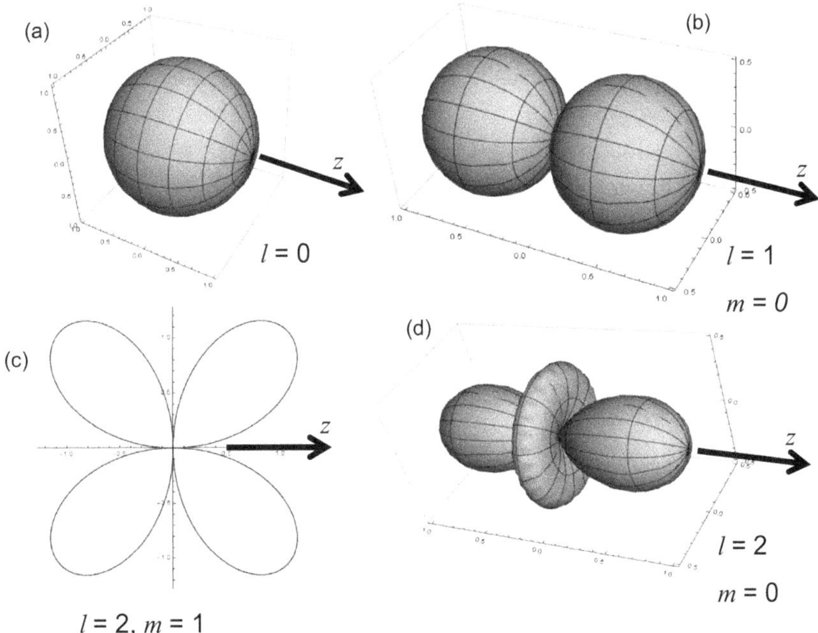

Figure 7.2 Examples of the angular structure of hydrogen wavefunctions $|Y(\theta,\phi)|$ for (a) $l = 0$, (b) $l = 1, m = 0$, (c) $l = 2, m = 1$ and (d) $l = 2, m = 0$. The direction of the z-axis is determined by, for example, an imposed magnetic field and is an axis of rotational symmetry in ϕ (see Figure 7.1). The $l = 2, m = 1$ wavefunction is shown (c) as a cross-section through the yz plane.

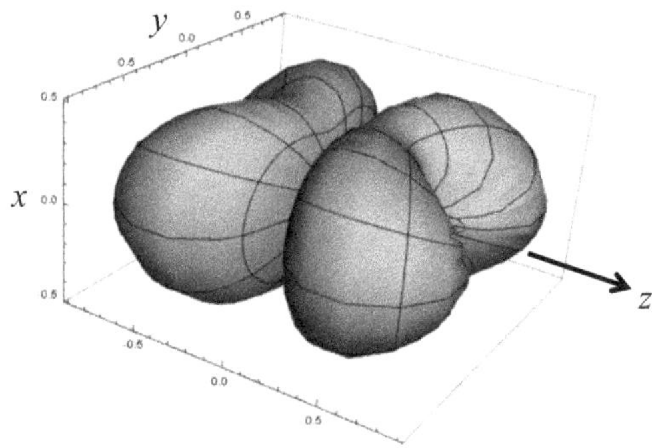

Figure 7.3 The angular structure of hydrogen wavefunction probability distribution $(1/2)(Y_{21} + Y_{2-1})^*(Y_{21} + Y_{2-1})$ for the $l = 2$, $m = \pm 1$ mixed state. The mixed state is not rotationally symmetric around the z-axis.

(Section 7.2.1) show that the energy levels are only determined by a principal quantum number as predicted by the Bohr model (Section 1.5). There is no difference in energy between $\pm m$ quantum numbers in the absence of an external electric or magnetic field (even including fine-structure effects discussed in Section 7.4). Consequently, an isolated atom or ion can exhibit a 'mixing' of the $\pm m$ quantum states so that there is no effective angular-momentum component along the z-axis. Electron probability distributions are often represented as the values of $(1/2)(Y_{lm} + Y_{l-m})^*(Y_{lm} + Y_{l-m})$. A linear combination of solutions is also a solution to the Schrodinger equation, though the added solutions are not eigenfunctions of the operator $\hat{L}_z$ for the angular-momentum component along the z-axis. To illustrate the typical probability distributions found for quantum states with $m \neq 0$, the angular distribution of the hydrogen wavefunction for the mixed states $l = 2$, $m = \pm 1$ is shown in Figure 7.3.

7.2.1 The Radial Wavefunctions of Hydrogen

The solutions of the radial wavefunctions $R(r)$ depend on the value of l (the orbital angular-momentum quantum number). However, the energy eigenvalues E of Equation 7.6 are independent of l, but depend on an additional quantum number which, following our earlier treatment of the Bohr model, we label as n and call the principal quantum number. The principal quantum number can only have integral values $n = 1, 2, 3, \ldots$ and must be greater than l. The values of the angular-momentum quantum number can thus vary up to $n - 1$, i.e. $l = 0, 1, 2, \ldots, n - 1$. The energies of the orbiting electron are as found assuming the simple Bohr model

$$E_n = -\frac{Z^2 e^4}{(4\pi\epsilon_0)^2} \frac{m_0}{2\hbar^2} \frac{1}{n^2}. \tag{7.11}$$

The radial wavefunctions R_{nl} vary as follows

$$R_{10} = 2\left(\frac{Z}{a_0}\right)^{\frac{3}{2}} e^{-Zr/a_0}$$

$$R_{21} = \frac{1}{\sqrt{3}}\left(\frac{Z}{2a_0}\right)^{\frac{3}{2}}\left(\frac{Zr}{a_0}\right) e^{-Zr/2a_0}$$

$$R_{20} = 2\left(\frac{Z}{2a_0}\right)^{\frac{3}{2}}\left(1 - \frac{Zr}{2a_0}\right) e^{-Zr/2a_0}$$

$$R_{32} = \frac{2\sqrt{2}}{27\sqrt{5}}\left(\frac{Z}{3a_0}\right)^{\frac{3}{2}}\left(\frac{Zr}{a_0}\right)^2 e^{-Zr/3a_0}$$

$$R_{31} = \frac{4\sqrt{2}}{3}\left(\frac{Z}{3a_0}\right)^{\frac{3}{2}}\left(\frac{Zr}{a_0}\right)\left(1 - \frac{Zr}{6a_0}\right) e^{-Zr/3a_0}$$

$$R_{30} = 2\left(\frac{Z}{3a_0}\right)^{\frac{3}{2}}\left(1 - \frac{2Zr}{3a_0} + \frac{2\,(Zr)^2}{27a_0^2}\right) e^{-Zr/3a_0}.$$

Here a_0 is the Bohr radius as introduced in our discussion of the Bohr model of the hydrogen atom. Surprisingly, we can see that the ground state radial wavefunction R_{10} and other wavefunctions with $l = 0$ have maximum values at the nucleus $r = 0$.

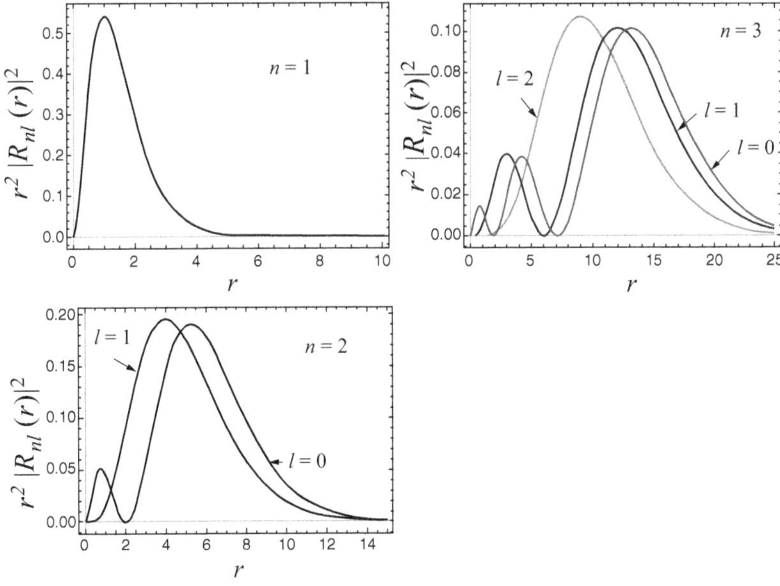

Figure 7.4 The probability density $r^2|R_{nl}(r)|^2$ of the radial hydrogen wavefunctions as a function of radius r (in units of the Bohr radius a_0).

However, the radius where the electron is most likely to be found is where $r^2 R_{nl}^* R_{nl}$ is maximum. For the ground state $n = 1, l = 0$, this can be shown to be at a radius $r = a_0/Z$ in agreement with the Bohr model. The radial-probability density as a function of radius of hydrogen energy levels up to $n = 3$ is shown in Figure 7.4.

7.3 Magnetic Moment, Electron Spin and Degeneracy

If a small current carrying wire encloses an area A in a plane and the current flowing is I, we can define a magnetic moment μ given by

$$\mu = I\mathbf{A}$$

where the vector directions are normal to the plane of the current carrying wire and we use the right-hand rule (point the fingers of the right hand in the direction of the current and the thumb will point in the direction of the vector **A**). The torque τ exerted on a particle with a magnetic moment μ in a magnetic field **B** is given by

$$\tau = \mu \times \mathbf{B} = \mu B \sin\theta$$

where θ is the angle between μ and **B**. Integrating in angle θ gives the work done to rotate the magnetic moment relative to the magnetic field. The work done relative to a starting angle perpendicular to the magnetic field is $-\mu \cdot \mathbf{B}$. The minimum energy configuration is when the magnetic moment is parallel to the magnetic field. An energy of $2\mu B$ is required to make the magnetic moment go to 180° (anti-parallel) to the magnetic field if it is initially parallel.

An electron rotating in a simple Bohr model of the hydrogen atom acts like a circuit with a current flow

$$I = \frac{-ev}{2\pi r}$$

where v is the electron-orbiting speed and r is the radius of the orbit. Multiplying the numerator and denominator for this expression for the current by m_0 the electron mass and r, we can quickly see that the magnetic moment is related to the electron angular momentum $L = m_0 v r$ by

$$\mu = -\frac{e}{2m_0}\mathbf{L}.$$

We saw in our treatment of the Bohr model that the angular momentum of orbiting electrons is supposedly given by $L = n\hbar$, so the magnetic moment associated with this concept would become

$$\mu = -\frac{e\hbar}{2m_0}n = -\mu_B n.$$

The value $e\hbar/2m_0$ is represented here by μ_B and is known as the Bohr magneton. It is a convenient atomic-scale unit for measuring magnetic moments even if the values of the magnetic moment for hydrogen are not $-\mu_B n$. As the component of electron angular momentum along the z-axis is actually given by $m\hbar$, the actual values of orbital magnetic moment for a hydrogen atom or hydrogen-like ion, for example, along the z-axis are given by

$$\mu = -m\mu_B \tag{7.12}$$

where m is the magnetic quantum number.

The energies of a magnetic moment in a magnetic field have an effect on quantum-state energies for atoms in magnetic fields. Imposing an external magnetic field on a hydrogen atom or hydrogen-like ion changes the energies of the m quantum states. Using Equation 7.12, quantum states with different m quantum numbers have slightly different energies ΔE_B from our treatment in Section 7.2.1. The energies vary with an imposed magnetic field B as

$$\Delta E_B = m\mu_B B. \tag{7.13}$$

This splitting in energy of quantum states in the presence of an external magnetic field is known as the Zeeman effect.

Experiments have shown that an electron has an intrinsic 'spin', producing effects as if the negatively charged electron is rotating. Experiments such as the Stern–Gerlach experiment (see Section 8.2.1) also show that, in the presence of a magnetic field, an electron only orientates with a magnetic moment component associated with spin which is parallel or anti-parallel to the magnetic field [122]. The spin of the electron can be regarded as being 'quantised' in that only two states are allowed: a spin component parallel or anti-parallel to a magnetic field. In the presence of a magnetic field as explained for Equation 7.13, the two electron spin orientations (parallel or anti-parallel) also have different energies. We shall see that this energy split for different electron spins leads to an energy splitting of quantum states with different electron spins (see Section 7.4).

The intrinsic angular momentum $\mathbf{s}$ of an electron has a magnitude given by $\sqrt{s(s+1)}\hbar$ where s is a new quantum number associated with the electron spin. The electron magnetic moment is given by

$$\mu_{\mathbf{s}} = -g_s \frac{e}{2m_0}\mathbf{s} \tag{7.14}$$

where g_s is the unimaginatively titled g-factor for the electron with a value obtained by measurement as 2.0023. The quantum number s has value $\frac{1}{2}$ so that the electron-spin component along a magnetic field (the z-axis) is $\pm\frac{1}{2}\hbar$, the total electron spin magnitude is $\frac{\sqrt{3}}{2}\hbar$ and the electron spin has a magnetic moment $\pm\sqrt{3}\frac{e\hbar}{2m_0}$ (the last here only being accurate to two decimal places as we have taken $g_s = 2$).

The g-factor g_s for the magnetic moment of a rotating or spinning charge introduced in Equation 7.14 for the spin of an electron depends on the relationship between the distribution of charge and mass. A rotating point charge and point mass have $g_s = 1$ as discussed for the Bohr model for hydrogen. If a disc of uniformly distributed charge Q and mass M rotates about its centre of mass, the magnetic moment is $\frac{Q}{2M}L$ where L is the rotating disc angular momentum, but if the charge is distributed only on the circumference of the disc then the magnetic moment is $2\frac{Q}{2M}L$.

In the absence of an external magnetic field, we saw in Section 7.2.1 that the energy levels of hydrogen are dependent only on the principal quantum number n, whereas there are several different wavefunctions associated with different quantum states associated with each n. There is consequently an apparent multiplicity or 'degeneracy' of the number of states associated with each hydrogen energy level.

The degeneracy is defined as the number of quantum states associated with a particular energy level. For hydrogen and hydrogen-like ions there are two possible electron–spin orientations, values of the magnetic quantum number m ranging from $-l$ to $+l$ (i.e. $2l+1$ different m values as we must include zero) and then values of l ranging from zero to $n-1$. The degeneracy g_n of an energy level associated with the principal quantum number n is therefore given by

$$g_n = \sum_{l=0}^{n-1} \sum_{m=-l}^{+l} 2 = \sum_{l=0}^{n-1} 2(2l+1) = 2n^2.$$

7.4 Hydrogen Fine Structure

We mentioned that the hydrogen energy levels found by the solution of the Schrodinger equation have been observed to split, so that instead of a single spectral line being observed when a radiative transition occurs between the energy levels, a doublet (i.e. two closely spaced spectral lines) is seen. For transitions involving higher levels ($n > 2$) more complicated spectral structures are seen. The understanding of these fine-structure splittings of the hydrogen energy levels is the topic of this section.

The origin of the fine structure is partly due to (i) 'extra' energy associated with the interaction between the magnetic field created by electron orbital angular momentum and the intrinsic magnetic moment of the electron due to its spin and partly due to (ii) a small correction to the hydrogen atom Hamiltonian that allows for relativistic mass increase of the electron.

7.4.1 The Spin–Orbit Coupling Energy

In the Bohr model of hydrogen, we can readily imagine that the orbiting negatively charged electron generates a magnetic field. The orbiting electron is like a small circular electric circuit. The electron in its frame of reference actually sees the positive nucleus spinning around so that the magnetic field for the electron is directed parallel to, and with a magnitude proportional to, the orbital angular momentum. In quantum mechanical treatments, the magnetic field is similarly associated with the electron angular momentum. The intrinsic electron magnetic moment μ due to the electron spin has energy $-\mu \cdot \mathbf{B}$ in the magnetic field generated due to the orbiting nature of the electron (see the discussion in Section 7.3).

To undertake a quantum mechanical treatment of the energy associated with the interaction of the electron magnetic moment with the magnetic field that the electron experiences, we use quantum mechanical perturbation theory. The Hamiltonian for the atom is modified by the addition of an extra (small) term representing the spin–orbit interaction. We can write

$$\hat{H} = \hat{H}_0 + \hat{H}_{so}$$

where the spin–orbit Hamiltonian is written as a scalar product of the spin operator $\hat{S}$ and the angular momentum operator $\hat{L}$, so we have

$$\hat{H}_{so} = A(r)\hat{S} \cdot \hat{L}. \tag{7.15}$$

The quantity $A(r)$ is a proportionality constant dependent on the distance from the nucleus r. The expression for $\hat{H}_{so}$ can be justified on the basis that the spin operator acts to determine something proportional to the magnetic moment of the electron and the angular momentum operator acts to determine the total angular momentum, which is proportional to the magnetic field produced by a rotating charge. The scalar product of these two quantities is proportional to the energy of the electron magnetic moment in the magnetic field produced by the rotating nucleus (in the frame of reference of the electron).

An expression for $A(r)$ can be obtained by evaluating the generated magnetic field. The magnetic field of the moving nuclear charge $+Ze$ is found from the Biot-Savart law to be

$$\mathbf{B}_L = \frac{\mu_0}{4\pi}\frac{Ze}{r^3}\mathbf{v} \times -\mathbf{r} = -\frac{\mu_0}{4\pi}\frac{Ze}{r^3}\mathbf{v} \times \mathbf{r}.$$

As the magnetic field is being calculated in the frame of reference of the electron, we have replaced the position vector for the orbiting electron $\mathbf{r}$ by $-\mathbf{r}$. The angular momentum of the electron is given by $\mathbf{L} = m_0\mathbf{r} \times \mathbf{v}$, so the magnetic field is given by

$$\mathbf{B}_L = \frac{\mu_0}{4\pi}\frac{Ze}{m_0 r^3}\mathbf{L}.$$

The interaction energy ΔE_{so} for the electron spin in the magnetic field generated by the orbiting charge at a single radius r is then given by

$$\Delta E_{so} = -\mu_s \cdot \mathbf{B}_L = g_s \frac{e}{2m_0} \mathbf{s} \cdot \mathbf{B}_L = \frac{g_s}{2} \frac{\mu_0}{4\pi} \frac{Ze^2}{m_0^2 r^3} \mathbf{s} \cdot \mathbf{L}$$

upon using Equation 7.14 for the intrinsic electron magnetic moment μ_s due to its spin. A correction to ΔE_{so} due to Thomas of a factor $\frac{1}{2}$ is needed to account for relativistic time-dilation effects in moving from the rest frame of reference to the electron frame of reference in the calculation as the electron is not an inertial frame of reference. Noting that $g_s \cong 2$, we have for the spin–orbit energy at radius r

$$\Delta E_{so} = \frac{1}{2} \frac{\mu_0}{4\pi} \frac{Ze^2}{m_0^2 r^3} \mathbf{s} \cdot \mathbf{L}. \tag{7.16}$$

From this relationship, we can see that the value of $A(r)$ in Equation 7.15 can be written

$$A(r) = \frac{1}{2} \frac{\mu_0}{4\pi} \frac{Ze^2}{m_0^2 r^3}.$$

Using the identity that $\mu_0 = 1/(c^2 \epsilon_0)$ and differentiating the expression for the electron potential $V(r)$ in the field of the nucleus, the value of $A(r)$ can also be written as

$$A(r) = \frac{1}{2m_0^2 c^2 r} \frac{dV(r)}{dr}.$$

These expressions for $A(r)$ are valid in a quantum mechanical treatment because the rotating nucleus (in the frame of the electron) is close to being a 'point' charge.

The energy shifts on the atom due to spin–orbit coupling are obtained using perturbation theory, which predicts the quantum mechanical spin–orbit coupling energy to be given by

$$E_{so} = \int_{Volume} \psi^*(\mathbf{r}) \hat{H}_{so} \psi(\mathbf{r}) dV. \tag{7.17}$$

As the spin–orbit energy is a small perturbation of the total hydrogen atom energy, the spin–orbit Hamiltonian $\hat{H}_{so}$ operates on the electron wavefunction $\psi(\mathbf{r})$ (as listed in Sections 7.2 and 7.2.1), the result of which is multiplied by the complex conjugate of the wavefunction $\psi^*(\mathbf{r})$. This result is integrated over all volume space.

Equation 7.16 shows that energy needs to be supplied to make the electron spin orient anti-parallel to the orbital angular momentum. To delineate the two possible scenarios: **s** parallel and anti-parallel to **L**, a new quantum number j is introduced to represent the total angular momentum of the atom – the electron intrinsic spin angular momentum added to the electron orbital angular momentum.

The quantum number j takes the value of l plus or minus $\frac{1}{2}$ to represent the two possible orientations of the electron spin $\frac{1}{2}$ in the direction of the orbital angular momentum (i.e. plus if the spin is in the direction of the orbital angular momentum and minus if the spin is in the opposite direction to the orbital angular momentum).

The total angular momentum represented by the quantum number j has an associated operator and similarly to the operator $\hat{L}$ for angular moment, we can write that

$$\hat{J}^2\psi = j(j+1)\hbar^2\psi.$$

To evaluate E_{so}, we need to determine $\hat{S}\cdot\hat{L}$. This can be evaluated by noting that $\hat{J} = \hat{L} + \hat{S}$ so that

$$\hat{J}^2 = \hat{L}^2 + \hat{S}^2 + 2\hat{S}\cdot\hat{L}$$

which means that

$$\hat{S}\cdot\hat{L}\psi = \frac{1}{2}\hbar^2[j(j+1) - l(l+1) - s(s+1)]\psi.$$

The integral for E_{so} can then be written

$$E_{so} = \frac{1}{2}\hbar^2[j(j+1) - l(l+1) - s(s+1)]\int_{Volume}\psi^*(\mathbf{r})A(r)\psi(\mathbf{r})dV. \tag{7.18}$$

This integral now represents an 'expectation value' for equivalently $1/r^3$ or $1/r\frac{dV(r)}{dr}$ (depending on which expression for $A(r)$ is considered). The mathematics for this is not too onerous, but we quote a solution later also containing a correction to the hydrogen energy levels due to relativistic electron mass increases.

7.4.2 The Relativistic Mass Correction to the Hydrogen Atom Energies

The kinetic energy operator $\hat{T}$ used in our original formulation of the Schrodinger equation was such that

$$\hat{T} = \frac{\hat{p}^2}{2m_0}$$

where $\hat{p}$ is the momentum operator and m_0 is the electron rest mass. For completeness, we need to use a relativistic expression for the kinetic energy and a relativistic form of the Schrodinger equation. A useful approximate treatment will be used first, before we consider the more complete solution in Section 7.4.4. At velocities approaching the speed of light, the correct expression for kinetic energy is

$$T = \sqrt{p^2c^2 + m_0^2c^4} - m_0c^2 = m_0c^2\left(\sqrt{1 + \frac{p^2}{m_0^2c^2}} - 1\right) \cong \frac{p^2}{2m_0} - \frac{p^4}{8m_0^3c^2} \tag{7.19}$$

upon expanding the square root expression to three terms. This means that the kinetic energy operator is more accurately represented by

$$\hat{T} = \frac{\hat{p}^2}{2m_0} - \frac{1}{8}\frac{\hat{p}^4}{m_0^3c^2}. \tag{7.20}$$

It is again possible to use perturbation theory to evaluate the effect of the extra term here for the kinetic energy operator. The operator $\hat{p}^4 = \hbar^4\nabla^4$ leads to an extra energy E_{rel} for the hydrogen atom quantum states such that

$$E_{rel} = -\frac{1}{8}\frac{\hbar^4}{m_0^3c^2}\int_{Volume} \psi^*(\mathbf{r})\nabla^4\psi(\mathbf{r})dV.$$

In practise, it is easier to solve this integral by noting that

$$\frac{\hat{p}^2}{2m_0} = E_n - V(r)$$

where E_n is the Bohr energy level and $V(r)$ is the central potential. Squaring the expression for $\hat{p}^2$ gives a Hamiltonian $\hat{H}_{rel}$ for relativistic effects of

$$\hat{H}_{rel} = -\frac{1}{2m_0c^2}(E_n^2 - 2E_nV(r) + V^2(r)) \tag{7.21}$$

so that

$$E_{rel} = \int_{Volume} \psi^*(\mathbf{r})\hat{H}_{rel}\psi(\mathbf{r})dV. \tag{7.22}$$

The solution of the integral now involves finding the expectation values of expressions with $1/r$ (for the $V(r)$ dependence) and $1/r^2$ (for the $V^2(r)$ dependence).

The expression for the spin–orbit coupling energy (Equation 7.18) shows for $l = 0$ states that $E_{so} = 0$ as $j = s$, but the relativity correction (Equation 7.22) indicates that energies for $l = 0$ states with non-zero wavefunction at the origin ($r = 0$) will have a large relativity correction factor E_{rel} due to the relativistic Hamiltonian $\hat{H}_{rel}$ involving terms with $1/r$ and $1/r^2$. However, including another effect known as the Darwin correction (discussed in Section 7.4.3) causes $l = 0$ energy states to have a higher energy (E_D), which compensates the extra relativity correction and brings the $l = 0$ energy up to that of the $l \neq 0$ state with the same j quantum number.

The total energy shift E_{fs} from the Bohr energy levels due to the spin–orbit coupling energy and relativistic energy correction found using perturbation theory plus the Darwin correction can be conveniently added together. The fine-structure energy level values can be written in terms of the total angular momentum quantum number j. The correction energy to the Bohr energy levels is given by

$$E_{fs} = E_{so} + E_{rel} + E_D = -\frac{E_n\alpha^2}{n}\left(\frac{1}{j+\frac{1}{2}} - \frac{3}{4n}\right)Z^2 \tag{7.23}$$

where α is the fine-structure constant. Given that $\alpha = e^2/(4\pi\epsilon_0\hbar c) = 1/137$, it can be seen that E_{fs} is much smaller than the Bohr energy levels E_n.

7.4.3 The Darwin Correction and Lamb Shift

The Darwin correction compensates for the extra relativity correction for the $l=0$ states. The Darwin correction arises because the full treatment (Dirac [26]; see Section 7.4.4) of a relativistic electron predicts an oscillation in the electron position (the oscillation is generally known by the German *zitterbewegung*). For a stationary electron, the oscillation has a period $\hbar/(2m_0c^2) = 6\times 10^{-22}$ seconds so cannot be directly observed, but it effectively blurs the electron position. It is only important if the electric field (or another field) gradient is large such as near the nucleus of an atom, where the gradient of potential energy (varying approximately as $1/r$) becomes large. A full treatment shows that the magnitude of the Darwin correction is opposite in sign to the relativistic correction for $l=0$ states introduced in Section 7.4.2. However, *zitterbewegung* has a completily negligible effect on energies for states $l \neq 0$ where the wavefunctions have a value of zero at the nucleus.

In quantum electrodynamics, electromagnetic fields are themselves quantised and the electron can be regarded as continually emitting a virtual photon and then re-absorbing the photon and also producing electron/positron pairs which subsequently quickly annihilate to produce virtual photons again. The effect is again prominent only at the nucleus ($r=0$), where the gradient of potential energy (varying approximately as $1/r$) becomes sufficiently high. Quantum electrodynamic effects change the energy levels, but only states with $l=0$ exhibit a significant effect because only these states have non-zero values of wavefunction at the nucleus. The energy shift is proportional to the value of the electron-probability density at the nucleus ($\psi^*(\mathbf{r}=0)\psi(\mathbf{r}=0)$), which causes $j=1/2$, $l=0$ states to have slightly higher energy (closer to the ionization limit) than $j=1/2$, $l=1$ states. This energy difference can be seen experimentally and is known as the Lamb shift, after Willis Lamb (1913–2008), who first observed the small energy difference. A schematic energy-level diagram for the fine structure of the hydrogen atom is given in Figure 7.5.

7.4.4 Relativistic Effects and the Dirac Equation

Our treatment of relativistic effects on the hydrogen atom and hydrogen-like ions in Section 7.4.2 is approximate and, as we have seen, requires a correction (the Darwin correction) for $l=0$ quantum states. A complete theory of relativistic effects in

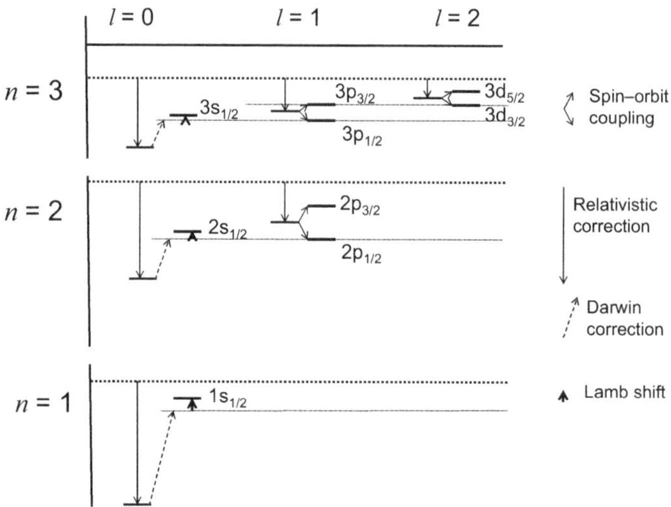

Figure 7.5 The schematic fine-structure energy levels of hydrogen and hydrogen-like ions. The down arrows show the energy deviation E_{rel} from the Bohr energy levels determined by relativistic corrections. Arrows going up and down show the splitting of energies due to spin–orbit coupling E_{so}. The Darwin corrections E_D and Lamb shift for $l = 0$ states are shown by respectively upward broken arrows and upward thick arrows. The spectroscopic notation discussed in Section 7.5 is used to designate the quantum states.

quantum mechanics has been developed and is briefly outlined here without mathematical detail.

The theory of special relativity gives the total energy of a particle of mass m_0 and momentum $\mathbf{p}$ as

$$E = \sqrt{p^2c^2 + (m_0c^2)^2}. \tag{7.24}$$

We have used this expression in our approximate treatment of relativity in Section 7.4.2 (Equation 7.19). Forming an operator in quantum mechanics is difficult with a square root, so we may suggest that the relativistic form of the equation for energy implies that the operator for the square of the particle energy is such that

$$\hat{E}^2\psi = [c^2\hat{\mathbf{p}} \cdot \hat{\mathbf{p}} + (m_0c^2)^2]\psi. \tag{7.25}$$

This equation has some validity and is known as the Klein-Gordon equation. The more complete treatment was found by Dirac, [26] who factorised the equation into a form

$$(\hat{E} - c\alpha \cdot \hat{\mathbf{p}} - \beta m_0c^2)(\hat{E} + c\alpha \cdot \hat{\mathbf{p}} + \beta m_0c^2)\psi = 0.$$

Only the first factor is needed for the left-hand side to equal zero, leading to the Dirac equation

$$(\hat{E} - c\alpha \cdot \hat{\mathbf{p}} - \beta m_0 c^2)\psi = 0 \tag{7.26}$$

with an expression for the relativistic energy operator:

$$\hat{H} = c\alpha \cdot \hat{\mathbf{p}} + \beta m_0 c^2. \tag{7.27}$$

The Dirac equation is deceptively simple, but constraints on α and β mean they need to be 4×4 matrices and the wavefunction ψ is no longer a single complex quantity, but a four-component complex quantity. However, electron spin is correctly included in the four-component wavefunction and evaluated energy eigenvalues can fully include the effects of relativity and spin–orbit coupling. The Dirac equation predicts the hydrogen and hydrogen-like fine structure given by Equation 7.23. Only quantum electrodynamic effects (as discussed in Section 7.4.3) leading to the Lamb shift need to be added for a complete theory of the hydrogen and hydrogen-like ion fine structure.

7.5 Spectroscopic Notation

The hydrogen quantum states have an associated historical system of designation. The angular momentum quantum number l is represented by a letter symbol such that $l = 0, 1, 2, 3, 4$ are referred to by the letters s, p, d, f, g (higher l values are alphabetical starting at h). The principal quantum number n is then used before the letter symbol, with the j quantum number added as a subscript to the angular quantum number symbol. The ground state of hydrogen thus has an electronic configuration of $1s_{1/2}$ and the different energy fine-structure states in the $n = 2$ energy level are $2p_{1/2}$, $2s_{1/2}$ and $2p_{3/2}$ (see Figure 7.5). This system of designating the quantum states of electrons is carried over to multi-electron atoms.

7.6 Hyperfine Structure: The Effect of Nuclear Spin

The effects on hydrogen and hydrogen-like ion energy levels due to the energy associated with electron spin in the magnetic field created by electron orbital angular momentum has been discussed. The spin/orbital angular momentum interaction leads to the fine-structure splitting of energy levels (see Section 7.4.1). Atomic nuclei also have spin and produce an additional energy term due to the interaction energy between the nuclear magnetic moment and the magnetic field arising from the angular momentum of the electron. The energies due to nuclear magnetic moment/electron orbit effects are approximately 1,000 times smaller than the energies due to electron magnetic moment/electron orbit effects and so the small variations of energy levels due to the nuclear spin are known as hyperfine structure.

Like the electron, protons and neutrons have spin $(1/2)\hbar$ and the nucleus can have a net magnetic moment. The nuclear magneton μ_N is analogous to the Bohr magneton (used for the electron) with value given by

$$\mu_N = \frac{e\hbar}{2m_p} \tag{7.28}$$

where m_p is the mass of the proton (and approximately the mass of the neutron). Due to the mass difference between the electron and proton, the nuclear magneton is 1,837 times smaller than the Bohr magneton. The mass difference between the electron and the proton is the main cause of the $\approx$1,000 $\times$ smaller effect on atomic energies due to the nuclear spin/electron angular momentum interaction (hyperfine structure) compared to the electron spin/electron angular momentum interaction (fine structure).

The magnetic moment of a proton, neutron or the nucleus as a whole is given by

$$\mu_{nuc} = g_{nuc}\mu_N \mathbf{I} \tag{7.29}$$

where g_{nuc} is the g-factor dependent on the relative distribution of charge and mass of a proton, neutron or nucleus and $\mathbf{I}$ is the proton, neutron or nucleus spin. A rotating object with charge and mass distributed proportionally to each other has a g-factor of unity. It is interesting to examine the g-factors needed in Equation 7.29. For a proton $g_{nuc} = 5.586$, while for a neutron $g_{nuc} = -3.826$. As a neutron has no charge, the existence of a non-zero magnetic moment is evidence for some 'internal' structure of the neutron where the charge and mass are not equally distributed; the quark model for particles explains the existence of the neutron magnetic moment. The g-factor for the nuclei of different elements also depends on the distribution of charge and mass in the nucleus and requires detailed nuclear physics calculation.

Nuclear spins $\mathbf{I}$ are quantised so that

$$|\mathbf{I}^2| = I(I+1)\hbar^2 \tag{7.30}$$

where I is the nuclear spin quantum number. The additional energy $-\mu_{nuc} \cdot \mathbf{B}$ resulting from the interaction between the nuclear magnetic moment μ_{nuc} and the magnetic field $\mathbf{B}$ associated with the total electron angular momentum $|\mathbf{J}| = \sqrt{J(J+1)}\hbar$ gives rise to the hyperfine structure. For hydrogen or hydrogen-like ions, the total angular momentum quantum number is j, but our treatment here is also appropriate for multi-electron atoms where the j quantum numbers for individual electrons can be added to give a net electron angular momentum J. As a direct analogy to the coupling between the electron orbital angular momentum and electron spin, a new quantum number F is introduced for the combination of the total electron angular momentum J and the nuclear spin I. We have

$$F = J + I \tag{7.31}$$

with an absolute value of the total angular momentum of $|\mathbf{F}| = \sqrt{F(F+1)}\hbar$. The quantum number F can have values

$$F = J - I, J - I + 1, \ldots, J + I.$$

The additional energy term leading to hyperfine structure can be constructed as for our treatment of LS coupling (see Equation 7.18). The hyperfine energy contribution $E_{HF} = -\mu_{nuc} \cdot \mathbf{B}$ can be represented by

$$E_{HF} = C_a g_{nuc} \mu_N \, \mathbf{I} \cdot \mathbf{J}$$

as the nuclear magnetic moment $\mu_{nuc} = g_{nuc}\mu_N \mathbf{I}$ (Equation 7.29) and the magnetic field is proportional to the electron angular momentum $\mathbf{J}$. Here C_a is a constant of proportionality dependent on the value of the magnetic field at the nucleus produced by the electron angular momentum. We have that

$$\mathbf{F}^2 = \mathbf{I}^2 + \mathbf{J}^2 + 2\mathbf{I} \cdot \mathbf{J}$$

which enables a convenient expression for $\mathbf{I} \cdot \mathbf{J}$ so that we have

$$E_{HF} = \frac{\hbar^2}{2} \left[F(F+1) - I(I+1) - J(J+1) \right] g_{nuc} \mu_N C_a. \tag{7.32}$$

The splitting due to hyperfine structure of the $n = 1$ and $n = 2$ states of hydrogen is illustrated in Figure 7.6a. Due to overlap of the wavefunctions of s-states ($l = 0$) with the nucleus (see Equation 7.12), the magnetic field created by the electron spin at the nucleus is much greater for s electrons than for other electron states, so the energy splitting due to the hyperfine interaction is much greater for s-states.

The hyperfine splitting of the ground state of hydrogen is particularly important in astrophysics and astronomy. If the electron and proton spins are aligned (parallel), the electron angular momentum j and proton spin I are parallel and we have $F = 1$ so that the ground 1s state is 5.9×10^{-6} eV higher in energy than if the proton and electron spins are anti-parallel and $F = 0$. It is possible to have a radiative transition between the two possible ground states if the spin of the electron flips from parallel to anti-parallel. The frequency of this transition is well into the radio-frequency spectrum at 1,420 MHz with a wavelength for the radiation of 21 cm. Radio-frequency emission and absorption at 21 cm is used by radio astronomers to map hydrogen density. Radio astronomy at 21 cm established the evidence for our galaxy being a spiral [10] enables measurements of the early universe during the

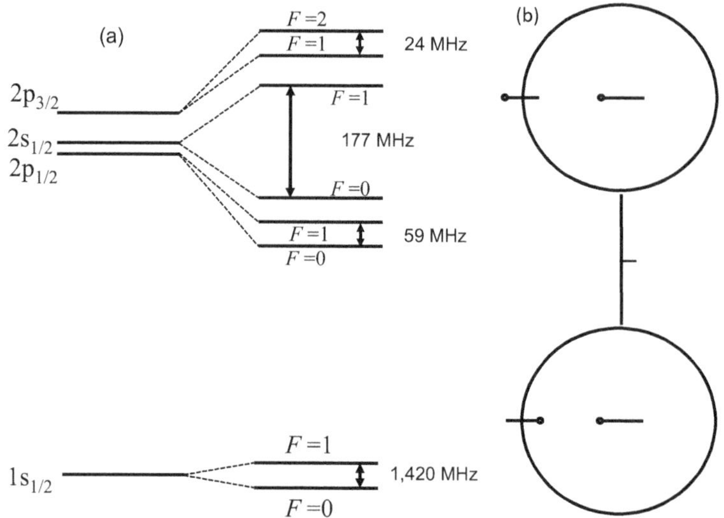

Figure 7.6 (a) Examples of hyperfine splitting in hydrogen. The energies of the hyperfine states are schematically illustrated along with the frequencies of transition between the states. The hyperfine energy scale (right side) is much smaller than the fine-structure scale (left side). (b) A copy of part of the *Voyager* record cover [121] and *Pioneer 10* and *Pioneer 11* plaques [94] constructed to illustrate in an abstract way the two hyperfine states of the ground state of hydrogen: presented in a way that, we hope, an alien intelligence could interpret.

formation of the first galaxies [74] and established the existence of dark matter in rotating galaxies [119].

The *Voyager 1* probe is the farthest human-made object from Earth. Launched in 1977, it has now left the solar system. A gold-plated phonograph record encoding sounds and images was placed aboard the spacecraft. This was intended to be decoded by any intelligent extraterrestrial life form or perhaps future humans who came across the equipment. A committee led by the famous popularising physicist Carl Sagan (1934–1996) decided on the sounds and images to be incorporated and on the coding system to enable an extraterrestrial to play the record. An image on the aluminium cover of the *Voyager* record conveyed the information that the 21 cm wavelength of the transition between the hyperfine states of hydrogen is used as a fundamental length scale for the images (see Figure 7.6b). The time for light to travel 21 cm corresponds to 0.704 ns and was included as a measure of time to be used in decoding the record. Similar images to Figure 7.6b were placed on the earlier *Pioneer 10* (launched 1972) and *Pioneer 11* (launched 1973) spacecraft [94]. In both of the *Pioneer* launches, a naked man and woman were also illustrated, with the woman's height shown in binary as | – –– meaning 1,000 = 8 (in decimal) implying 8×21 cm = 168 cm (5 feet 6 inches).

7.7 Summary for Hydrogen and Hydrogen-Like Ions

A quantum mechanical description of the hydrogen atom and the equivalent hydrogen-like ion has been presented. Several different quantum numbers have been introduced with different possible values:

- The principal quantum n determines the energy of states in the hydrogen atom or hydrogen-like ion (of atomic number Z) to good accuracy. The principal quantum number can have postive non-zero values – $n = 1, 2, 3, \ldots$. The energy levels are given by

$$E_n = -\frac{13.6Z^2}{n^2}\text{eV}.$$

- Other quantum numbers have values: $l = 0, 1, 2, 3, \ldots n - 1$, $m = -1, -l + 1, \ldots 0 \ldots l - 1, l$, and $s = \pm\frac{1}{2}$.
- A relativistic correction to the hydrogen atom Hamiltonian and the addition of the coupling energy between the magnetic moment associated with the electron spin and the magnetic field associated with an orbital angular momentum gives the fine structure of the hydrogen energy levels with the fine-structure energies only dependent on the j quantum number, defined by $j = l \pm 1/2$. A Lamb shift correction to $l = 0$ states causes them to have a slightly different (higher) energy to other states with the same j.

Exercises

7.1 Considering the hydrogen-like ion wavefunctions, show that the most probable radius to find an electron in the ground state of the hydrogen-like ion is a_0/Z, where a_0 is the Bohr radius and Z is the atomic number of the hydrogen-like ion.

7.2 The expectation value for radius r for an electron with wavefunction ψ is given by

$$< r >= \int \psi^* \mathbf{r} \psi dV$$

where the integration is over all space. Show that $< r > = \frac{3}{2}a_0/Z$ for the hydrogen-like ion ground state.

7.3 The angular variation of the hydrogen and hydrogen-like quantum states for an isolated atom or ion are given by mixed states of form $Y_{mix} = C(Y_{lm} + Y_{l-m})$ for $m \neq 0$. Using the wavefunctions given in Equation 7.10, show for $l = 1$, $m = \pm 1$ that the necessary normalisation $\int Y^*_{mix} Y_{mix} d\Omega = 1$ requires $C = 1/\sqrt{2}$.

7.4 The angular variation of the hydrogen and hydrogen-like quantum states are such that

$$Y_{lm} = |Y_{lm}|e^{im\phi}$$

and

$$Y_{l-m} = (-1)^m|Y_{lm}|e^{-im\phi}$$

for $m \geq 0$. Show that the mixed state

$$\begin{aligned} Y_{lm} + Y_{l-m} &= 2i\sin(m\phi)\,|Y_{lm}| \quad (m = 1, 3, \ldots) \\ Y_{lm} + Y_{l-m} &= 2\cos(m\phi)\,|Y_{lm}| \quad (m = 0, 2, \ldots). \end{aligned} \tag{7.33}$$

7.5 Consider a disc rotating about its centre of mass with a uniform distribution of charge Q and mass M. Show that the magnetic moment of the rotating disc is given by $\mu = Q/(2M)\mathbf{L}$, where $\mathbf{L}$ is the angular momentum of the disc. If the charge is only found on the circumference of the disk, show that $\mu = 2Q/(2M)\mathbf{L}$. [The electron spin has an experimentally measured magnetic moment $\mu = -2.023e/(2m_0)\mathbf{L}$, which suggests that, in some sense, the electron has a rotating-disk geometry with the charge mainly distributed on the edge of the disk.]

7.6 Given that the fine-structure energy levels for hydrogen and hydrogen-like ions depart from the Bohr values by E_{fs} where

$$E_{fs} = -\frac{E_n\alpha^2}{n}\left(\frac{1}{j+\frac{1}{2}} - \frac{3}{4n}\right)Z^2,$$

show that the Lyman alpha emission for transitions 2p–1s consists of a doublet with a wavelength separation of 0.53×10^{-3} nm for hydrogen and all hydrogen-like ions.

7.7 Spin–orbit coupling produces an energy perturbation ΔE to the hydrogen quantum state energies which is related to the relevant quantum numbers j, l and s by the equation

$$\Delta E = \frac{1}{2}\hbar^2[(j(j+1) - l(l+1) - s(s+1)]C$$

where C is a constant dependent on the n and l quantum numbers. For $l \geq 1$, show that spin–orbit coupling in hydrogen produces two different energy states with an energy separation of $\hbar^2(l + \frac{1}{2})C$.

7.8 A radio telescope at Ohio State University detected a single, 72-second burst on August, 15, 1977, at a frequency 50 kHz higher than the $F = 1$ to $F = 0$ hyperfine transition of the ground 1s state of hydrogen. Astronomer

Jerry R. Ehman wrote the exclamation ‘Wow!’ on the computer print-out of the signal as he thought it was possibly communication from an alien civilisation. However, in 2015 it was verified that a comet passed in the direction of the signal at the observed time [80]. Assuming the radio signal arose from the hydrogen tail of the comet, calculate the relative velocity of the comet to Earth along the Earth-comet line of sight. [10.56 km s^{-1} towards Earth]

8

Discrete Bound States: Many-Electron Atoms and Ions

Much of the physics determined for hydrogen and hydrogen-like ions is relevant to atoms and ions with many electrons. In a multi-electron atom or ion, the quantum state energies and wavefunctions are dominated by the central potential arising from the nuclear charge as is the case for hydrogen and hydrogen-like ions. Consequently, quantum wavefunctions similar in form to those found for hydrogen or hydrogen-like ions are produced. There are some corrections needed to the exact electron energies associated with the nuclear electric field experienced by each electron due to shielding effects by other electrons – and some perturbing energy effects due to electron–electron Coulomb repulsion. The electron wavefunctions have similar orbital or angular shapes as for hydrogen and hydrogen-like ions (see Figures 7.2 and 7.3). The same electron-configuration system (i.e. 1s, 2s, 2p, ...) can be used and the same degeneracies are associated with different values of principal quantum number n (degeneracy $2n^2$) and orbital angular momentum quantum number l (degeneracy $2 \times 2l(l+1)$) as found for hydrogen and hydrogen-like ions.

As multi-electron atoms and ions have quantum states following the 'rules' established for hydrogen and hydrogen-like ions, the quantum state energy values are primarily determined by a principal quantum number n with values $n=1,2,3,\ldots$. Angular quantum numbers l with $l=0,1,2,\ldots n-1$ are found to have a small effect on energy values due to shielding of the nuclear charge. As for hydrogen, the number of individual quantum states at a particular energy involves the different magnetic quantum numbers m with values $m=-l,-l+1,\ldots,0,1\ldots+l$ and the two possible spin orientations (relative to the orbital angular momentum). The states which the electrons occupy are called the electronic configuration for the atom or ion.

An important feature of multi-electron atoms is established by the Pauli exclusion principle which requires that only one electron can occupy an individual quantum state. In the ground state of an atom or ion, the required number of electrons effectively fills the lowest energy quantum states designated by the n, l, m and s values according to the hydrogen degeneracies.

8.1 Exchange Parity and the Pauli Exclusion Principle

Before examining the energy levels of multi-electron atoms, it is worthwhile to examine the Pauli exclusion principle, which is responsible for the electron configuration of atoms. We examine the fundamental issue of the number of identical particles that can occupy the same quantum state.

We suppose that a wavefunction $\psi(\mathbf{r}_1,\mathbf{r}_2)$ represents two identical particles with positions $\mathbf{r}_1$ and $\mathbf{r}_2$. If the two particles are identical, then there should be no measurable difference between $\psi(\mathbf{r}_1,\mathbf{r}_2)$ and $\psi(\mathbf{r}_2,\mathbf{r}_1)$. Now consider an (imaginary) 'exchange' operator $\hat{P}$ which has the action of interchanging the positions of the two particles. We can write

$$\hat{P}\psi(\mathbf{r}_1,\mathbf{r}_2) = \psi(\mathbf{r}_2,\mathbf{r}_1). \tag{8.1}$$

If the eigenvalues of $\hat{P}$ are p, we can also write that

$$\hat{P}\psi(\mathbf{r}_1,\mathbf{r}_2) = p\psi(\mathbf{r}_1,\mathbf{r}_2).$$

The possible values of the eigenvalues p can be found by applying the exchange operator twice. Thinking of the eigenvalue equation, this gives

$$\hat{P}(\hat{P}\psi(\mathbf{r}_1,\mathbf{r}_2)) = p^2\psi(\mathbf{r}_1,\mathbf{r}_2).$$

Applying the operator twice just returns the wavefunction to its original state i.e.

$$\hat{P}(\hat{P}\psi(\mathbf{r}_1,\mathbf{r}_2)) = \psi(\mathbf{r}_1,\mathbf{r}_2).$$

For both these last two equations to be valid requires that $p^2=1$ or $p=\pm 1$. It is consequentially a fundamental property that any particle has an eigenvalue to the exchange operator of either $+1$ or -1. This property is so fundamental that different names are given for particles with different exchange operator eigenvalues. Particles with $p=+1$ are known as bosons, while particles with $p=-1$ are known as fermions.

We can cite examples of these different types of particle. Most of the familiar particles making up nuclei and atoms are fermions. Electrons, protons and neutrons are fermions and all have half-integer spin (units of $\frac{1}{2}\hbar$). The particles which are bosons are most familiarly represented by photons. These can be shown to have integer spin (units of $\hbar$). For some applications, whole atoms can be regarded as particles which are either bosons or fermions. If the total atom has an integer spin, the atom as a whole has the nature of a boson, while if the total atom has a half-integer spin, the atom as a whole will act like a fermion.

It is possible to deduce some further properties of these two classes of particles. As we have assumed that the particles are indistinguishable, we can write

$$\hat{P}\psi(\mathbf{r}_1,\mathbf{r}_2) = \psi(\mathbf{r}_2,\mathbf{r}_1) = \psi(\mathbf{r}_1,\mathbf{r}_2). \tag{8.2}$$

This equation is consistent with the eigenvalue equation

$$\hat{P}\psi(\mathbf{r}_1,\mathbf{r}_2) = p\psi(\mathbf{r}_1,\mathbf{r}_2)$$

only for bosons with $p = +1$. For fermions with $p = -1$, it is not possible for both Equation 8.2 and the eigenvalue equation to be valid unless the wavefunction $\psi(\mathbf{r}_1,\mathbf{r}_2)$ is identically equal to zero. Having $\psi(\mathbf{r}_1,\mathbf{r}_2) = 0$ means that it is not possible to have two indistinguishable particles in the same quantum state – the only situation where this can apply is if there is no particle density (zero wavefunction). This enables us to state that no two fermions can occupy the same quantum state. This result is known as the Pauli exclusion principle. No two electrons in an atom or ion can have the same set of quantum numbers.

Identical bosons (with $p = 1$) can exist in the same quantum state as it is possible for Equation 8.2 and the eigenvalue equation both to apply and the wavefunction to be non-zero.

8.2 The Central Field Approximation

An approximation known as the central field approximation (sometimes referred to as the 'orbital approximation') can be used to calculate the energies and wavefunctions of a multi-electron atom or ion. It is assumed that individual electrons are affected predominantly by the Coulomb potential of the Z protons in the nucleus with the other electrons in the atom simply acting to shield this Coulomb potential from the electron under investigation. The central potential of a multi-electron atom or ion is assumed to depend only on the distance from the nucleus so that it can be written as

$$V(r) = -\frac{Z(r)e^2}{4\pi\epsilon_0 r} \tag{8.3}$$

where $Z(r)$ reflects the effective charge seen by an electron at a distance r from the nucleus. Close to the nucleus, we would expect $Z(r) \approx Z$ the charge on the nucleus, while at large distance from the nucleus all the other electrons shield the nucleus causing $Z(r) \approx 1$ (for an atom) or $Z(r) \approx Z_i + 1$ (for an ion of charge Z_i). The total electronic energy of the atom or ion is determined by adding up the individual electron energies calculated using the Schrodinger equation (Equation 7.3) solved separately for each electron.

A method to determine $Z(r)$ is to assume that the electron wavefunctions are the same as those of hydrogen and use the calculated probability densities for electrons (i.e. the square of the wavefunction amplitudes) to evaluate $Z(r)$. A revised set of wavefunctions can then be numerically calculated using the central potential given by Equation 8.3 and a revised value of $Z(r)$ calculated as well. The process can continue iteratively until a required level of accuracy in the wavefunctions and individual energy level values is found. This iterative method is known as the Hartree self-consistent field method of evaluating the energy and wavefunctions of multi-electron atoms.

Without undertaking detailed numerical calculations, we can start to think about the ordering of the energy levels that an electron can occupy in multi-electron atoms. Due to shielding of the nucleus by other electrons, the energy of an electron in a multi-electron atom depends on the average proximity of its wavefunction to the nucleus.

The wavefunctions of electrons associated with different quantum states (i.e. different quantum numbers) are close to those for hydrogen-like ions as the central potential approximation (Equation 8.3) will yield wavefunctions which are hydrogen-like. From the radial wavefunctions for hydrogen-like ions (listed in Section 7.2.1), we see that the radial electron probability distribution varies with the orbital angular quantum number l. For example, electrons with $l=0$ (s-states) have a finite probability distribution at the nucleus ($r=0$). This means that on average an s-electron sees more of the nuclear charge than a p-orbital ($l=1$) or higher l orbital, which all have wavefunctions of zero at the nucleus. The s-orbital electrons are consequently more tightly bound. There is in fact an ordering of the proximity of electron-probability distributions from the nucleus – $l=0$ is always tightly bound and the degree of nuclear charge seen by an electron generally decreases as l increases.

The effect of the different radial variation of wavefunctions associated with different orbital angular momentum quantum numbers l causes the energies in multi-electron atoms to be ordered not only due to the principal quantum number n, but also with l. The principal quantum number n still dominates the energy ordering so that different n-values are referred to as 'shells'. However, within each shell, there is an ordering dependent on l. Associated with our argument regarding shielding of the nuclear charge, the ordering within a shell is such that $l=0$ electrons have lower energy (farther from the ionisation limit), and the energies generally increase (get closer to the ionisation limit) as l increases. Electrons with the same n and l quantum numbers are referred to as being in the same sub-shell. Due to the importance of n in determining the electron energies, electron shells are often designated by symbols K, L, M and N to represent n-values of 1, 2, 3 and 4, respectively.

For a multi-electron atom, the Pauli exclusion principle requires that only one electron is in each quantum state. The ground electronic configuration of each atom is therefore determined by the energy ordering of the shells and sub-shells

Atomic number → $_{79}$Au ← Atomic symbol
197.0 ← Atomic weight
Electronic configuration → $5d^{10}$ $6s$

$_{1}$H 1.00 $1s$																	$_{2}$He 4.00 $1s^2$
$_{3}$Li 6.94 $2s$	$_{4}$Be 9.01 $2s^2$											$_{5}$B 10.81 $2s^2 2p$	$_{6}$C 12.01 $2s^2 2p^2$	$_{7}$N 14.01 $2s^2 2p^3$	$_{8}$O 16.00 $2s^2 2p^4$	$_{9}$F 19.00 $2s^2 2p^5$	$_{10}$Ne 20.18 $2s^2 2p^6$
$_{11}$Na 22.99 $3s$	$_{12}$Mg 24.21 $3s^2$											$_{13}$Al 26.98 $3s^2 3p$	$_{14}$Si 28.09 $3s^2 3p^2$	$_{15}$P 30.97 $3s^2 3p^3$	$_{16}$S 32.07 $3s^2 3p^4$	$_{17}$Cl 35.45 $3s^2 3p^5$	$_{18}$Ar 39.95 $3s^2 3p^6$
$_{19}$K 39.10 $4s$	$_{20}$Ca 40.08 $4s^2$	$_{21}$Sc 44.96 $3d$ $4s^2$	$_{22}$Ti 47.88 $3d^2$ $4s^2$	$_{23}$V 50.94 $3d^3$ $4s^2$	$_{24}$Cr 52.00 $3d^5$ $4s$	$_{25}$Mn 54.94 $3d^5$ $4s^2$	$_{26}$Fe 55.85 $3d^6$ $4s^2$	$_{27}$Co 58.93 $3d^7$ $4s^2$	$_{28}$Ni 58.69 $3d^8$ $4s^2$	$_{29}$Cu 63.55 $3d^{10}$ $4s$	$_{30}$Zn 65.39 $3d^{10}$ $4s^2$	$_{31}$Ga 69.72 $4s^2 4p$	$_{32}$Ge 72.61 $4s^2 4p^2$	$_{33}$As 74.92 $4s^2 4p^3$	$_{34}$Se 78.96 $4s^2 4p^4$	$_{35}$Br 79.90 $4s^2 4p^5$	$_{36}$Kr 83.80 $4s^2 4p^6$
$_{37}$Rb 85.47 $5s$	$_{38}$Sr 87.62 $5s^2$	$_{39}$Y 88.91 $4d$ $5s^2$	$_{40}$Zr 91.22 $4d^2$ $5s^2$	$_{41}$Nb 92.91 $4d^4$ $5s$	$_{42}$Mo 95.94 $4d^5$ $5s$	$_{43}$Tc 97.9 $4d^6$ $5s$	$_{44}$Ru 101.1 $4d^7$ $5s$	$_{45}$Rh 102.9 $4d^8$ $5s$	$_{46}$Pd 106.4 $4d^{10}$ -	$_{47}$Ag 107.9 $4d^{10}$ $5s$	$_{48}$Cd 112.4 $4d^{10}$ $5s^2$	$_{49}$In 114.8 $5s^2 5p$	$_{50}$Sn 118.7 $5s^2 5p^2$	$_{51}$Sb 121.8 $5s^2 5p^3$	$_{52}$Te 127.6 $5s^2 5p^4$	$_{53}$I 126.9 $5s^2 5p^5$	$_{54}$Xe 131.6 $5s^2 5p^6$
$_{55}$Cs 132.9 $6s$	$_{56}$Ba 137.3 $6s^2$	$_{57}$La 138.9 $5d$ $6s^2$	$_{72}$Hf 178.5 $4f^{14}$ $5d^2$ $6s^2$	$_{73}$Ta 180.9 $5d^3$ $6s^2$	$_{74}$W 183.8 $5d^4$ $6s^2$	$_{75}$Re 186.2 $5d^5$ $6s^2$	$_{76}$Os 190.2 $5d^6$ $6s^2$	$_{77}$Ir 192.2 $5d^9$ -	$_{78}$Pt 195.1 $5d^9$ $6s$	$_{79}$Au 197.0 $5d^{10}$ $6s$	$_{80}$Hg 200.6 $5d^{10}$ $6s^2$	$_{81}$Tl 204.4 $6s^2 6p$	$_{82}$Pb 207.2 $6s^2 6p^2$	$_{83}$Bi 209.0 $6s^2 6p^3$	$_{84}$Po 209 $6s^2 6p^4$	$_{85}$At 210 $6s^2 6p^5$	$_{86}$Rn 222 $6s^2 6p^6$
$_{87}$Fr 223 $7s$	$_{88}$Ra 226.0 $7s^2$	$_{89}$Ac 227.0 $6d$ $7s^2$	$_{104}$Rf 267	$_{105}$Db 268	$_{106}$Sg 271	$_{107}$Bh 272	$_{108}$Hs 270	$_{109}$Mt 276	$_{110}$Un 281	$_{111}$Uu 272	$_{112}$Ub 285						

Lanthanides	$_{58}$Ce 140.1 $4f^2$ $6s^2$	$_{59}$Pr 140.9 $4f^3$ $6s^2$	$_{60}$Nd 144.2 $4f^4$ $6s^2$	$_{61}$Pm 145 $4f^5$ $6s^2$	$_{62}$Sm 150.4 $4f^6$ $6s^2$	$_{63}$Eu 152.0 $4f^7$ $6s^2$	$_{64}$Gd 157.3 $4f^7$ $5d$ $6s^2$	$_{65}$Tb 158.9 $4f^8$ $5d$ $6s^2$	$_{66}$Dy 162.5 $4f^{10}$ $6s^2$	$_{67}$Ho 164.9 $4f^{11}$ $6s^2$	$_{68}$Er 167.3 $4f^{12}$ $6s^2$	$_{69}$Tm 168.9 $4f^{13}$ $6s^2$	$_{70}$Yb 173.0 $4f^{14}$ $6s^2$	$_{71}$Lu 175.0 $4f^{14}$ $5d$ $6s^2$
Actinides	$_{90}$Th 232.0 - $6d^2$ $7s^2$	$_{91}$Pa 231.0 $5f^2$ $6d$ $7s^2$	$_{92}$U 238.0 $5f^3$ $6d$ $7s^2$	$_{93}$Np 237.0 $5f^4$ $7s^2$	$_{94}$Pu 244 $5f^6$ $7s^2$	$_{95}$Am 243 $5f^7$ $7s^2$	$_{96}$Cm 247 $5f^7$ $6d$ $7s^2$	$_{97}$Bk 247	$_{98}$Cf 251	$_{99}$Es 252	$_{100}$Fm 257	$_{101}$Md 258	$_{102}$No 259	$_{103}$Lr 260

Figure 8.1 The periodic table of the ground state of un-ionised elements. The electronic configuration of outer and un-filled sub-shells alongside the atomic number, symbol and atomic weight are listed. Taken from a table of constants available for examination at the Department of Physics, University of York, UK.

with electrons 'filling up' each state in energy order starting with $n=1, l=0$. The degeneracy of each sub-shell, discussed in Chapter 7, now reflects the number of electrons required to fill the sub-shell. The periodic table of elements is built up in this way (see Figure 8.1 for the periodic table listing). Each element is defined by the number of protons in the nucleus (the atomic number), which implies that, for a neutral atom, there will be the same number of electrons occupying electronic quantum states.

After placing electrons into sub-shells of increasing energy, the last un-filled sub-shell largely controls the chemical behaviour of the element as the wavefunction(s) of the electrons(s) in this sub-shell extend farther away from the nucleus so that these electrons are more likely to interact with neighbouring atoms. The number of electrons in the chemically active sub-shell determines the 'valency' of the element. The valency of an element is the number of hydrogen atoms an element can combine with or displace, so is typically the number of electrons in an un-filled sub-shell or the number of electrons required to fill the sub-shell.

The un-filled sub-shells also determine the angular momentum and spin of an atom or ion. Filled sub-shells have no net angular momentum as the different electron values of the m-component of angular momentum along the z-axis cancel ($m = -l, \ldots 0, \ldots +l$). Filled sub-shells also have no net spin as each of the two opposite spin states cancel. It is possible to show that adding the probability distributions of all the electrons in a filled sub-shell produces a spherically symmetric electron-probability distribution. This last is important as it means that electrons in un-filled sub-shells have energies determined by a spherically symmetric potential determined by the point nuclear charge and the spherically symmetric spread of the charge of the electrons in filled sub-shells. An electron in an un-filled sub-shell with most of its wavefunction at radii beyond the radii of the filled sub-shell electron wavefunctions sees a central potential similar to hydrogen or a hydrogen-like ion with a charge comprising the nuclear charge minus the charge of the filled sub-shell electrons. Such a central potential causes the un-filled sub-shell electron energies and wavefunctions to be similar to those for hydrogen or a hydrogen-like ion.

There are some peculiarities of the periodic table for the ground state of un-ionised atomic elements (Figure 8.1). For example, after filling up electrons in the $n = 3$ and $l = 0$ and 1 sub-shells to produce argon (atomic number $Z = 18$), the next-higher atomic element (potassium with $Z = 19$) has the additional electron in the 4s sub-shell, rather than the 3d sub-shell. The s-orbitals have a large value of wavefunction close to the nucleus and are less shielded from the nuclear charge by other electrons. The s-orbitals are consequently more tightly bound (and hence have lower energy) than the higher l, but lower n quantum states for principal quantum number $n \geq 4$.

With ionisation, the energy ordering of ionic quantum states depends more strongly on the principal quantum number n as there is a larger net central positive charge near the nucleus. The shells of principal quantum number n for ions are filled when the number of bound electrons equals the degeneracy $2n^2$. For example, ions lying on the nickel isoelectronic series with twenty-eight electrons (known as nickel-like ions) have a ground state configuration of $1s^2 2s^2 2p^6 3s^2 3p^6 3d^{10}$. The neutral nickel atom has a different electronic configuration: $1s^2 2s^2 2p^6 3s^2 3p^6 3d^8 4s^2$ (see Figure 8.1).

8.2.1 The Stern–Gerlach Experiment

An experiment was undertaken in 1922 by Stern and Gerlach to measure the magnetic dipole moment of silver atoms. The experiment provided evidence that electrons have an intrinsic magnetic moment or spin, but requires an understanding of the multi-electron atom structure to properly interpret [122]. Silver has an atomic number of 47 so that all the electrons are in filled sub-shells, except for one electron in a 5s orbital (see Figure 8.1). This electron has zero orbital angular momentum $\sqrt{l(l+1)}\hbar$ as $l=0$ and all the filled sub-shells have zero angular momentum and zero spin. The nuclear spin can be neglected as the magnetic moment of the silver nucleus is much smaller than that of electrons due to the much greater nuclear mass (see Section 7.6). Consequently, the magnetic moment of the atom is determined by the one 5s electron.

A magnetic moment μ with component μ_z oriented parallel to a non-uniform magnetic field $\mathbf{B}$ has a minimum energy $-\mu \cdot \mathbf{B} = -\mu_z B$ at the maximum of the magnetic field, while a magnetic moment with component $-\mu_z$ oriented in the opposite direction to the magnetic field has a minimum energy $-\mu \cdot \mathbf{B} = +\mu_z B$ at the minimum magnetic field. In the Stern–Gerlach experiment, a beam of silver atoms from an oven passed through a non-uniform magnetic field produced by two curved pole pieces (Figure 8.2). The experiment should cause the atoms to minimise their potential energy by moving to a higher or lower magnetic field depending on the direction of the component of magnetic moment to the magnetic field. It was found that the detected atoms either moved to higher or lower magnetic field strength as they passed along the non-uniform magnetic field, but the atoms were only detected with two values of the component of the atom magnetic moment to the magnetic field: one value parallel and one value anti-parallel. If the electrons could orient their magnetic moment at any angle to the magnetic field, the

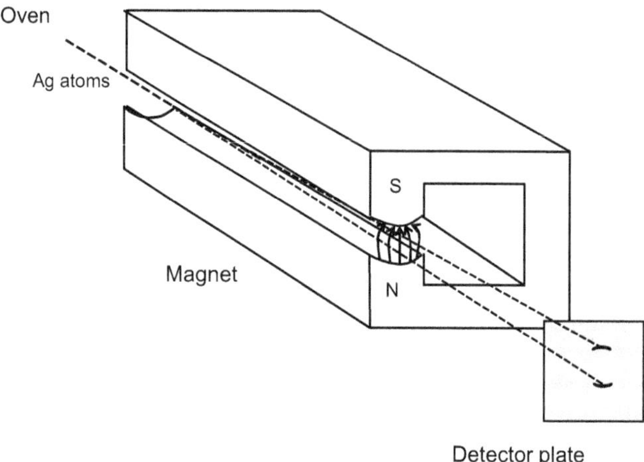

Figure 8.2 A schematic of the Stern–Gerlach experiment. A non-uniform magnetic field causes a beam of silver atoms to deflect towards increasing magnetic field or decreasing magnetic field depending on the spin of an electron in the 5s orbital of silver.

experiment would produce a broadened beam on detection, but only two discrete magnetic moments were detected, indicating only two orientations of the electron magnetic moment to the magnetic field.

The Stern–Gerlach experiment showed that the electron-intrinsic magnetic moment is oriented with the same component either parallel or anti-parallel to a magnetic field and not at any other angle. As the electron is charged, the intrinsic electron magnetic moment can be said to occur due to rotation or spin of the electron. The electron is negatively charged, so the spin is in the opposite direction to the magnetic moment.

In 1927 Phipps and Taylor used the Stern–Gerlach technique with a hydrogen beam and obtained a similar result as Stern and Gerlach. At the time, this was a more satisfactory result as only a single electron was involved and there was no need to think about the (now-known) zero magnetic moment contribution from the electrons in the filled sub-shells.

8.3 The Coulomb and Spin–Orbit Interactions

To obtain the detailed energy structure of many electron atoms and ions, we need to consider some interactions that we have so far ignored. We have ignored the energy associated with the Coulomb repulsion that electrons will experience with each other and there are spin–orbit coupling interactions which are significant and apply in a similar way to our treatment for hydrogen. These two effects are small compared to the energy-ordering arguments of Section 8.2, so can be treated using quantum mechanical perturbation theory. We can identify an applicable Hamiltonian for the Coulomb energy $\hat{H}_c$ such that

$$\hat{H}_c = \Sigma_{i,j} \frac{e^2}{4\pi\epsilon_0 r_{ij}}$$

where the sum is over all electrons i and j, and r_{ij} represents the distance between the electron wavefunctions. The spin–orbit coupling is treated in a similar way as for hydrogen, but we need to sum over all electrons. The Hamiltonian $\hat{H}_{so}$ is given by

$$\hat{H}_{so} = \Sigma_i A(r_i)\hat{S}\,.\,\hat{L}_i.$$

It is possible to verify by checking expressions for the orbital wavefunctions for hydrogen (see Equation 7.10) that the addition of the electron probability distributions for all electrons in a sub-shell gives rise to a spherical distribution of electron probability (see, for example, Exercise 8.1). In addition, the sum of all electron orbital angular momentum and electron spin for a full sub-shell is zero. The orbital angular momenta and spin for different electron quantum states within a sub-shell are oriented in different directions so that they cancel when added together. Remember that the component of orbital angular momentum along the z-axis varies as $m\hbar$ with $m = -l, -l+1, \ldots, 0, \ldots, l-1, l$. Orbital angular momentum is equally spread in the minus and plus directions to the z-axis. With the electron spin, only two states are allowed $\pm\frac{1}{2}\hbar$ and these clearly cancel for a full sub-shell with equal number of electrons with $\pm\frac{1}{2}$ spin. For these reasons, full shells and sub-shells are referred to as 'closed'. There is little interaction with electrons in un-filled shells, apart from the shielding of the nuclear potential. As the closed shells and sub-shells have spherically symmetric electron distributions, the central field assumption of Equation 8.3 is also accurate.

We first consider a treatment where the Coulomb interaction Hamiltonian has a bigger effect than the spin–orbit interaction. For historical reasons, the treatment is known as LS or Russell-Saunders coupling. The energy ordering and a designatory symbol (known as the term symbol) for quantum states in atoms and ions where LS coupling applies is determined by some guidelines known as Hund's rules, described in Section 8.3.1.

8.3.1 LS Coupling

The detailed energy ordering of multi-electron atom and ion quantum states depends on the total atomic or ionic spin and the total orbital angular momentum of all the electrons. The total atomic orbital angular momentum is found by adding the angular orbital momentum for electrons in un-filled sub-shells. We have already seen that filled sub-shells will have zero total angular momentum, so do not need to be considered further. The total angular momentum is given by a quantum number L (written in upper-case letters) where L can have positive integer values $L = |\Sigma_i \mathbf{l}_i|$, where the summation is such that the individual electron orbital angular momentum $\mathbf{l}_i$ in an un-filled shell is treated as a vector, but only allowing the result that L is an integer. For example, if there are two p-electrons in an un-filled shell, the possible values of L are 0, 1 and 2. This is a further example of quantisation on the atomic scale. The orientation of angular momenta of individual electrons is such that there is a limited range of orientations. They all orient with integer components of angular momentum along the z-axis and this ensures that L is an integer. Similar to hydrogen, the total angular momentum for an atom is designated by a symbol for different L values, but with the symbol being written in upper-case letters. We write S, P, D, F, G, H, ... for L values of respectively 0, 1, 2, 3, 4, 5, ...

The electron spin of all electrons in un-filled shells is similarly added to give the total spin S for the atom (not to be confused with the use of S to indicate a value of $L = 0$ as just explained). The addition of electron spin is easier to understand than the addition of orbital angular momentum as there are only two spin orientations, so for example the total spin with two electrons in an un-filled shell is either 0 or 1.

In LS coupling, the Coulomb Hamiltonian $\hat{H}_c$ dominates the spin–orbit coupling Hamiltonian $\hat{H}_{so}$, so we need to take account of the Coulomb repulsion energy first. It is possible to invoke arguments (Hund's rules) to indicate the order of atomic energy levels that arise due to the different effects. We consider a situation where two electrons are in an un-filled sub-shell, but the treatment can be extended to three or more electrons.

Different values of total electron spin cause the dominant effect on the atom energies. For two electrons, the total atomic spin $S=0$ if the individual electron spins are opposite to each other. If this is the case, the two electrons can have wavefunctions which overlap considerably as the Pauli exclusion principle does not apply because they are in different quantum states. However, the two electrons will repel each other so that it takes less energy to remove them from the atom. The atomic energy level is higher (closer to the ionisation limit). If $S=1$, then the two electrons have the same spin and they must have wavefunctions which are separated spatially, otherwise they would be trying to occupy the same quantum state, which is not allowed by the Pauli exclusion principle. With spatially separated wavefunctions, the Coulomb repulsion is less and the electrons are more tightly bound so the atomic energy as a whole is lower (farther from the ionisation limit).

Similar arguments apply for the energy ordering associated with different total electron angular momentum L. High values of L require that the individual electrons have the same values of individual orbital angular momentum quantum number l. Again, the Pauli exclusion principle says that these electron wavefunctions must be spatially distinct for the same l quantum numbers, so the Coulomb repulsion is less. For the atom as a whole, the energy ordering is such that high L values have lower energy (farther from the ionisation limit).

The division of atomic energies depending on the total spin S and total angular momentum L gives rise to atomic energies which are slightly different depending on the value of S and L. These different energy states are referred to as 'terms'. The terms are designated by the symbol for the L value (S, P, D, ...) with a superscript before the symbol designating the value of $2S+1$, where S is the total spin. The reason for this possibly obscure use of $2S+1$ is that $2S+1$ gives the degeneracy or number of 'multiplets' associated with each term arising due to spin–orbit coupling (which is discussed immediately below). For example, if there are two electrons in an un-filled p-sub-shell, the possible values of S are 0 or 1 and the possible values of L are 0, 1 or 2. The highest energy term would be designated ${}^1\mathrm{S}$.

In determining the terms that can exist, we must make sure that a term does not arise with two identical electrons as this violates the Pauli exclusion principle. With the example just discussed of two electrons in an un-filled p-sub-shell, this would mean that the lowest energy term is not ^{3}D as this would require two identical electrons with the same spin and orbital angular momentum. The lowest energy term is in fact ^{3}P as here the orbital angular momentum of the two electrons is different.

The remaining effect to consider is that of the spin–orbit interaction, which is an energy associated with the orientation of the total atomic spin in the magnetic field associated with the total orbital angular momentum for the electrons. The summation of the spin–orbit coupling for individual electrons can be subsumed into a calculation involving operators for the total atomic angular momentum $\hat{L}_a$ and total spin $\hat{S}_a$. We write

$$\hat{H}_{so} = \Sigma_i A(r_i)\hat{S} \cdot \hat{L}_i = A(L,S)\hat{L}_a \cdot \hat{S}_a.$$

Just as in the treatment of spin–orbit coupling for hydrogen, the effect of the spin-orbit operator on the atomic energy levels is best represented by introducing a new quantum number J for the total atomic angular momentum including the atomic orbital angular momentum and the atomic spin. The operator $\hat{J}^2$ for the square of total angular momentum has eigenvalues $\hbar^2[J(J+1)]$ so that

$$\hat{J}^2\Psi = \hbar^2[J(J+1)]\Psi. \tag{8.4}$$

J can take integer (if S is an integer) or half-integer (if S is a half-integer) values from a maximum of $L+S$ to $|L-S|$. Similarly, the square of the orbital angular momentum and spin eigenvalues are respectively $\hbar^2[L(L+1)]$ and $\hbar^2[S(S+1)]$.

Following the treatment of spin–orbit coupling for hydrogen (Equation 7.18), the spin–orbit energy becomes

$$E_{so} = \frac{1}{2}C(L,S)[J(J+1) - L(L+1) - S(S+1)] \tag{8.5}$$

where $C(L,S)$ is a constant with the dimensions of energy. States with the same values of L and S are split according to the value of J. The terms divide into slightly different energy levels (mulitplets). There are $2S+1$ multiplets for each term. Lower values of J produce a multiplet state of lower energy (farther from the ionisation limit) just as is the case for hydrogen spin–orbit coupling.

The spectral designation for the terms is modified to reflect a particular multiplet by adding a subscript to the term (after the S, P, D, ... designation), which is the value of the J quantum number. For example, with two electrons in an un-filled p-sub-shell, the lowest-energy multiplet is designated $^3P_0$. An example of the term and multiplet structure of an atom with two electrons in an un-filled shell is shown in Figure 8.3. For this example, we assume that all other sub-shells are full except for a 3p and 3d electron. Other details of the atom are not needed to determine the

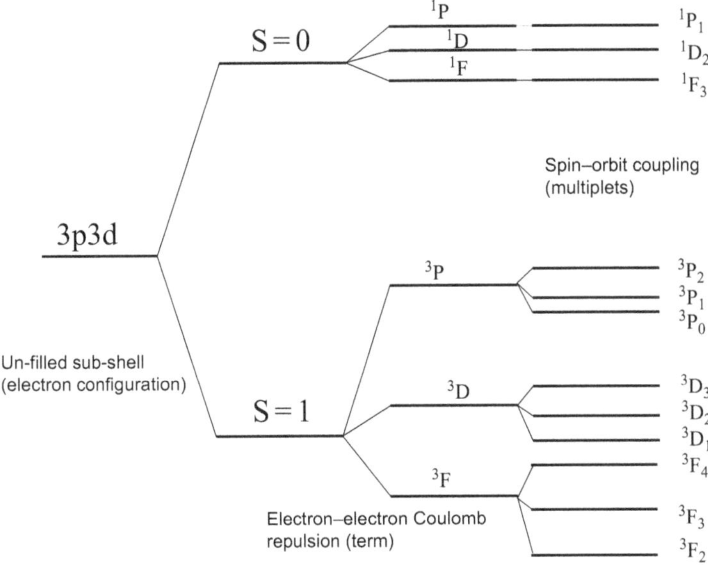

Figure 8.3 Schematic of the term structure due to electron–electron Coulomb interaction and the multiplet structure due to spin–orbit interaction in a multi-electron atom with electrons in un-filled sub-shells of electron configuration 3p3d exhibiting LS coupling (i.e. the Coulomb interaction energy is larger than the spin–orbit interaction energy). Additional processes affecting the energies of the quantum states are added going from left to right: the effect of total spin S on electron–electron Coulomb repulsion, the effect of total angular momentum L on electron–electron Coulomb repulsion and finally the effects of spin–orbit coupling.

term and multiplet structure, but we can say that the atom must be in an excited state and that all other electrons are in filled sub-shells.

8.3.2 j-j Coupling

If the spin–orbit coupling effect is greater than the Coulomb repulsion between electrons, the atomic energies of an electronic configuration split according to the possible values of the J quantum number in an approximately similar way to the hydrogen atom. States are designated simply by the notation $(j_1,j_2,\ldots)J$, where j_1,j_2, etc., are the j-quantum numbers of the individual electrons in the un-filled sub-shell.

The atomic states of elements with atomic number Z in the range from helium $Z=2$ to sodium $Z=11$ are given by LS coupling (with the Coulomb effect dominating the spin–orbit interaction). Many ions also follow the LS coupling rules. However, for neutral atoms with $Z > 11$, j-j coupling rules for details of the energy levels often apply.

8.4 Summary for Multi-Electron Atoms and Ions

- Multiple electron atoms have electronic configurations similar to the hydrogen atom, but shielding of the nuclear potential by the other electrons causes electrons with low l quantum number to be more tightly bound – to have lower energy.
- Due to the Pauli exclusion principle, only one electron can exist in each electronic quantum state. The ground state of the elements is determined by filling each electronic quantum state in energy order with the number of electrons associated with the element (i.e. the atomic number).
- Multiple electron atoms with filled values of n and l quantum numbers, known as 'sub-shells', have spherical electron probability distributions, zero angular momentum and zero magnetic moment.
- Coulomb repulsion between electrons causes a 'term' structure of finer-detail energy levels. If the electron probability distributions are farther apart, the Coulomb repulsion is lower, the electrons are more tightly bound and the overall atom energy is lower. Higher total spin S and higher total angular momentum L have individual electron probability distributions which are further apart – hence they have lower energy.
- The spin–orbit coupling energy between the magnetic field associated with the total angular momentum of the atom and the total electron spin causes further quantum state energy division into 'multiplets'.
- An individual term is designated by a symbol (S, P, D, F, G, ...) for the total angular momentum $L = 0, 1, 2, 3, 4, \ldots$. A superscript before the symbol gives the value of $2S + 1$ where S is the total spin. To designate an individual multiplet, the total angular momentum quantum number J is added as a subscript. For example, the electronic configuration for the ground state of helium is $1s^2$ and the term is $^1S_0$. The first shell of excited states of helium include, for example, 1s2s $^3S_0$, 1s2p $^3P_1$ and 1s2p $^1P_1$ states.

Exercises

8.1 Use the hydrogen-like spherical wavefunctions listed in Equation 7.10 to show that the sum of the probability distribution of all m quantum number electrons in a p-state is such that

$$Y_{10}^* Y_{10} + Y_{11}^* Y_{11} + Y_{1-1}^* Y_{1-1} = \frac{3}{2\pi}.$$

8.2 In the Stern–Gerlach experiment the electron spin of the $5s$ orbital in silver is shown to have a component either parallel or anti-parallel to a magnetic field. Given that the total electron spin is $\sqrt{s(s+1)}\hbar$ and the component of spin parallel to the magnetic field is $\pm s\hbar$ where $s = 1/2$ is the spin quantum number, determine the two angles that the electron spin subtends to a magnetic field. [Answer: $\cos(1/\sqrt{3}) \approx 54.7°$ and 125.3°]

8.3 The next elements in the periodic table above silver are cadmium and indium (see Figure 8.1). If cadmium and indium atoms were used in the Stern–Gerlach experiment, describe and explain the signal that you would expect to see on the detector plate. [Cadmium: – no splitting; indium: – splitting into two according to whether the lowest energy 5p $j = 1/2$ state has angular momentum aligned parallel or anti-parallel to the magnetic field]

8.4 Given that spin–orbit coupling for multi-electron atoms causes energy shifts according to

$$E_{so} = \frac{1}{2} C(L, S)\, [J(J+1) - L(L+1) - S(S+1)],$$

show that adjacent multiplets J and $J + 1$ are separated by an energy of $(J+1)\, C(L, S)$. This result is known as the Landé interval rule.

8.5 Consider two electrons in an un-filled sub-shell of an atom with total angular momentum quantum number $L \geq 1$. Using the equation for the spin-orbit coupling energy given in the above question, show that if the electrons have total spin quantum number $S = 0$ then $E_{so} = 0$. If $S = 1$, show that there are three possible values of E_{so}, such that $E_{so} = +LC(L, S)$, $-C(L, S)$ and $-2L\, C(L, S)$. If $L = 1$, write down the total angular momentum quantum number J associated with each of these multiplet energies. [In ascending energy order $J = 0$, 1 and 2]

9

Discrete Bound States: Molecules

A molecule can be defined as a group of two or more atoms held together by sharing one or more electrons. Approximately equal electron sharing between atoms is said to produce a covalent bond between the atoms. The shared electron(s) in a covalent bond attract the positively charged atomic nuclei sufficiently to overcome the mutual electrostatic repulsion between the nuclei. When the atoms in a molecule are not the same element, the electron(s) may not be shared equally around the nuclei. Depending on the level of electron sharing, the bond or attraction between atoms can have a contribution due to a net negative charge near one atom (associated with that atom attracting more of the electron wavefunction) and a net positive charge near another, leading to an ionic bond.

We discussed the energy levels of atoms and ions comprising a single element in Chapters 7 and 8. In discussing the periodic table of the pure elements, we saw that, after placing electrons into sub-shells of increasing energy, the last un-filled sub-shell largely controls the chemical behaviour of the element as the wavefunction(s) of the electrons(s) in this sub-shell usually extend farther away from the nucleus so that these electrons are more likely to interact with neighbouring atoms.

Neutral and ionised molecules occur in plasmas if the particle temperatures are sufficiently low that vibrations and collisions do not have sufficient energy to cause the molecules to dissociate. In 'plasma chemistry', plasma ions react with surfaces or other plasma constituents to form molecules [36]. Temporary molecules are sometimes formed in plasmas. Charge exchange occurs when a neutral atom such as hydrogen interacts with a highly stripped ion (with no or few electrons) to form a temporary molecule. It has been shown that to achieve good accuracy the Rosseland mean opacity (discussed in Chapter 6) should include molecular transitions at temperatures below 5,000 K [2].

We consider a simple molecular ion comprising two protons and an electron: the hydrogen ion molecule. The physics associated with this simple H_2^+ molecule illustrates the behaviour of more complex molecules and is relevant to plasma physics. Ionisation of neutral molecular hydrogen gas creates the hydrogen ion molecule. For example, molecular H_2^+ is produced in sources used for generating neutral beams of atomic hydrogen used, for example, to heat tokamak plasmas. The magnetosphere of Saturn is known to have H_2^+ ions [113].

9.1 The Hydrogen Molecule Ion H_2^+

The hydrogen molecule ion has a total energy made up of the Coulomb repulsion energy of the two positively charged nuclei and an electron energy associated with the solution of the Schrodinger equation in the field of both nuclei. If the nuclei are separated by a distance r_{ab}, the energy E_{nuc} required to bring the two nuclei from at rest at an infinite distance to that distance has a positive value of

$$E_{nuc} = \frac{e^2}{4\pi\epsilon_0 r_{ab}} = 2\frac{R_d}{r_{ab}/a_0}$$

where R_d is the Rydberg energy and a_0 is the Bohr radius.

Schrodinger's equation for the electron energy E can be written as the electron kinetic energy plus the potential energy in the field of nucleus a and nucleus b. We have

$$\hat{H}\psi = \left(-\frac{\hbar^2}{2m_0}\nabla^2 - \frac{e^2}{4\pi\epsilon_0 r_a} - \frac{e^2}{4\pi\epsilon_0 r_b}\right)\psi = E\psi \tag{9.1}$$

where r_a and r_b are distances from respectively nucleus a and b in the molecule. We see that the Schrodinger equation for the molecule is essentially the same as that for the hydrogen atom, but with the radial variation of the potential energy represented by $1/r_a + 1/r_b$. Due to this similarity, we could expect that the molecule electron wavefunctions can be expressed in terms of the hydrogen atom wavefunctions.

A trial solution for the electron wavefunction ψ of the hydrogen ion molecule can be written as

$$\psi = C_a\psi(\mathbf{r}_a) + C_b\psi(\mathbf{r}_b)$$

where C_a and C_b are constants to be determined and $\psi(\mathbf{r}_a)$ and $\psi(\mathbf{r}_b)$ are wavefunctions for the isolated hydrogen atom. Substitution into Equation 9.1 using hydrogen wavefunctions verifies that this addition of the wavefunctions for hydrogen are solutions of the molecule Schrodinger equation.

We are interested in finding the minimum energy configuration for H_2^+. When the two nuclei are identical this occurs when the electron probability distribution $\psi^*\psi$ is equally distributed between the two nuclei. We can see almost intuitively that, for identical nuclei, energy input is required to move from a symmetric distribution of the electron-probability distribution to a distribution where there is a greater probability of the electron being close to one nucleus. Imagine if initially $|C_a| = |C_b|$ and we slowly reduce, say, $|C_a|$ and increase $|C_b|$. This is equivalent to moving the electron charge from around nucleus a, which means the electric field attracting the remaining charge to a increases, requiring more work to remove the charge. Meanwhile, the electron shielding around nucleus b increases as negative electron charge is added around nucleus b (decreasing the electric field around b) and less energy is gained than was required to remove the charge from a.

Having $C_a = \pm C_b$ gives the minimum energy configuration for a system with one electron shared between two identical nuclei. The electron then has a probability distribution $\psi^*\psi$, which is symmetric about a plane perpendicular to and midway between the line joining the two nuclei (the electron is 'covalent', i.e. shared between the two nuclei). We can write for the electron wavefunction

$$\psi_\pm = C(\psi(\mathbf{r}_a) \pm \psi(\mathbf{r}_b)) \tag{9.2}$$

where the constant $C = 1/\sqrt{2(1 \pm S)}$ with

$$S = \int \psi^*(\mathbf{r}_a)\psi(\mathbf{r}_b)dV \tag{9.3}$$

which is known as the overlap integral. The constant C has this complicated expression because we need to normalise the wavefunction ψ so that $\int \psi^*\psi dV = 1$. Remember that, with this normalisation, the quantity $\psi^*\psi dV$ gives the probability of finding the electron in the volume dV. We designate the electron wavefunction ψ by a $\pm$ subscript depending on whether the hydrogen wavefunctions add or subtract. We allow $\pm$ values of the wavefunction addition because the minimum energy configuration only requires $\psi^*\psi$ to be symmetrical.

The energies of the electron in the molecule are given by the normal quantum mechanics evaluation

$$E_\pm = \int \psi_\pm^* \hat{H} \psi_\pm dV. \tag{9.4}$$

There are analytic solutions for $E_\pm$ treated in the next section. Combining the solutions with E_{nuc} gives a total energy E_{mol} for the molecule; see Figure 9.1 for the ground state solution. Two energy values are at each internuclear distance depending on whether the wavefunctions in Equation 9.2 are added or one is subtracted from the other. The subtraction condition is known as 'antibonding' as the minimum energy is for an internuclear distance of infinity. The addition of the wavefunctions leads to a minimum at approximately 2.2 Bohr radii and represents a stable molecule with this distance between the nuclei. The attraction between the electron wavefunction distribution and each of nucleus a and b is sufficient to overcome the repulsion between the nuclei as there is a significant probability density for the electron between the two nuclei (see Figure 9.2).

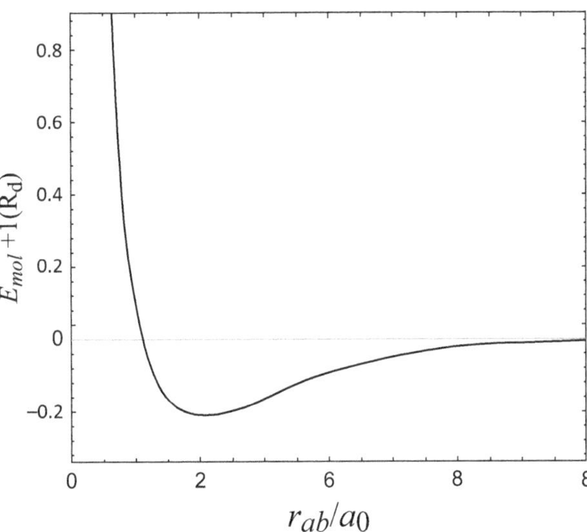

Figure 9.1 The total ground state energy E_{mol} of the H_2^+ ion showing the bonding molecule energies as a function of the internuclear distance. Internuclear distances are in units of the Bohr radius, while the energy scale $E_{mol}+1$ is the binding energy in Rydberg (1 R_d = 13.6 eV). At large r_{ab} the molecule energy E_{mol} relative to the energies of infinitely separated particles is that of one hydrogen atom (– 1 R_d).

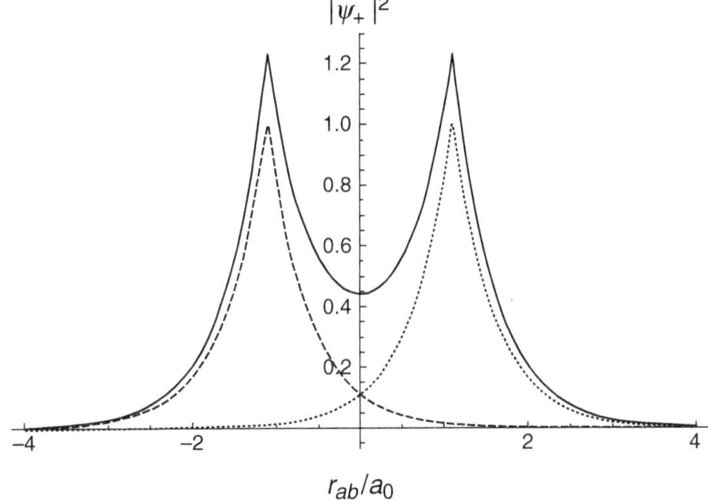

Figure 9.2 The probability distribution of the electron wavefunction $|\psi_+|^2$ along a line through the two nuclei in the H_2^+ ion electronic ground state (solid curve) and the individual hydrogen wavefunctions (broken curves). The distance scale r_{ab} is in units of the Bohr radius a_0 with the two protons separated by $2.2a_0$.

The H_2^+ molecule electronic states are designated by the hydrogen wavefunction taken as the solution of the Schrodinger equation for the molecule. The German word *gerade* (abbreviated to 'g') is used to designate the bonding E_+ solution where the wavefunction contributions around each nucleus have the same sign. The German for odd, *ungerade* (abbreviated as 'u') is used for the anti-bonding E_- solution. A symbol is used to designate the shape of the wavefunction when viewed along the direction of a line joining the two nuclei. If the wavefunction shape is circular viewed along the line joining the nuclei, the symbol is σ. The ground state of H_2^+ is consequently written as $1s\,\sigma_g$.

Excited electronic states of H_2^+ exist if instead of using the ground hydrogen wavefunction in, for example, Equation 9.5, the excited quantum wavefunctions of hydrogen along with the appropriate excited principal quantum number n are employed. The energies $E_\pm$ of excited electron states are higher (less negative) and the total molecular energy E_{mol} for the bonding solution has a minimum at an increasing nuclear distance r_{ab} as n increases. As in atomic hydrogen, the excited molecular electronic states are also increasingly close in energy as n increases. The $n = 2$ bonding solutions are designated $2s\,\sigma_g$ and $2p\,\sigma_g$.

9.1.1 Analytic Solutions for H_2^+

The molecule Hamiltonian $\hat{H}$ for H_2^+ can be expanded as an isolated hydrogen atom Hamiltonian plus the potential of the 'other' nucleus. We can write

$$\hat{H} = \hat{H}_a - \frac{e^2}{4\pi\epsilon_0}\frac{1}{r_b} = \hat{H}_b - \frac{e^2}{4\pi\epsilon_0}\frac{1}{r_a}$$

where $\hat{H}_a$ and $\hat{H}_b$ are the Hamiltonians associated with isolated nuclei a and b respectively. Using Equation 9.2 the energies $E_\pm$ are given by

$$\begin{aligned} E_\pm = C^2 \bigg[& \int \psi^*(\mathbf{r}_a)\hat{H}_a\psi(\mathbf{r}_a)dV - \frac{e^2}{4\pi\epsilon_0}\int \psi^*(\mathbf{r}_a)\frac{1}{r_b}\psi(\mathbf{r}_a)dV \\ & + \int \psi^*(\mathbf{r}_b)\hat{H}_b\psi(\mathbf{r}_b)dV - \frac{e^2}{4\pi\epsilon_0}\int \psi^*(\mathbf{r}_b)\frac{1}{r_a}\psi(\mathbf{r}_b)dV \\ & \pm \int \psi^*(\mathbf{r}_a)\hat{H}\psi(\mathbf{r}_b)dV \pm \int \psi^*(\mathbf{r}_b)\hat{H}\psi(\mathbf{r}_a)dV \bigg]. \end{aligned}$$

The first and third integrals give energies E_n as found for hydrogen, while the second and fourth plus fifth and sixth integrals are respectively equal to each other. We can write

$$E_\pm = 2C^2\left[-\frac{R_d}{n^2} - \frac{e^2}{4\pi\epsilon_0}\int \psi^*(\mathbf{r}_a)\frac{1}{r_b}\psi(\mathbf{r}_a)dV \pm \int \psi^*(\mathbf{r}_a)\hat{H}\psi(\mathbf{r}_b)dV\right] \quad (9.5)$$

where $R_d = (1/2)e^2/(4\pi\epsilon_0 a_0)$ is the Rydberg energy (13.6 eV).

The first integral in Equation 9.5 for $E_\pm$ represents the 'expectation value' for the quantity $1/r_b$ around nucleus a. We designate this integral by F, so define

$$F = \int \psi^*(\mathbf{r}_a) \frac{1}{r_b/a_0} \psi(\mathbf{r}_a) dV. \tag{9.6}$$

We can also expand $\hat{H}\psi(\mathbf{r}_b) = (E_n - e^2/(4\pi\epsilon_0 r_a)\psi(\mathbf{r}_b)$ to give for the second integral in Equation 9.5:

$$\int \psi^*(\mathbf{r}_a)\hat{H}\psi(\mathbf{r}_b)dV = -\frac{R_d}{n^2}S - 2R_d G$$

where

$$G = \int \psi^*(\mathbf{r}_a) \frac{1}{r_a/a_0} \psi(\mathbf{r}_b) dV. \tag{9.7}$$

Using $C = 1/\sqrt{2(1 \pm S)}$ where S is the overlap integral defined by Equation 9.3, we can write for the electronic energies

$$E_\pm = -R_d \left(\frac{1}{n^2} + 2\frac{F \pm G}{1 \pm S} \right). \tag{9.8}$$

The total energy E_{mol} for the molecule is made up of the energy E_{nuc} associated with the repulsive force between the two nuclei and $E_\pm$. We have $E_{mol} = E_{nuc} + E_\pm$ so that

$$E_{mol} = -R_d \left(\frac{1}{n^2} + 2\frac{F \pm G}{1 \pm S} \right) + 2\frac{R_d}{r_{ab}/a_0}. \tag{9.9}$$

Values of E_{mol} can be evaluated in a fairly straightforward way as the wavefunctions $\psi(\mathbf{r}_a)$ and $\psi(\mathbf{r}_b)$ here are the hydrogen atom wavefunctions. Values of $E_\pm$ are negative – energy is released when an electron is brought from infinity into the molecule, while E_{nuc} is positive. Energy needs to be applied to bring the two nuclei to within the internuclear distance. We see that the resultant energies E_{mol} are negative and that the solution with the wavefunctions for hydrogen added together (labelled '+' on Figure 9.1) has a minimum at an internuclear distance of approximately 2.2 Bohr radii. The other solution (labelled '–' in Equation 9.8) has minimum energy when the internuclear distance is infinite, i.e. this solution does not result in a stable molecule with the nuclei separated by a finite distance; it is 'antibonding'.

When r_{ab} becomes large, the integrals F, G and S in Equation 9.9 approach zero (see Equations 9.3, 9.6 and 9.7). The electron energy becomes $-R_d/n^2$ as for an isolated hydrogen atom. When r_{ab} is small, the integrals F and G become approximately equal, while the overlap integral $S = 1$. The electron energies

E_+ for the bonding solution given by Equation 9.8 are equal to $-4R_d/n^2$, which is the energy of a helium ion. This is the result expected as joining the two nuclei together forms the central potential for a helium ion. The electron energies E_- for the antibonding solution are given by $-R_d/n^2$ as $F \approx G$.

9.2 Covalent and Ionic Molecular Bonds

The energies of all stable covalently bonded molecules have a similar form to that shown for the bonding solution for $H_2{}^+$ of Figure 9.1. Ground and excited electron states have the same form, with the minimum energy tending to move to larger internuclear distance for higher energy excited states.

If the atoms making up a molecule are not identical, the minimum energy configuration is not necessarily found when the wavefunction is evenly distributed between the two nuclei. There is an ionic bond if the electron wavefunction has lower energy when localised close to one atom (leading to a net negative charge) with the creation of a net positive charge in the vicinity of the other atom.

Following Equation 9.2, we can write that the wavefunction distribution around the two nuclei a and b for a covalent bond is given by

$$\psi_+ = C(\psi(\mathbf{r}_a) + \psi(\mathbf{r}_b)) \tag{9.10}$$

where C is a normalisation factor. For an ionic bond, we have

$$\psi_{ion} = C_a\psi(\mathbf{r}_a) + C_b\psi(\mathbf{r}_b)) \tag{9.11}$$

where C_a and C_b are constants determined by the requirement that the energy of the molecule is minimised and that ψ_{ion} is normalised.

9.3 Molecular Vibrational and Rotational States

At non-zero temperatures, the internuclear distances between the atoms of a molecule are not fixed and tend to oscillate in time. Molecules are also free to rotate in space. As we are dealing with small masses, the quantum mechanical quantisation of these motions is significant, leading to discrete vibrational and rotational energy states. We assume the Born–Oppenheimer approximation that the nuclear motion is decoupled from the electrons due to the much-greater nuclei mass: the electrons quickly adjust to any new nuclear positions. The Born–Oppenheimer approximation allows the electron wavefunctions and energy eigenvalues to be independent of the nuclear velocites. The total energy of a molecule becomes a summation of the electron energy and the (smaller) quantised vibrational and rotational energies of the nuclear masses.

9.3.1 Vibrational States

The nuclei in a molecule are subject to a potential dependent on the internuclear distance and the particular electronic state. Molecular vibrations are constrained by the potential energy E_{mol} variation with internuclear distances (see e.g. Figure 9.1) so that oscillations are centred around the minimum energy distance. The previous treatment of the ground electronic states of the $H_2{}^+$ molecule shows a minimum energy at an internuclear distance $r_{ab} = r_{min} = 2.2a_0$. The potential energy E_{mol} at internuclear distances smaller and larger than r_{min} for $H_2{}^+$ and other molecules can be approximated as having a parabolic shape of form

$$E_{mol} = E_{min} + \frac{1}{2}k_s r^2 \tag{9.12}$$

where $r = r_{ab} - r_{min}$ is the internuclear distance from the minimum energy position E_{min} and k_s is a constant determined by the shape of the potential variation ($k_s = d^2E_{mol}/dr^2$). A parabolic potential gives rise to internuclear oscillations described by solutions for a harmonic oscillator (see Appendix A.3). Solving the one-dimensional Schrodinger equation with a potential of the form of Equation 9.12 gives rise to energy eigenvalues of the form

$$E_p = \left(p + \frac{1}{2}\right)\hbar\omega_{HO} \tag{9.13}$$

where ω_{HO} is a characteristic angular frequency of the oscillation and p is an integer with value $p = 0, 1, 2, \ldots$. The total energy of the molecule is $E_{min} + E_p$. The value of ω_{HO} is given by

$$\omega_{HO} = \left(\frac{k_s}{m_{red}}\right)^{1/2}$$

where $m_{red} = m_1m_2/(m_1 + m_2)$ is a reduced mass for the molecule with two atoms of mass m_1 and m_2. The key point given by Equation 9.13 is that the vibrational energy levels are evenly spaced with energy separation $\hbar\omega_{HO}$.

Although the harmonic oscillator potential is only an approximation to the electronic molecular potential energy variation with internuclear distance, it is often observed that the vibrational energy levels of molecules are spaced evenly, as predicted here. The typical spacing between vibrational energy levels is ≈ 0.1 eV so that transitions between molecular vibrational levels are in the infra-red. The vibrational wavefunctions for a harmonic oscillator alternate in 'parity' as p increases in steps of one, so that vibrational radiative transitions in molecules are allowed with unit changes in p ($\Delta p = \pm 1$). The issue of parity in determining allowed radiative transitions is discussed in Chapter 10.

9.3.2 Rotational States

Besides vibrations of the internuclear distance, the rotational motion of molecules about the common centre of mass of the nuclei occurs. In both classical and quantum rotating systems, the rotational angular momentum R is related to the rotational energy E by

$$E = \frac{R^2}{2I} \tag{9.14}$$

where I is the moment of inertia made up of the sum of the nuclei moments of inertia:

$$I = m_p \sum_i A_i r_i^2$$

where A_i is the atomic mass of nucleus i, r_i is the distance of nucleus i from the molecule centre of mass, m_p is the mass of a proton and the summation is over the number of atoms in the molecule.

As we found for the angular momentum of atomic electrons, the quantum mechanical operator corresponding to the square of angular momentum R^2 has eigenvalues of form $q(q+1)\hbar^2$, where $q = 0, 1, 2, 3, \ldots$ is an integer. The energy is quantised as

$$E_q = \frac{q(q+1)\hbar^2}{2I}. \tag{9.15}$$

The spacing of the rotation energy levels increases with q and E_q. We have that

$$E_{q+1} - E_q = \frac{\hbar^2}{I}(q+1). \tag{9.16}$$

The increase in energy spacing of molecular rotational states with increasing q contrasts with the approximately constant spacing of energy states associated with vibrational molecular states. The typical spacing between molecular rotational energy levels is 10^{-3} eV. Allowed radiative transitions between rotational states occur when $\Delta q = \pm 1$ or 0.

Exercises

9.1 Two vibrating nuclei of mass m_1 and m_2 in a diatomic molecule are distances r_1 and r_2 from the centre of mass of the molecule so that the internuclear distance $r_{ab} = r_1 + r_2$. The force $-k_s r$ between the nuclei depends on their separation $r = r_{ab} - r_{min}$, where r_{min} is the internuclear separation at the molecular energy minimum E_{mol}. Here k_s is a constant of proportionality. The equation of motion of each nucleus can be written as

$$m_1 \frac{d^2 r_1}{dt^2} + k_s r = 0,$$

$$m_2 \frac{d^2 r_2}{dt^2} + k_s r = 0.$$

Show that the equation of motion of the vibrating nuclei can be written as

$$m_{red} \frac{d^2 r}{dt^2} + k_s r = 0$$

where the reduced mass is given by

$$m_{red} = \frac{m_1 m_2}{m_1 + m_2}.$$

[Hint: the position of the centre of mass is determined by $m_1 r_1 = m_2 r_2$.]

9.2 The approximately parabolic shape of the potential energy E_{mol} of the $H_2{}^+$ molecular ion is given by

$$E_{mol} = E_{min} + \frac{1}{2} k_s r^2$$

where r is the internuclear distance variation from the minimum energy E_{min} value. If the parameter k_s fits the potential energy curve with

$$k_s = \frac{2.53e}{a_0^2}\ \mathrm{Jm}^{-2},$$

estimate the energy spacing of the vibrational energy levels. [0.27 eV]

9.3 There is much experimental and theoretical interest in the hydrogen ion molecule HD^+ composed of hydrogen and the isotope deuterium. Assuming that the molecular potential energy is unchanged from that given in Exercise 9.2, estimate the energy spacing of the vibrational energy levels of HD^+. [0.23 eV. Wing et al. [123] measured an energy of $1{,}869\ \mathrm{cm}^{-1} = 0.23$ eV for the transition from the $p = 1$ to 0 vibrational state of HD^+.]

9.4 Show that the wavefunction for the ground state of the $H_2{}^+$ molecule vibrational quantum state given by

$$\psi = \left(\frac{\hbar}{\pi}\right)^{1/4} \left(\frac{1}{k_s m_{red}}\right)^{1/8} \exp\left(-\frac{(k_s m_{red})^{1/2}}{2\hbar} r^2\right)$$

is a solution of the one-dimensional Schrodinger equation. Verify the normalisation $\int_{-\infty}^{\infty} \psi^* \psi \, dr = 1$ for this wavefunction.

9.5 Assuming the k_s value used in Exercise 9.2 and wavefunction for the ground state given in Exercise 9.4, determine the distance from the minimum of the $H_2{}^+$ potential curve in units of the Bohr radius a_0, where the probability density for the vibration wavefunction drops to e^{-1} of the peak value. [$0.33a_0$]

9.6 Estimate the moment of inertia of the $H_2{}^+$ molecule and then the energy spacing between the ground and first excited rotational state. [6.1×10^{-3} eV]

10

Radiative Transitions between Discrete Quantum States

Radiation can change the energy state of matter via three optical processes: spontaneous emission, absorption and stimulated emission. In spontaneous emission, an excited atomic or molecular state decays to a lower energy state with the emission of a photon of energy equal to the difference in the quantum state energies. Photon absorption is the reverse process, with a photon causing excitation from a lower to an excited quantum state. Stimulated emission was first proposed by Einstein in a paper published in 1917. He attempted to balance spontaneous emission and photon absorption, but could not get the correct balance without the postulated process of stimulated emission (see Section 4.2). With stimulated emission, an excited quantum state is stimulated to decay to a lower quantum state by an impinging photon. The existing photon survives with another identical photon being created (so that energy is conserved).

The rates at which radiative transitions occur can be calculated using quantum mechanics. The time-dependent Schrodinger equation is used with the system wavefunction comprising time-varying linear combinations of the two wavefunctions for the lower and upper quantum states. In semi-classical treatments, an atom is treated quantum mechanically with the radiation field treated classically. There is a perturbation in energy associated with, say, the electric field of the radiation interacting with the electric dipole of the atom, though other interactions can be significant if the electric dipole interaction energy is small. In absorption, for example, a differential equation representing the change in 'weighting' of the wavefunction from the lower to upper state is obtained. A 'time constant' for the speed of change represents the rate of the radiative process. We examine the quantum mechanical nature of radiation absorption in Section 10.1 before discussing (Section 10.3) another important parameter arising with radiative transitions between discrete bound quantum states: the lineshape function.

10.1 Quantum Theory of the Atom–Radiation Interaction

An expression for the Einstein B-coefficient for photo-absorption is derived using quantum mechanics in this section. Einstein determined the detailed balance relationship between the rates of spontaneous emission, photo-absorption and stimulated emission when light interacts with an atom by considering an atomic system of two energy levels which are in equilibrium with a black-body radiation field. We use a semi-classical treatment where the atom is treated quantum mechanically and the radiation field is treated as a classical oscillating electric field. We determine the Einstein B-coefficient for photo-absorption, which enables the B-coefficient for stimulated emission and the transition probability for spontaneous emission to be calculated using the relationships between them found by considering detailed balance. The photo-absorption B-coefficient is the easiest to evaluate as it is not necessary to consider the quantum mechanics of the radiation field.

10.1.1 Time-Dependent Quantum Mechanics

The Einstein B-coefficient for photo-absorption describes the response of an atom when there is an external field in the vicinity of the atom. The time-dependent Schrodinger equation allows a quantum mechanical treatment of this system. As outlined in Section 7.1, the time-dependent Schrodinger equation is given by

$$\hat{H}\Psi = i\hbar\frac{d\Psi}{dt} \tag{10.1}$$

where $\hat{H}$ is the quantum-mechanical Hamiltonian and Ψ is the time-dependent wavefunction.

An isolated atom has a Hamiltonian comprising expressions for the kinetic energy and potential energy of a bound electron. We have for an isolated atom

$$\hat{H}_0 = -\frac{\hbar^2}{2m_0}\nabla^2 + V \tag{10.2}$$

where m_0 is the electron mass and V is the potential energy of the electron in the field of the nucleus.

It is possible to express the time-dependent wavefunction Ψ as a product of two functions, one of time and the other of spatial co-ordinates. Showing the explicit dependence on spatial co-ordinates by using a vector $\mathbf{r}$, we write that

$$\Psi(\mathbf{r}, t) = \psi(\mathbf{r})\Phi(t). \tag{10.3}$$

If we substitute this expression for the time-dependent wavefunction (Equation 10.3) into the time-dependent Schrodinger equation (Equation 10.1), we can write that

$$\frac{1}{\psi(\mathbf{r})}\left[-\frac{\hbar^2}{2m_0}\nabla^2+V(\mathbf{r})\right]\psi(\mathbf{r})=-\frac{\hbar}{i\Phi(t)}\frac{\partial\Phi(t)}{\partial t} \tag{10.4}$$

upon dividing throughout by $\Psi(\mathbf{r},t)$.

The left-hand side of Equation 10.4 is a function of spatial co-ordinates only, while the right-hand side is a function of time only. The two sides must both be equal to a constant (independent of $\mathbf{r}$ and t). We denote this constant by the symbol E and so can write

$$\left[-\frac{\hbar^2}{2m_0}\nabla^2+V(\mathbf{r})\right]\psi(\mathbf{r})=E\psi(\mathbf{r}) \tag{10.5}$$

$$i\hbar\frac{\partial\Phi(t)}{\partial t}=E\Phi(t). \tag{10.6}$$

Equation 10.5 is the time-independent Schrodinger equation. We can recognise that Equation 10.5 is an eigenvalue equation using the Hamiltonian operator with E the energy eigenvalue. Equation 10.6 is a relatively simple first-order differential equation which can be readily integrated. We obtain

$$\Phi(t)=\exp\left(-i\frac{Et}{\hbar}\right) \tag{10.7}$$

if we assume the normalisation that $\Phi^*(t)\Phi(t)=1$. Normalising the time-varying component of the time-dependent wavefunction in this way ensures that the time-independent component $\psi^*(\mathbf{r})\psi(\mathbf{r})dV$ represents a measure of the probability of finding the electron in an element of volume dV.

10.1.2 The Interaction Hamiltonian

In order to evaluate the photo-absorption of light, we consider the effect on the atom of electromagnetic radiation. The interaction of the electromagnetic radiation with the atom causes the atom to have an additional energy component which can be treated using quantum mechanical perturbation theory. We add an 'interaction' Hamiltonian $\hat{H}_1$ to the isolated atom Hamiltonian so that

$$\hat{H}=\hat{H}_0+\hat{H}_1. \tag{10.8}$$

We consider two 'generic' energy levels of energy E_1 and E_2 in an atom ($E_2>E_1$) and calculate the effect of the interaction of electromagnetic radiation of frequency close to ω_0 where $\hbar\omega_0=E_2-E_1$. Only the two atomic states E_1 and E_2 need to be considered, so the time-dependent wavefunction $\Psi(\mathbf{r},t)$ of the atom can be represented by a linear superposition of the isolated atom time-dependent wavefunctions ($\Psi_1(\mathbf{r}, t)$ and $\Psi_2(\mathbf{r}, t)$) associated with the two states (1 and 2). We have for the total atom time-dependent wavefunction that

$$\Psi(\mathbf{r}, t) = C_1(t)\Psi_1(\mathbf{r}, t) + C_2(t)\Psi_2(\mathbf{r}, t) \tag{10.9}$$

where $C_1(t)$ and $C_2(t)$ are time-varying proportionality constants for the contribution of states 1 and 2. We are considering photo-absorption and have delineated that $E_1 < E_2$, so we can expect that, as photo-absorption proceeds, the value of $C_2(t)$ will increase and $C_1(t)$ will decrease; we discuss this further in Section 10.1.4.

We require that the total atom wavefunction is normalised with

$$\int \Psi^*(\mathbf{r}, t)\Psi(\mathbf{r}, t)dV = 1$$

with the integration over all space. If the isolated atom wavefunctions for energies E_1 and E_2 are similarly normalised, we must have that

$$|C_1(t)|^2 + |C_2(t)|^2 = 1. \tag{10.10}$$

If we substitute the wavefunction summation of the two quantum states (Equation 10.9) into the time-dependent Schrodinger equation (Equation 10.1), we obtain

$$\hat{H}_1(C_1\Psi_1 + C_2\Psi_2) = i\hbar\left(\Psi_1\frac{dC_1}{dt} + \Psi_2\frac{dC_2}{dt}\right) \tag{10.11}$$

upon cancelling

$$\hat{H}_0\Psi = i\hbar\frac{\partial\Psi}{\partial t}$$

from both sides for both Ψ_1 and Ψ_2. Multiplication of Equation 10.11 from the left by Ψ_1^* followed by an integration over all volume gives

$$\begin{aligned} C_1 \int \psi_1^*\hat{H}_1\psi_1 dV + C_2\exp(-i\omega_0 t)\int \psi_1^*\hat{H}_1\psi_2 dV \\ = i\hbar\left(\frac{dC_1}{dt} + \exp(-i\omega_0 t)\frac{dC_2}{dt}\int \psi_1^*\psi_2 dV\right) \end{aligned} \tag{10.12}$$

upon using

$$\Phi_1^*(t)\Phi_2(t) = \exp(-i\omega_0 t).$$

Similarly, multiplication of Equation 10.11 from the left by Ψ_2^* followed by an integration over all volume gives

$$\begin{aligned} C_2 \int \psi_2^*\hat{H}_1\psi_2 dV + C_1\exp(-i\omega_0 t)\int \psi_2^*\hat{H}_1\psi_1 dV \\ = i\hbar\left(\frac{dC_2}{dt} + \exp(i\omega_0 t)\frac{dC_1}{dt}\int \psi_2^*\psi_1 dV\right). \end{aligned} \tag{10.13}$$

These two Equations (10.12 and 10.13) can be simplified if we use some of the properties of the interaction Hamiltonian $\hat{H}_1$.

10.1.3 Form of the Interaction Hamiltonian

The usual dominant contribution to the energy of interaction between an atom or ion and an electromagnetic field arises due to the effect of the radiation electric field on the atom or ion with the atom or ion treated as an electric dipole. The atom forms an electric dipole due to the separation of the negative electron charge from the positive nuclear charge.

An electric dipole moment can be represented by a vector **d** such that

$$\mathbf{d} = q\mathbf{r}$$

where **r** is a vector from a position with a charge of $-q$ to the position of a charge of $+q$. There is an energy associated with moving the $+q$ charge (conventionally electric effects, of course, assume movement of a positive charge, though in practise the much less massive electrons usually move). Assuming movement of the positive charge at an angle θ to the electric field direction, the force on the positive charge in an electric field E is $qE\cos\theta$. Moving a distance r at angle θ is associated with an energy of $-qEr\cos\theta$ (with the minus sign indicating a release of energy for $\theta < 90^0$). The energy of the electric dipole **d** in an electric field **E** thus has an energy which can be written as $-\mathbf{d}\cdot\mathbf{r}$.

We assume that the electric field of the radiation has a wavelength much greater than the size of an atom so that we only need to consider the temporal change of electric field which oscillates as $\mathbf{E}_0\cos(\omega t)$ for incident radiation with a frequency ω. Associated with the energy of a dipole moment in an electric field, the interaction Hamiltonian can be written as

$$\hat{H}_1 = -\mathbf{d}\cdot\mathbf{E}_0\cos(\omega t) = e\mathbf{r}\cdot\mathbf{E}_0\cos(\omega t) \tag{10.14}$$

where **r** is a vector from the nucleus to spatial positions around the atom.

Isolated atom wavefunctions have symmetry, but the wavefunction values can be positive and negative. A useful and important property of integrals of wavefunction over all space is the parity of the wavefunction. A function $f(\mathbf{r})$ has even parity if

$$f(\mathbf{r}) = f(-\mathbf{r})$$

and odd parity if

$$f(\mathbf{r}) = -f(-\mathbf{r}).$$

Functions lacking symmetry have an indeterminate parity as neither of the above apply. However, the concept of parity is particularly useful in simplifying integrals of wavefunctions as wavefunctions are typically highly symmetric. If the integrand has odd parity, the integral over all space is equal to zero. The interaction Hamiltonian (Equation 10.14) has odd parity, which means that if the wavefunctions ψ_1 and ψ_2 have different parity, integrals of the form

$$\int \psi_1^* \hat{H}_1 \psi_2 dV$$

have non-zero value. If the wavefunctions have different parity, we also then have that

$$\int \psi_1^* \psi_2 dV = 0.$$

If the wavefunctions ψ_1 and ψ_2 have the same parity, Equations 10.12 and 10.13 both have the left side equal to zero and the equations are not mutually consistent unless

$$\frac{dC_1}{dt} = \frac{dC_2}{dt} = 0.$$

Assuming the electric dipole Hamiltonian of Equation 10.14, wavefunctions of the same parity, consequently, cannot have radiative transitions between them as the rate of change of the wavefunction proportionality is always zero.

If the wavefunctions ψ_1 and ψ_2 have different parity, Equations 10.12 and 10.13 simplify considerably. We can respectively write that

$$i\frac{dC_1}{dt} = C_2 \exp(-i\omega_0 t) I_{12} \cos(\omega t) \tag{10.15}$$

and

$$i\frac{dC_2}{dt} = C_1 \exp(i\omega_0 t) I_{21} \cos(\omega t) = C_1 \exp(i\omega_0 t) I_{12}^* \cos(\omega t) \tag{10.16}$$

where we use a symbol I_{21} and I_{12} such that

$$I_{12} = \frac{eE_0}{\hbar} \int \psi_1^* \, x \, \psi_2 dV. \tag{10.17}$$

The integral in the expression for I_{12} represents an overlap integral of the two wavefunctions multiplied by the component of the position vector **r** along the direction of the electric field of the radiation (assumed for convenience to be the x-direction in Cartesian co-ordinates).

10.1.4 The Transition Rate

We have set out to consider the photo-absorption of light with the atom starting in energy state E_1 and finishing in energy state E_2. We can consequently set up initial conditions for solving the differential Equations 10.15 and 10.16 by requiring that the total wavefunction comprises solely the state 1 wavefunction at time $t = 0$. We have

$$C_1(0) = 1, \tag{10.18}$$
$$C_2(0) = 0. \tag{10.19}$$

The rate of change of the population N_2 of the upper state is given by

$$\frac{dN_2}{dt} = N\frac{d|C_2|^2}{dt} \tag{10.20}$$

where $N = N_1 + N_2$ is the total population of atoms (N_1 is the population of the lower energy state). The wavefunction proportionality constant C_2 is modulus squared because the modulus square of the wavefunction gives the probability density of the quantum state.

The Einstein B-coefficient for absorption (B_{12}) is defined in terms of the rate of change of the population due to photo-absorption. We have

$$\frac{dN_2}{dt} = B_{12}W(\omega)N_1 \tag{10.21}$$

where $W(\omega)$ is the radiation energy density at the atom. Using Equations 10.20 and 10.21, we can write that

$$\frac{d|C_2|^2}{dt} = B_{12}W(\omega)\frac{N_1}{N_1 + N_2}.$$

Comparing this equation to Equation 10.21, we see that the Einstein B_{12}-coefficient can be determined by considering a time t approaching zero when N_2 is negligible. We have

$$\left(\frac{|C_2|^2}{t}\right)_{t=0} = B_{12}W(\omega). \tag{10.22}$$

We can integrate Equation 10.16 assuming a time t close to zero so that $C_1 = 1$. Expanding

$$\cos(\omega t) = \frac{1}{2}(\exp(i\omega t) + \exp(-i\omega t)),$$

we find that an integration from time zero to time t yields

$$C_2(t) = -i\frac{I_{12}^*}{2}\left(\frac{1 - \exp(i(\omega_0 - \omega)t)}{\omega_0 - \omega}\right) \tag{10.23}$$

after dropping a solution involving $1/(\omega_0+\omega)$, which will be very small as the two frequencies are large. We can write for the modulus squared of C_2 that

$$|C_2(t)|^2 = \frac{1}{4}t^2|I_{12}|^2\frac{\sin^2((\omega-\omega_0)t/2)}{((\omega-\omega_0)t/2)^2}. \tag{10.24}$$

The frequencies of the radiation impinging on the atom will not all be in reality at one frequency ω. We need to integrate Equation 10.24 over all frequencies. The 'sinc' function represented by Equation 10.24 is quite peaked around frequency ω_0 so assuming that the radiation electric field magnitude E_0 is constant in frequency is usually very accurate. Using

$$\int_{-\infty}^{\infty} \frac{sin^2 x'}{x'^2} dx' = \pi$$

and the relationship between the radiation energy density $W(\omega)$ and electric field E_0 (viz. $W(\omega) = (1/2)\epsilon_0 E_0^2$), we can integrate Equation 10.24 and re-arrange to get

$$|C_2(t)|^2 = \frac{\pi e^2}{\epsilon_0 \hbar^2} W(\omega)\, t \left| \int \psi_1^* x \psi_2 dV \right|^2. \tag{10.25}$$

The integral in this equation assumes a particular orientation of the atom to the electric field. Remember that x is the component of the position vector $\mathbf{r}$ along the direction of the radiation electric field. In reality, atoms will be randomly oriented relative to the electric field direction, so we must consider an angle average of the dipole moment. Assuming that the dipole moment has so far been oriented at an angle θ to the electric field, we need to average the square of the component of the dipole moment along the electric field direction. This average is 1/3 of the dipole moment (see Appendix A.4). Using Equation 10.22, we can now evaluate the Einstein B_{12}-coefficient. Our expression for the Einstein B_{12}-coefficient becomes

$$B_{12} = \frac{\pi e^2}{3\epsilon_0 \hbar^2} \left| \int \psi_1^* \mathbf{r} \psi_2 dV \right|^2. \tag{10.26}$$

The integral here is known as a matrix element [69].

The Einstein A-coefficient for spontaneous emission is related to the photo-absorption coefficient B_{12} by

$$A_{21} = \frac{\hbar \omega_0^3 g_1}{\pi^2 c^3 g_2} B_{12} \tag{10.27}$$

where g_1 and g_2 are the degeneracies of the two quantum states. This relationship can be demonstrated by considering a balance between photo-absorption, spontaneous emission and stimulated emission for an atom with equilbrium populations immersed in an equilibrium black-body radiation field (see Section 4.2). We can consequently write for the spontaneous transition probability that

$$A_{21} = \frac{e^2\omega_0^3}{3\pi\epsilon_0\hbar c^3}\frac{g_1}{g_2}\left|\int \psi_1^*\mathbf{r}\psi_2 dV\right|^2. \tag{10.28}$$

Equation 10.28 shows that the transition probability A_{21} has a non-zero value if the two wavefunctions ψ_1 and ψ_2 have different parity. The position vector $\mathbf{r}$ associated with the electric-dipole interaction Hamiltonian has odd parity, so the total integrand of the overlap integral has an even parity when the two wavefunctions have different parity. A transition between two quantum states is said to be electric-dipole allowed if the states have opposite parity. This parity rule for electric dipole transitions is known as the Laporte rule.

As an example of the Laporte rule, consider a p-orbital ($l = 1$) electronic wavefunction comprising two wavefunction 'lobes' with odd parity (see Section 7.2). Transitions to s-orbitals ($l = 0$) with spherically symmetric wavefunctions and even parity have a non-zero A_{21}. We have that p–s transitions are allowed. More generally the Laporte rule for an allowed transition requires $\Delta l = \pm 1$. Another example was briefly examined in discussing vibrational energy levels in molecules which to a first approximation have wavefunctions as for a harmonic oscillator (Section 9.3.1). The vibrational wavefunctions for a harmonic oscillator alternate in parity as the vibrational quantum number p increases in steps of one, so that vibrational radiative transitions in molecules are allowed with unit changes in p ($\Delta p = \pm 1$).

10.1.5 Thomas-Reiche-Kuhn Sum Rule

Using Equation 4.29 for the relationship of the transition probability to the absorption oscillator strength, we can determine upper bounds for electron dipole transition probabilities using the Thomas-Reiche-Kuhn sum rule introduced in Section 3.3. The Thomas-Reiche-Kuhn sum rule states that the sum of the absorption oscillator strengths for transitions from a quantum state is equal to the number of electrons in the state. We can write for the sum of the oscillator strengths from a state 1 to all other states i in an atom or ion:

$$\sum_i f_{1i} = \frac{2m_0}{\hbar^2}\left(\frac{1}{3}\right)\sum_i (E_i - E_1)\left|\int \psi_1^*\mathbf{r}\psi_i dV\right|^2 = 1 \tag{10.29}$$

following Exercise 10.1. Here $E_i - E_1$ is the energy difference between states i and the quantum state labelled 1 and ψ_i is the wavefunction of the ith state. In atomic units, energies are measured in Rydbergs (R_d) and distances in units of the Bohr radius (a_0). If

$$\sum_i (E_i - E_1) \left| \int \psi_1^* \mathbf{r} \psi_i dV \right|^2 = X R_d a_0^2,$$

the summation in Equation 10.29 has a value

$$\sum_i f_{1i} = \frac{2m_0}{\hbar^2} \frac{1}{3} \sum_j (E_i - E_1) \left| \int \psi_1^* \mathbf{r} \psi_i dV \right|^2 = \frac{1}{3} X \qquad (10.30)$$

which means that $X = 3$. Absorption oscillator strengths where the i level is higher than state 1 are positive, but Equations 10.29 and 10.30 apply also if state 1 is an excited state. If 1 is an excited state, then the oscillator strengths to lower i states are negative (as suggested by the form of Equations 10.29 and 10.30). A demonstration of the proof of the Thomas-Reiche-Kuhn sum rule (Equation 10.29) is given in Appendix A.4.

10.1.6 Sample Calculation of a Transition Probability

Hydrogen atoms comprise 74% of the observable mass of the universe (another 24% of the total observable mass is helium[1]). The most intense radiative transition in hydrogen is between the principal quantum number $n = 2$ and $n = 1$ states of hydrogen: a transition known as the Lyman alpha transition. It is the 'resonance transition' for hydrogen (see the discussion in Section 3.3). As a sample calculation of a radiative transition probability, in this section we evaluate the transition probability for the hydrogen Lyman alpha transition. In addition to its astrophysical importance, the Lyman alpha transition probability is also fortunately amenable to an exact analytical evaluation.

The allowed radiative transitions for the Lyman alpha $n = 2$ and $n = 1$ line in hydrogen are of form 2p–1s, so that the two degeneracies of the upper and lower states are $g_2 = 6$ and $g_1 = 2$. From the Bohr theory of the atom, we know that the frequency of the $n = 2 \gg 1$ transition is given by

$$\hbar\omega_0 = R_d \left(\frac{1}{1^2} - \frac{1}{2^2} \right)$$

where the Rydberg energy R_d is 13.6 eV. The angular frequency of this light, $\omega_0 = 1.55 \times 10^{16}\,\mathrm{s}^{-1}$ If we assume that the value of the matrix element integral in Equation 10.28 is given by the Bohr radius $a_0 = 5.3 \times 10^{-11}$ m, we find that

[1] I have been careful here and at the beginning of Chapter 1 to refer to the 'observable' mass of the universe. If we include energy converted to mass using $E = mc^2$, cosmological evidence suggests that only 4.6% of the universe's mass is in the form of 'baryonic' mass, primarily made up of protons and neutrons. The rest of the mass comprises dark energy (68%) and dark matter (27%). Dark energy is evidenced by the measured increasing rate of expansion of the universe and dark matter by, for example, the rotation behaviour of galaxies.

$A_{21} = 4 \times 10^8\ \mathrm{s}^{-1}$. This value is close to the value obtained ($6.265 \times 10^8\ \mathrm{s}^{-1}$) when a proper integration over the s- and p-shaped wavefunctions is undertaken.

A proper evaluation of the square of the matrix element $|D_{12}|$ in Equation 10.28 can be made. We have for the square of the matrix element

$$|D_{12}|^2 = \left[\int \psi_1^* \mathbf{r} \psi_2 dV\right]^2.$$

We need to evaluate

$$|X_{12}|^2 = \left[\int \psi_1^* x \psi_2 dV\right]^2$$

where x is the component of a position vector along the direction of the electric field of the electromagnetic wave. Here $|X_{12}|^2 = \frac{1}{3}|D_{12}|^2$ (see the discussion in going from Equation 10.25 to 10.26 in Section [10.1.4]). For a hydrogen atom, the wavefunctions for the ground 1s and excited 2p quantum states are given by:

$$\psi_{1s} = R_{10}Y_{00} = 2\exp(-r)\frac{1}{\sqrt{2}}\frac{1}{\sqrt{2\pi}} \tag{10.31}$$

$$\psi_{2p} = R_{21}Y_{10} = \frac{1}{2\sqrt{6}}\,r\,\exp(-r/2)\sqrt{\frac{3}{2}}\cos\theta\frac{1}{\sqrt{2\pi}} \tag{10.32}$$

using Equations 7.10 and 7.12. For brevity all distances r are given in units of the Bohr radius a_0 and we have used the spherical 2p wavefunction for $m = 0$ as an identical result is obtained for $m = \pm 1$ (with then a $\sin\theta$ dependence rather than $\cos\theta$). We can explicitly calculate

$$\int \psi_1^* x \psi_2 dV = \int_0^\infty 2\exp(-r)\frac{1}{2\sqrt{6}}\,r^2\exp(-r^2/2)r^2 dr$$
$$\times \int_0^\pi \frac{1}{\sqrt{2}}\cos\theta\sqrt{\frac{3}{2}}\cos\theta\sin\theta d\theta \int_0^{2\pi}\frac{1}{\sqrt{2\pi}}\frac{1}{\sqrt{2\pi}}d\phi,$$

where $dV = r^2\sin\theta dr d\theta d\phi$ and $x = r\cos\theta$. The integrations on the righthand side can be carried out independently. The integral over ϕ is trivial: $\int_0^{2\pi} 1/2\pi d\phi = 1$. The integral over θ reduces to

$$\frac{\sqrt{3}}{2}\int_0^\pi \cos^2\theta\sin\theta d\theta.$$

We introduce a change of variable $u = \cos\theta$ and $du = -\sin\theta d\theta$ and the integral becomes

$$-\frac{\sqrt{3}}{2}\int_{-1}^{1} u^2 du = \frac{1}{\sqrt{3}}.$$

The radial integral can be evaluated as

$$\frac{1}{\sqrt{6}}\int_{0}^{\infty} r^4 \exp(-3r/2)dr = \frac{2^7}{3^4}\sqrt{\frac{2}{3}}$$

using definite integrals. The total transition integral for the transition 2p–1s now becomes

$$\int \psi_1^* x \psi_2 dV = \frac{1}{\sqrt{3}}\frac{2^7}{3^4}\sqrt{\frac{2}{3}} = \frac{2^{15/2}}{3^5}. \tag{10.33}$$

As this integral is evaluated in units of a_0, the square of the matrix element $|X_{12}|^2$ becomes

$$|X_{12}|^2 = a_0^2 \left| \int \psi_1^* x \psi_2 dV \right|^2 = \frac{2^{15}}{3^{10}} a_0^2 \tag{10.34}$$

and the square of the matrix element $|D_{12}|^2$ is

$$|D_{12}|^2 = 3|X_{12}|^2 = \frac{2^{15}}{3^9} a_0^2 \approx 1.665 a_0^2. \tag{10.35}$$

We can now calculate the spontaneous emission rate

$$A_{21} = \frac{g_1 e^2 \omega_0^3 |D_{12}|^2}{3\pi \epsilon_0 g_2 \hbar c^3} = 6.268 \times 10^8 \text{ s}^{-1}. \tag{10.36}$$

The accepted value for the 2p–1s spontaneous emission rate according to the NIST database is 6.265×10^8 s^{-1} [63].

The form of Equation 10.36 shows how transition probabilities A_{21} vary for hydrogen-like ions of nuclear charge Z. We saw in Section 1.5 that the radii of electron orbits scale as a_H/Z, where a_H is the hydrogen orbital radius. All hydrogen-like wavefunctions have the same form as hydrogen with a simple change of scale from a_0 to a_0/Z (see Equations 7.10 and 7.12). The matrix elements D_{12} then scale as $1/Z$ (see Equation 10.34), and the radiation frequency ω_0 scales proportionally to Z^2 so that all A_{21} for hydrogen-like ions scale $\propto Z^4$.

10.2 Selection Rules

We have seen that the Laporte selection rule discussed in Section 10.1.4 for a dipole-allowed transition requires that the atomic orbital angular momentum quantum number l changes such that

$$\Delta l = \pm 1. \tag{10.37}$$

There is no selection-rule restriction on the change in the principal quantum number n. The overlap integral (Equation 10.28) between the lower and upper quantum state wavefunctions is larger, leading to larger transition probabilities if the wavefunctions coincide spatially. The spatial overlap of wavefunctions is larger for smaller changes of principal quantum number. However, transition probabilities scale with frequency as ω_0^3 (see Equation 10.28), so tend to be small for $\Delta n = 0$ as ω_0 is small. Typically transition probabilities for $\Delta n = 1$ transitions have the largest values.

The electron total angular momentum j quantum number is given by the summation of the orbital angular momentum and electron spin quantum numbers: $j = l \pm 1/2$. If the electron spin stays the same during an optical transition, then the change of total angular momentum $\Delta j = \Delta l$. If the electron spin changes during the transition (so that $\Delta s = 1$), then it is possible for $\Delta j = 0$. We can summarise that the selection rule for the j quantum number change during an optical transition is that

$$\Delta j = 0, \pm 1. \tag{10.38}$$

In addition to the Laporte rule, constraints on radiative transitions arise due to conservation of angular momentum. A photon has a spin of one corresponding to an angular momentum of $\hbar$. Consequently, in a radiative transition, the angular momentum of the atom or ion must change by $\hbar$ in the opposite direction to the photon spin.

Multiple electron atoms and ions have similar selection rules to those for single electrons given by the selection rule shown in Equation 10.38 for the total atomic angular momentum J. We have that

$$\Delta J = 0, \pm 1 \tag{10.39}$$

but with 0–0 not allowed as it is impossible to balance the atomic momentum change with the photon momentum. Multi-electron atoms and ions tend to have stronger transitions if the following rule is satisfied:

$$\Delta L = 0, \pm 1 \tag{10.40}$$

with 0–0 not allowed. The selection rule shown in Equation 10.40 arises from the selection rule 10.39 if the total electron spin S does not change. We can write another selection rule implying stronger transitions if $\Delta S = 0$. The usually weaker transitions where the spin changes ($\Delta S \neq 0$) are known as 'intercombination' transitions.

10.2.1 Forbidden Transitions

If a transition is not allowed so that Equation 10.28 predicts $A_{21} = 0$, the transition is referred to as a 'forbidden' transition. However, a non-zero but typically small A_{21} value can be obtained assuming a different interaction Hamiltonian to the electric dipole Hamiltonian of Equation 10.14. The total energy of an atom or ion in an electromagnetic field has other (usually smaller) terms than the electric dipole energy of Equation 10.14 and these can result in a non-zero A_{21} value when the electron dipole transition rate is zero. An example of a different interaction energy and Hamiltonian is the magnetic moment of the atom or ion interaction in the magnetic field of the electromagnetic wave. Radiative transitions where the magnetic moment interaction of an atom gives the dominant contribution to the transition probability are known as magnetic dipole transitions.

The 21 cm transition between the ground state hyperfine states of hydrogen discussed in Section 7.6 is an example of a magnetic dipole transition. Calculations show that the transition probability is incredibly small ($2.85 \times 10^{-15}\,\text{s}^{-1}$) implying a lifetime of eleven million years. We show in Section 10.1.6 that the 2p–1s electronic dipole transition in hydrogen has a transition probability 2.2×10^{23} times larger (at $6.268 \times 10^{8}\,\text{s}^{-1}$).

Other forbidden transitions arise when an interaction Hamiltonian involving the energy of an electric quadrupole in an electric field dominates. An electric quadrupole has alternate positive and negative charges distributed in a four-leaf clover pattern and can be represented as two electric dipoles oriented anti-parallel to each other. The energy of interaction with an electric field is much reduced compared to a single electric dipole, as the two opposing electric dipoles have opposite signs in their interaction energy with an electric field.

The selection rules for forbidden transitions are different to those given above (the start of Section 10.2) for electric dipole transitions. A listing of selection rules for transitions where the interaction Hamiltonian is determined by considering the atom or ion as a magnetic dipole or electric quadrupole are given, for example, in the textbook by Tennyson [111].

10.3 Lineshapes

The lineshape function $f(\omega)$ allows the evaluation of the emission coefficient or absorption coefficient for a bound-bound transition over a narrow frequency range. We have defined the lineshape $f(\omega)$ such that $f(\omega)d\omega$ is the probability of the radiative process taking place in the frequency range ω to $\omega + d\omega$, assuming that the radiative process for the particular transition does take place at some frequency. This last assumption requires that

$$\int_0^\infty f(\omega)d\omega = 1. \tag{10.41}$$

Line profiles have a region near the peak value known as the line centre, whereas the line profile at higher or lower frequencies is known as the line wing.

The expressions for the line profiles presented here are developed in terms of the angular frequency ω which has units of radians/second. This is particularly useful for some theoretical developments, for example, when Fourier transforms are used as in the discussion of natural broadening (Section 10.3.1). However, in experimental spectroscopy, the spectral content of radiation is often measured in terms of wavelength λ in units of length (such as nanometre) or frequency ν in units of Hertz or s^{-1}. At short wavelengths (high frequencies), the spectral content is measured in terms of the photon energy $E = \hbar\omega = h\nu$.

It is possible to readily convert expressions for line profiles from ω to ν using $\omega = 2\pi\nu$ and $d\omega = 2\pi d\nu$. The lineshape functions expressed in terms of angular frequency $f(\omega)$ are related to lineshape functions $f_\nu(\nu)$ expressed in terms of frequency ν measured in Hertz by

$$f_\nu(\nu)d\nu = f(\omega)d\omega.$$

The conversion of small increments of frequency ($\Delta\omega$ or $\Delta\nu$) to small increments of wavelength ($\Delta\lambda$) or photon energy (ΔE) as being considered in line profiles proceeds by noting that the proportional changes in wavelength, frequency and photon energy are equal. We have that

$$\frac{\Delta\omega}{\omega} = \frac{\Delta\nu}{\nu} = -\frac{\Delta\lambda}{\lambda} = \frac{\Delta E}{E}.$$

Using $\lambda\nu = c$ gives, for example, that

$$\Delta\nu = -\frac{c\Delta\lambda}{\lambda^2}.$$

Before considering line profiles associated with real plasma emission or absorption, note that there are two line profiles sometimes used in theoretical considerations. The Dirac-delta function $\delta(\omega - \omega_0)$ is employed to represent an infinitely narrow line profile. This profile has a value of zero for all frequencies ω except at the frequency ω_0, while simultaneously having the property that an integral over all frequencies of the function has a value of one. Similar but useful when a finite spectral width is required is the Milne profile. For a Milne profile $f_M(\omega)$, we can write

$$\begin{aligned} f_M(\omega) &= \frac{1}{\Delta\omega} \; (-\Delta\omega/2 < \omega - \omega_0 < \Delta\omega/2) \\ &= 0 \text{ (otherwise)} \end{aligned} \tag{10.42}$$

where $\Delta\omega$ represents the spectral line width. The Milne profile has a flat top (that is a top-hat or mesa shape) centred on the frequency ω_0 which enables some calculations, particularly in radiative transfer, to be simplified. When integrated over all frequencies the Milne profile as written here gives unity as required by Equation 10.41. The Milne profile is identical to the Dirac-delta profile when $\Delta\omega$ approaches a value of zero.

10.3.1 Natural Line Broadening

The most fundamental factor affecting a lineshape is the natural broadening associated with the finite lifetime of the upper quantum state. The Heisenberg uncertainty principle implies that there is a broadening of the measured energy of the upper quantum state due to the finite lifetime.[2] The spectral range of frequencies observed from the radiative exponential decay of the population of the upper quantum state gives rise to the natural broadening frequency lineshape, which can be interpreted as a corresponding upper-state energy uncertainty. We first examine the electromagnetic line-profile shape expected as the population of an upper quantum state radiatively exponentially decays and then obtain the (same) lineshape by adding an extra spontaneous decay term to the differential Equation 10.16 obtained earlier in our quantum mechanical treatment of the absorption of light.

From our treatment of the Einstein A-coefficient (the radiative transition probability), we know that the electric field due to spontaneous emission from an ensemble of excited quantum states decays at a rate $E(t) = E(0)\exp(-A_\Sigma t/2)$, where A_Σ is the sum of the transition probabilities for all transitions to lower quantum states from the upper quantum state. The factor 1/2 arises because the radiative intensity decays as $\exp(-A_\Sigma t)$ and the square of the electric field is proportional to the intensity. The Fourier transform of the exponential temporal decay of the radiation electric field gives the electric field variation in frequency ω. We have the electric field variation in frequency given by the Fourier transform in time of the following form:

[2] The Heisenberg uncertainty principle itself is more general and arises due to the relationship between the effective widths of Fourier transform variable pairs: pairs of operators such as position and momentum, energy and time can be measured only to an accuracy proportional to the magnitude of the commutator of the operators (see Appendix A.4). For position/momentum and energy/time, the multiplication of the standard deviation of the two measurements has a minimum of $\hbar/2$.

$$E(\omega) \propto \int_0^\infty \exp\left(\frac{-A_\Sigma t}{2}\right) \exp(-i(\omega - \omega_0)t)dt = \frac{1}{A_\Sigma/2 + i(\omega - \omega_0)} \tag{10.43}$$

where the central line frequency is ω_0. The intensity $I(\omega)$ of the radiation is then given by

$$I(\omega) \propto E(\omega)E^*(\omega) \propto \frac{1}{1 + 4(\omega - \omega_0)^2/A_\Sigma^2}$$

where the asterisk implies complex conjugation. Taking account of our requirement that the integral over all frequency is equal to unity, we then have for the lineshape function associated with natural broadening

$$f_L(\omega) = \frac{2/(\pi A_\Sigma)}{1 + 4(\omega - \omega_0)^2/A_\Sigma^2}. \tag{10.44}$$

This profile shape $f_L(\omega)$ is known as a Lorentzian line profile. We see that the full width at half maximum of the lineshape is equal to A_Σ. If only one transition from the upper state 2 to lower state 1 is possible, then $A_\Sigma = A_{21}$. Natural broadening is often small compared to other line-broadening mechanisms, but it is a fundamental broadening process that is always present.

Quantum Mechanics of Natural Broadening

We can derive an expression for the natural line broadening of a transition based on the semi-classical quantum mechanics method used previously in this chapter. Equation 10.16 predicts that the coefficient C_2 will increase in time as a result of absorption, but we did not take spontaneous emission into account. We can include spontaneous emission by adding an extra spontaneous decay term $-i\frac{\gamma}{2}C_2$ to the left-hand side of Equation 10.16, viz.

$$I_{12}^* \cos(\omega t) \exp(i\omega_0 t)C_1 - i\frac{\gamma}{2}C_2 = i\frac{dC_2}{dt}. \tag{10.45}$$

In the absence of an (applied) electromagnetic wave, the equation can be integrated readily to give

$$C_2(t) = C_2(0) \exp\left(-\frac{\gamma}{2}t\right). \tag{10.46}$$

For a plasma of N atoms or ions, the number excited at a given time is described by

$$N_2(t) = N|C_2|^2 = N_2(0) \exp(-\gamma t). \tag{10.47}$$

We know from our discussions on the Einstein A-coefficients that the decay of an excited state is described by an exponential decay of the form $\exp(-A_{21}t)$. This means that the additional term we introduced in Equation 10.45 gives the correct fluorescence decay rate if

$$\gamma = A_{21}. \tag{10.48}$$

The decay rate γ is also equivalent to the dissipation rate for a bound accelerated electron associated with radiation, as discussed in Section 3.2.

The electric dipole moment of an atom parallel to x (the direction of the electric field) is given by

$$d(t) = -\int \Psi^*(t) eX\Psi(t) dV, \tag{10.49}$$

where

$$-eX = -\sum_{j=1}^{Z} ex_j, \tag{10.50}$$

with Z the number of electrons. As before, the general form of the atomic wavefunction is

$$\Psi(t) = C_1(t)\psi_1 \exp(-iE_1 t/\hbar) + C_2(t)\psi_2 \exp(-iE_2 t/\hbar). \tag{10.51}$$

Substitution of Equation 10.51 into Equation 10.49 yields

$$d(t) = -e\left\{C_1^* C_2 X_{pq} \exp(-i\omega_0 t) + C_2^* C_1 X_{21} \exp(i\omega_0 t)\right\} \tag{10.52}$$

where $X_{12} = \left[\int \psi_1^* x \psi_2 dV\right]$.

The method to solve this is similar to the determination of the Einstein B-coefficient, in that solutions for C_1 and C_2 are required. The initial conditions before the field is applied are that all atoms or ions are in their ground states. Application of the oscillating electric field causes each atom to undergo transitions so that there is some probability of finding the atom in its excited state. At early time C_1 can be set equal to unity on the left of Equation 10.45. Integration of this equation then leads to a solution for C_2. It is appropriate to take the indefinite integral of Equation 10.45, and the result is

$$C_2(t) = -\frac{1}{2} I_{12}^* \left[\frac{\exp\{i(\omega_0 + \omega)t\}}{\omega_0 + \omega - i\gamma/2} + \frac{\exp\{i(\omega_0 - \omega)t\}}{\omega_0 - \omega - i\gamma/2}\right], \tag{10.53}$$

using the relation $\cos(\omega t) = \frac{1}{2}[\exp(i\omega t) + \exp(-i\omega t)]$. Again, we assume that ω is close to ω_0 which means we can neglect the first term in Equation 10.53 and write

$$C_2(t) \simeq -\frac{1}{2} I_{12}^* \left[\frac{\exp\{i(\omega_0 - \omega)t\}}{\omega_0 - \omega - i\gamma/2}\right]. \tag{10.54}$$

The single-atom dipole moment $d(t)$ is found by substituting Equation 10.53 and $C_1(t) = 1$ in Equation 10.52. We find

$$d(t) = \frac{e^2|X_{12}|^2 E_0}{2\hbar} \left\{ \frac{\exp(-i\omega t)}{\omega_0 - \omega - i\gamma/2} + \frac{\exp(i\omega t)}{\omega_0 - \omega - i\gamma/2} \right\} \tag{10.55}$$

where $|X_{12}|^2 = \left[\int \psi_1^* x \psi_2 dV\right]^2$. The dipole moment of a single atom must now be related to the macroscopic polarisation of the medium. Allowing for the random orientation of the atoms, we make the replacement $|X_{12}|^2 = \frac{1}{3}|D_{12}|^2$ where $|D_{12}|^2 = \left[\int \psi_1^* \mathbf{r} \psi_2 dV\right]^2$. The macroscopic polarisation can be written as

$$P(t) = \frac{Nd(t)}{V} = \frac{Ne^2|D_{12}|^2 E_0}{6\hbar V} \left\{ \frac{\exp(-i\omega t)}{\omega_0 - \omega - i\gamma/2} + \frac{\exp(i\omega t)}{\omega_0 - \omega - i\gamma/2} \right\}, \tag{10.56}$$

where N is the number of atoms or ions, $d(t)$ is the single atom or ion dipole moment, and V is the volume.

The macroscopic polarisation (see Equation 2.34) can also be written as

$$P = \epsilon_0 \chi E = \epsilon_0 \chi E_0 cos(\omega t) = \frac{1}{2}\epsilon_0 E_0 \left\{ \chi(\omega) \exp(-i\omega t) + \chi(\omega) \exp(i\omega t) \right\}. \tag{10.57}$$

Comparing Equations 10.56 and 10.57 we find

$$\chi(\omega) = \frac{Ne^2|D_{12}|^2}{3\epsilon_0 \hbar V} \left(\frac{1}{\omega_0 - \omega - i\gamma/2} \right) = \frac{Ne^2|D_{12}|^2}{3\epsilon_0 \hbar V} \frac{\omega_0 - \omega + i\gamma/2}{(\omega_0 - \omega)^2 + (\gamma/2)^2}. \tag{10.58}$$

The absorption coefficient K is related to the imaginary part of the susceptibility by $K = \omega_0/(c\eta)\chi''$ (Equation 2.45). For a dilute plasma we can assume $\eta \approx 1$ which leads to

$$K = \frac{\pi N e^2 |D_{12}|^2 \omega_0}{3\epsilon_0 \hbar c V} \frac{\gamma/(2\pi)}{(\omega_0 - \omega)^2 + (\gamma/2)^2}. \tag{10.59}$$

The frequency dependence of the absorption coefficient is determined by the function

$$f_L(\omega) = \frac{\gamma/(2\pi)}{(\omega_0 - \omega)^2 + (\gamma/2)^2}, \tag{10.60}$$

which has the form of the Lorentzian lineshape. Using $\gamma = A_{21}$ (Equation 10.48) we find

$$f_L(\omega) = \frac{1}{2\pi} \frac{A_{21}}{(\omega_0 - \omega)^2 + (A_{21}/2)^2}. \tag{10.61}$$

This expression for the natural broadening lineshape is equivalent to Equation 10.44 if only two energy levels are considered.

10.3.2 Doppler Line Broadening

Gaseous or plasma media are affected by a line-broadening process caused by the Doppler shift in frequency of the atomic or ionic response due to the atom or ion velocity. Relativity theory gives the relationship of the frequency of radiation in the frame of reference of an observer to that of the frame of reference of the emitter. We have for the frequency ω in the frame of reference of the observer

$$\omega = \omega_0 \left(\frac{1 + v/c}{1 - v/c} \right)^{1/2} \tag{10.62}$$

where ω_0 is the frequency in the frame of reference of the emitter and v is the velocity of the emitter relative to the observer. For velocities v much smaller than the speed of light c, the denominator $1/(1 - v/c) \approx 1 + v/c$ and we have

$$\omega - \omega_0 \approx \omega_0 \frac{v}{c}. \tag{10.63}$$

Along a line of sight, a gas or plasma has a Maxwellian distribution f_v of velocities in any one direction given by the proportionality

$$f_v \propto \exp\left(-\frac{Mv^2}{2kT} \right)$$

where T is the atomic or ionic temperature and M is the mass of the atom or ion. Substituting for v using the Doppler relationship (Equation 10.63), gives an expression for the form of a Doppler line profile:

$$f_D(\omega) \propto \exp\left(-\frac{Mc^2(\omega - \omega_0)^2}{2kT\omega_0^2} \right). \tag{10.64}$$

The Doppler line profile has a Gaussian form – it varies as the exponential of the square of the frequency difference from line centre. We want the integral over all frequencies to be equal to unity, so we normalise the profile and write

$$f_D(\omega) = \frac{2(\ln 2)^{1/2}}{\pi^{1/2}\Delta\omega_D} \exp\left(-4\ln 2 \left(\frac{\omega - \omega_0}{\Delta\omega_D} \right)^2 \right) \tag{10.65}$$

where $\Delta\omega_D$ is the full width at half maximum of the line width. We can check that $\Delta\omega_D$ is the full width at half maximum by substituting $\omega - \omega_0 = (1/2)\Delta\omega_D$ to see that $f_D(\omega_0 + (1/2)\Delta\omega_D) = (1/2)f_D(\omega_0)$.

By comparing Equations 10.64 and 10.65 we can obtain an expression for $\Delta\omega_D$. We have

$$\Delta\omega_D = 2\sqrt{2\ln 2}\frac{\omega_0}{c}\sqrt{\frac{kT}{M}}. \tag{10.66}$$

The spectral width of the Doppler line profile is proportional to the square root of the temperature. If Doppler broadening dominates, measuring the widths of spectral lines gives a measure of the temperature of a gas or plasma.

10.3.3 Pressure Broadening

At high densities, the energy levels of an ion in a plasma or the radiation process itself are perturbed by the environment. The effect on spectral line profiles is referred to as pressure broadening, though the term Stark broadening or collisional broadening is also used if the dominant contribution to the broadening process arises due to respectively (i) quasistatic energy level shifts due to the Stark effect arising from micro-electric fields associated with other ions near the emitting ion in the plasma or (ii) ions or electrons colliding with emitting ions. In practise these effects occur together, though as with most line broadening, a single process can dominate for a particular spectral line. A general theory of pressure broadening where many processes can contribute to the line profile is introduced first, followed by an examination of the lineshapes produced when (i) Stark broadening or (ii) collisions dominate.

A General Theory of Pressure Broadening

Pressure broadening of spectral lines arises due to the effects of the plasma environment on the emitting ions. The local environment for an ion in a plasma varies, so the ensemble average effect is a broadening of line-profile emission. The emission process or the energies of quantum states are shifted due to the perturbing influence of nearby ions or by electrons.

It is possible to formulate a general approach to line broadening which, in principle, allows consideration of all the pressure-broadening effects together. An expression for the emission coefficient for a spectral line was given in Equation 4.23 with the necessary transition probability A_{21} determined by Equation 10.28. We can write for the emission coefficient ϵ for spectral lines associated with atoms or ions which are effectively isolated from each other:

$$\epsilon = N_2 A_{21} \hbar \omega_{21} f_L(\omega, \omega_{21}) = N_2 \frac{e^2 \omega_{21}^4}{3\pi \epsilon_0 c^3} \frac{g_1}{g_2} \left| \int \psi_1^* \mathbf{r} \psi_2 dV \right|^2 f_L(\omega, \omega_{21}). \quad (10.67)$$

The line profile $f_L(\omega, \omega_{21})$ for isolated atoms or ions is the Lorentzian lineshape function for natural broadening with central frequency ω_{21} and N_2 is the number density of the upper quantum state involved in the transition. In a plasma where pressure broadening is important, the emission from an ensemble of ions is obtained by summing up the emission from all the individual perturbed states ($\psi_{1,j}$ and $\psi_{2,j}$) taking particular care to allow for the frequency shifts in emission so that

ω_j represents the central frequency of ions with a degree of perturbation j due to the plasma environment. We can write

$$\epsilon = N_2 \frac{e^2}{3\pi\epsilon_0 c^3} \frac{g_1}{g_2} \sum_j \left[\omega_j^4 \left| \int \psi_{1,j}^* \mathbf{r} \psi_{2,j} dV \right|^2 f_L(\omega, \omega_j) \, P_j^* \right] \tag{10.68}$$

where P_j^* represents the probability of a particular perturbation j occuring due to the plasma environment.

Two simplifying assumptions to this general expression are almost always made [38]. The frequency range of emission is assumed small so that the ω_j^4 term can be moved outside the summation (we replace it with ω_0^4) and the Lorentzian lineshape function $f_L(\omega, \omega_{21})$ is replaced with the Dirac-delta function $\delta(\omega - \omega_j)$ (which has an integral over all frequency of unity, but has non-zero value only at frequency $\omega = \omega_j$). Replacing the Lorentzian lineshape function with the Dirac-delta function is accurate if the natural broadening is much smaller than the frequency shifts $\omega_j - \omega_0$. We can write that

$$\epsilon = N_2 \frac{e^2 \omega_0^4}{3\pi\epsilon_0 c^3} \frac{g_1}{g_2} \sum_j \left[\left| \int \psi_{1,j}^* \mathbf{r} \psi_{2,j} dV \right|^2 \delta(\omega - \omega_j) \, P_j^* \right]. \tag{10.69}$$

The lineshape function $f_P(\omega)$ for a pressure-broadened line is then written as

$$f_P(\omega) = \sum_j \left[\left| \int \psi_{1,j}^* \mathbf{r} \psi_{2,j} dV \right|^2 \delta(\omega - \omega_j) \, P_j \right] \tag{10.70}$$

where the probability P_j for the j degree of perturbation is now normalised so that

$$\sum_j \left[\left| \int \psi_{1,j}^* \mathbf{r} \psi_{2,j} dV \right|^2 P_j \right] = 1. \tag{10.71}$$

It is often sufficiently accurate to assume that the wavefunctions $\psi_{1,j}$ and $\psi_{2,j}$ are not greatly affected by the plasma environment so that the isolated atom or ion values can be used. The line-broadening calculation then reduces to determining the probability P_j of a frequency shift in emission to ω_j for the ensemble of atoms. We can write for the pressure-broadened line profile

$$f_P(\omega) = \sum_j \delta(\omega - \omega_j) \, P_j^* \tag{10.72}$$

with normalisation

$$\sum_j P_j^* = 1. \tag{10.73}$$

This approach can be used in the derivation of the profile for the line wings of a Stark-broadened line profile (see Equation 10.75).

Quasistatic Stark Broadening

If the emission process occurs from an ion that is effectively static during the period of emission, the most important perturbing effect is usually the electric field produced by nearby ions. The energy levels are shifted depending on the micro-electric field appropriate for each ion giving an ensemble line broadening. The effect is most pronounced where energy levels are degenerate (or near degenerate) in the absence of an electric field, but split in energy due to the Stark effect in the presence of a field.

The Stark effect arises due to an energy associated with an electric dipole moment present in an atom or ion. An electric dipole moment $\mathbf{d} = q\mathbf{r}$ is produced when a charge q is displaced at a distance $\mathbf{r}$ leaving a charge $-q$ at the original position. The energy ΔE_E associated with an electric dipole $\mathbf{d}$ in the presence of an electric field E is given by

$$\Delta E_E = -\mathbf{d} \cdot \mathbf{E}.$$

An approximate expression for Stark-broadened line profiles can be evaluated by considering the probability of experiencing an electric field E. Designating this probability by $P^*(E)dE$, we can relate it to the probability of finding another ion at a distance r which gives E. For hydrogen and hydrogen-like ions where electron degeneracy is lifted by an imposed electric field, the Stark shift in frequency is large and proportional to the electric field (this is known as the 'linear' Stark effect). We then have for the spectral intensity at frequency ω in a Stark-broadened line:

$$f_S(\omega)\, d\omega \;\propto\; P^*(E)\, dE \;\propto\; 4\pi r^2 dr \tag{10.74}$$

as $4\pi r^2 dr$ is the volume of a spherical shell at a distance r which is proportional to the number of perturbing ions at a distance r. The electric field E at a distance r from an ion is proportional to $1/r^2$ so $dE \propto -2/r^3 dr$ and:

$$f_S(\omega)\, d\omega \;\propto\; -r^2\, dr \;\propto\; E^{-5/2} dE \propto \delta\omega^{-5/2} d\omega \tag{10.75}$$

as the Stark shift in frequency from line centre is $\delta\omega \propto E$.

The functional form of Equation 10.75 for the line-profile shape is valid away from the line centre. The assumption of a small perturbation on the quantum state energy levels due to a nearby ion breaks down when the electric field due to a nearby ion approaches the electric field strength of the emitting ion. This will occur when r is such that:

$$\frac{4}{3}\pi r^3 = \frac{1}{n_i} \tag{10.76}$$

where n_i is the number density of emitting ions. The ion density is proportional to $1/r^3$ which is in turn proportional to $E^{3/2}$. As the frequency from line centre $\delta\omega \propto E$ the line profile changes shape from the dependence shown in Equation 10.75 $\left(f_S(\omega) \propto \delta\omega^{-5/2}\right)$ at a frequency from line centre close to the spectral line width $\Delta\omega \propto n_i^{2/3}$. This change of dependence effectively controls the width of the line profile. Indeed, the measurement of line widths of Stark-broadened lines is an accurate way to determine ion densities in dense plasmas where the broadening is dominated by the Stark effect. A typical, linear Stark-broadened line profile has a peak in value at frequencies above and below the line centre at a frequency from line centre proportional to $n_i^{2/3}$, with a drop in value at greater distances $\delta\omega$ from line centre decreasing proportionally to $1/\delta\omega^{5/2}$.

Collisional Broadening

In collisional broadening, the exponential decay in time of the electric field resulting from a transition between two bound quantum states is interrupted by another ion or an electron colliding with the emitting ion. With an electron–ion collision, the dominant effect is to cause a collisional transition from the ionic upper quantum state which reduces the effective lifetime of the upper state (electron–ion collisions are discussed in Chapter 11). With ion–ion collisions, the colliding ion perturbs the energy levels of the emitting ion during the collision and so interrupts the wavetrain of electromagnetic emission. The net effect with both electron–ion and ion–ion collisions is to reduce the lifetime of coherent emission from an atom undergoing a collision.

With both ion–ion and electron–ion collisions, the frequencies of the emitted radiation are given by Equation 10.43 with the electric field decay rate per unit time increased from $A_\Sigma/2$ to ν_{col}, where ν_{col} is the frequency of the collision being considered. The resulting lineshape is Lorentzian shaped as found for natural broadening with a variation given by

$$f_{col}(\omega) = \frac{1/(\pi\nu_{col})}{1 + (\omega - \omega_0)^2/\nu_{col}^2}. \tag{10.77}$$

The frequency of ion–ion or electron–ion collisions can be estimated as

$$\nu_{col} = v_{th} n_{i,e} \sigma_{col} \tag{10.78}$$

where $v_{th} = (2kT/m_{i,e})^{1/2}$ is the average ion or electron velocity, $m_{i,e}$ is the ion or electron mass, $n_{i,e}$ is the ion or electron density, and σ_{col} is the cross-section for the collisional process. We show in Section 10.3.5 that two line-broadening processes both producing Lorentzian profiles result in a Lorentzian profile with spectral width given by the addition of the two individual spectral widths associated with the separate line-broadening effects. Hence, the line-broadening full width at half maxima induced by ion–ion and electron–ion collisions will be added.

It is implicitly assumed in this treatment that the 'collision' causing the perturbation of the radiation emission occurs on a timescale much shorter than the time $1/\nu_{col}$ between collisions. However, more involved treatments show that often many small-angle collisions occur, producing, for example, small phase changes which can add to produce a sufficient disturbance in the electromagnetic wave of the ensemble of emitting ions to affect the line shape [38].

Ion–ion collisions can cause a spectral narrowing of the line width for Doppler-broadened lines (as treated in Section 10.3.2). This seemingly perverse narrowing occurs when ion–ion collisions occur much more rapidly than the radiative emission timescale (that is the $1/A_{21}$ value). The effective average velocity decrease due to the rapid changes in direction of the emitting ions due to collisions during the average time of emission means that the average Doppler shift in frequency is reduced below that calculated from the instantaneous ion velocity determined by the ion temperature. The effect is known as Dicke narrowing after Robert Dicke (1916–1997), who first showed how line profiles could be affected in this way [25].

10.3.4 Other Broadening Effects

Measurements of the spectral widths of spectral lines can be affected by other effects to those discussed above. The spectrometer used for a linewidth measurement will have a wavelength resolution and an instrument profile. If the plasma under observation has a relative velocity to the observer, a Doppler shift of the line profile occurs. A varying velocity of the plasma material to the observer is reflected in a broadened line profile.

An effect known as opacity broadening occurs as the centre of a line profile is absorbed more than the line wings. Referring to Chapter 6, we can use Equation 6.4 to examine changes in line-profile shape due to opacity. Explicitly adding a line profile $f(\omega)$ variation to the optical depth τ_0 at line centre ω_0, the intensity of a spectral line from an optically thick plasma at frequency ω is given by

$$I(\omega) = S_0 \left(1 - \exp\left(-\tau_0 \frac{f(\omega)}{f(\omega_0)}\right)\right) \tag{10.79}$$

where S_0 is the source function which is independent of the line profile if the emission and absorption profiles are the same. The ratio of line intensity at frequency ω to the intensity at line centre ω_0 is given by

$$\frac{I(\omega)}{I(\omega_0)} = \frac{1 - \exp\left(-\tau_0 f(\omega)/f(\omega_0)\right)}{1 - \exp\left(-\tau_0\right)}. \tag{10.80}$$

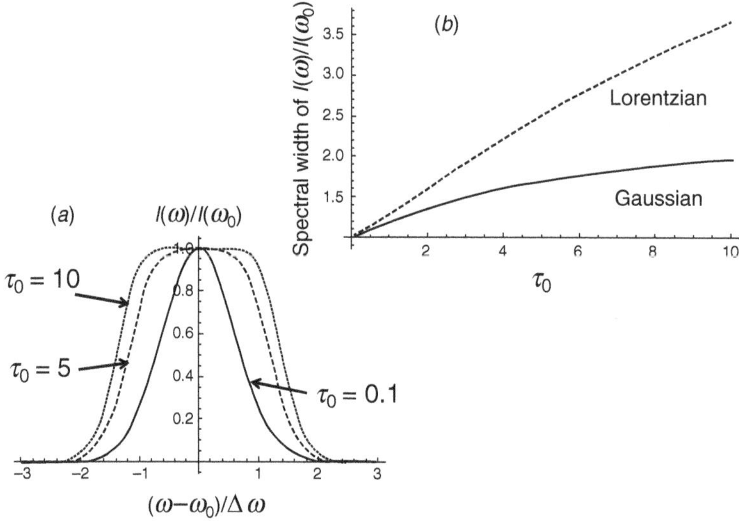

Figure 10.1 The effects of opacity broadening. (a) Spectral line intensities for a Gaussian emission and absorption profile $f(\omega)$ of $\Delta\omega$ full width at half maximum. Different values of the optical depth τ_0 at line centre are labelled. (b) The relative full width at half maximum of spectral line intensities as a function of optical depth τ_0 at line centre for Gaussian and Lorentzian line profiles $f(\omega)$ as labelled.

Examples of line profiles evaluated using Equation 10.80 are shown in Figure 10.1a. At small optical depth τ_0, the exponentials in Equation 10.80 can be accurately expanded to two terms and we have for the ratio of intensity at frequency ω to that at line centre ω_0:

$$\frac{I(\omega)}{I(\omega_0)} = \frac{f(\omega)}{f(\omega_0)}.$$

The spectral width of the line intensity at small optical depth τ is the same as the line-profile width $f(\omega)$ associated with the spectral line emission coefficient. However, for large optical depth τ_0, the spectral width of the ratio $I(\omega)/I(\omega_0)$ given by Equation 10.80 can be significantly broader than the emission coefficient profile width (see Figure 10.1b).

10.3.5 Combinations of Broadening Effects

Line profiles can be broadened by several mechanisms simultaneously. The spectrometer measuring the profile may also have a resolution which is close to that of the line width, resulting in the measured line profile having a profile shape determined by both the instrument resolution and the emitted profile shape. The

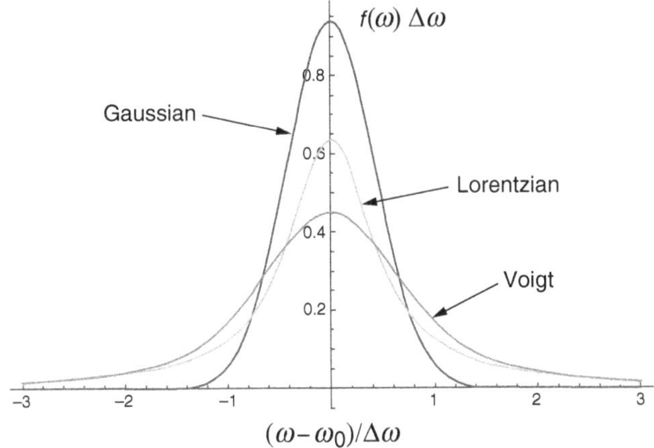

Figure 10.2 Gaussian, Lorentzian and example Voigt line profiles. For the Gaussian and Lorentzian line profles, the profile function value multiplied by the spectral width at half maximum $\Delta\omega$ is plotted as a function of the frequency from line centre $\omega - \omega_0$ in units of the full width at half maximum $\Delta\omega$. The Voigt profile is the convolution of the plotted Gaussian and Lorentzian profiles.

resultant profile $f(\omega)$ when there are two independent broadening mechanisms, say $f_1(\omega)$ and $f_2(\omega)$, is the convolution of the two profile functions:

$$f(\omega) = \int f_1(\omega - \omega')f_2(\omega')d(\omega'). \tag{10.81}$$

The convolution of two Gaussian profiles of width $\Delta\omega_1$ and $\Delta\omega_2$ gives a final width $\Delta\omega$ given by:

$$\Delta\omega^2 = \Delta\omega_1^2 + \Delta\omega_2^2 \tag{10.82}$$

while for two Lorentzian profiles of width $\Delta\omega_1$ and $\Delta\omega_2$, the final width $\Delta\omega$ is:

$$\Delta\omega = \Delta\omega_1 + \Delta\omega_2. \tag{10.83}$$

The profile arising from the convolution of Gaussian and Lorentizian profiles is termed a Voigt profile. It does not have a simple mathematical expression. Examples of line profiles are shown in Figure 10.2.

10.4 Transitions between States Affected by Zeeman and Stark effects

The Zeeman and Stark effects are the splitting of the energies of quantum states because of respectively imposed magnetic or electric fields. We saw in Section 7.3

that the energy of a quantum state is changed by a small amount ΔE_B in a magnetic field **B** due to the Zeeman effect such that

$$\Delta E_B = -\mu \cdot \mathbf{B}$$

where μ is the magnetic moment associated with the angular momentum of the quantum state. The hydrogen magnetic quantum numbers m determine the energy shift due to orbital angular momentum as

$$\Delta E_B = m \mu_B B \tag{10.84}$$

where $\mu_B = e\hbar/2m_0$ is the Bohr magneton.

We also need to take account of the electron spin contribution to the atomic angular momentum. A magnetic moment due to orbital angular momentum and electron spin can be obtained using the idea of g-factors introduced in Section 7.3. A component m_j of the total angular momentum in the direction of a magnetic field associated with a j quantum number can be employed. We then have an energy shift given by

$$\Delta E_B = m_j \, g_j \mu_B B \tag{10.85}$$

where g_j is known as the Landé g-factor. The pure orbital angular momentum effect on the energy in a magnetic field has $g_j = 1$, while the energy shift associated solely with electron spin is $g_j = 2.0023$ (as discussed in Section 7.3). Most quantum state energy shifts are associated with intermediate values of g_j, though hydrogen-like ions are dominated by the linear Stark effect with $g_j = 1$. For historical reasons, Zeeman splitting where the electron spin is important (and $g_j > 1$) is referred to as the 'anomalous' Zeeman effect.

The deduction of transition probabilities in Section 10.1 showed that electric dipole transitions between two quantum states are only possible if the upper and lower wavefunctions have different parity. States with different magnetic quantum numbers have wavefunctions aligned differently to a magnetic field (see, for example, Figure 7.2). The parity along the magnetic field axis (the z-axis) determines whether a transition is allowed and also the polarisation of the radiation involved in the transition. If the radiation is polarised in the z-direction, interaction with an atom only occurs with transitions where $\Delta m = 0$. If the polarisation is in the x or y directions, then $\Delta m = \pm 1$. Detecting the polarisation of the emitted spectral lines associated with Zeeman line splitting consequently enables the direction of a magnetic field to be deduced with the energy shift of such spectral lines giving a measure of the magnitude of the magnetic field.

The linear Stark effect on quantum state energies was introduced in Section 10.3.3 as part of a discussion of pressure broadening of spectral line profiles. A range of electric microfields in a plasma due to the presence of nearby ions causes an energy shift proportional to the electric field perturbation. The energy shift on the energy of a quantum state associated with an imposed electric field **E** results in a single value of the energy shift. We can write

$$\Delta E_E = -\mathbf{d} \cdot \mathbf{E}$$

where **d** is the induced electric dipole moment associated with a quantum state. The electron spin does not contribute to the shift in quantum state energy due to the Stark effect and the energy shifts only depend on the magnitude of the m quantum number (not the sign of m). The polarisation of emitted radiation is as outlined for transitions between states split by the Zeeman effect ($\Delta m = 0$ transitions are polarised in the direction of the electric field).

The motional Stark effect arises for a beam of fast-moving atoms or ions traversing a magnetic field. In the frame of reference of the atoms or ions with velocity $\mathbf{v}$, the $e\mathbf{v} \times \mathbf{B}$ force is equivalent to the force $e\mathbf{E}$ from an electric field $\mathbf{E}$ and the energies of the atom or ion are shifted by the Stark effect. Measuring the polarisation and energy of emission from the atoms or ions in the beam allows a measure of the orientation and magnitude of the local magnetic field. The motional Stark effect is employed as a diagnostic in tokamak plasmas [51].

Exercises

10.1 The radiative transition probability for a transition from quantum state 2 to 1 is given by Equation 10.28. We have

$$A_{21} = \frac{e^2 \omega_{21}^3}{3\pi \epsilon_0 \hbar c^3} \frac{g_1}{g_2} \left| \int \psi_1^* \mathbf{r} \psi_2 dV \right|^2$$

where ω_{21} is the frequency of the transition. Using Equation 4.29, show that the absorption oscillator strength for the transition can be written as

$$f_{12} = \frac{2}{3} \frac{\omega_{21} m_0}{\hbar} \left| \int \psi_1^* \mathbf{r} \psi_2 dV \right|^2 .$$

10.2 Use the oscillator strength expression from Exercise 10.1 and the evaluation of the matrix element given by Equation 10.35 for the $n = 2$ to $n = 1$ transition in hydrogen to determine the oscillator strength for absorption from $n = 1$ to $n = 2$ in hydrogen. [0.4162]

10.3 The power radiated by an accelerating charge was considered in Chapter 2. In the exercises at the end of Chapter 2 (see Exercise 2.13), the power P_n radiated by the centripetal acceleration of a classical electron orbiting in the nth Bohr orbit of hydrogen was determined, with the result that

$$P_n = \frac{2\, r_e R_d^2}{3\, c\, m_0\, a_0^2} \frac{1}{n^4}.$$

Relating this to the power $A_{21}\hbar\omega_{21}$ radiated by an electron in decaying to the ground state in a quantum mechanical calculation from the $n = 2$ level, find the absorption oscillator strength f_{12} for hydrogen, assuming the classical power emission is correct. [$f_{12} = 0.44$, provided we assume that the degeneracy of the $n = 2$ level is 6 (corresponding to the degeneracy of the 2p upper states which have allowed transitions to the ground 1s state). The correct quantum mechanical $n = 1$ to 2 oscillator strength is 0.4162; see Exercise 10.2.]

10.4 Considering a single atom, the Einstein coefficient B_{12} for the absorption of a photon producing a transition from a state of wavefunction ψ_1 to a state of wavefunction ψ_2 is given by

$$B_{12} = \frac{\pi e^2}{\epsilon_0 \hbar^2} \left| \int \psi_1^* x \psi_2 dV \right|^2$$

where x is the component of a vector $\mathbf{r}$ in the direction of the electric field. Assuming that atoms are randomly oriented relative to the electric field, show that the average value of B_{12} for a large number of atoms is given by

$$B_{12} = \frac{\pi e^2}{3\epsilon_0 \hbar^2} \left| \int \psi_1^* \mathbf{r} \psi_2 dV \right|^2.$$

10.5 Radiative transitions between quantum states which are not allowed, assuming the electromagnetic/atomic interaction energy arises from the atomic electric dipole interaction with the radiation electric field, can have a non-zero transition probability if a different interaction energy is considered. Assuming an atomic magnetic dipole interaction energy in the magnetic field of radiation results in a transition probability for the 2s–1s transition of a hydrogen-like ion of atomic number Z of $3.9 \times 10^{-7} Z^{10}\,\mathrm{s}^{-1}$ [78]. Another process for 2s–1s radiative decay results in the emission of two photons and has a transition probability of $8.23 Z^6\,\mathrm{s}^{-1}$. Determine the hydrogen-like ion atomic number for the 2s–1s transition where the transition probability of magnetic dipole emission exceeds the transition probability of two-photon emission. Can either magnetic dipole or two-photon emission from a 2s state have a greater transition probability than 2p–1s emission for a hydrogen-like ion? [$Z = 68$; for the 2s–1s magnetic dipole transition probability to exceed the electric dipole 2p–1s transition probability requires $Z > 345$, which does not exist.]

10.6 Evaluate (a) the average time $< t >$ for the radiative decay of an upper quantum state given by

$$<t>= \frac{\int_0^\infty t \exp(-A_{21}t)dt}{\int_0^\infty \exp(-A_{21}t)dt}$$

and (b) the standard deviation Δt for this radiative decay time given by

$$(\Delta t)^2 = \frac{\int_0^\infty (t- <t>)^2 \exp(-A_{21}t)dt}{\int_0^\infty \exp(-A_{21}t)dt}.$$

(c) Use the Heisenberg uncertainty relationship to determine the energy uncertainty (standard deviation) of the upper quantum state. [(a) $1/A_{21}$, (b) $1/A_{21}$, (c) $\hbar A_{21}/2$]

10.7 Using the line profile for natural broadening expressed in angular frequency (Equation 10.44), show that the line profile for natural broadening expressed in frequency ν (Hz) is given by

$$f_L(\nu) = \frac{8/A_\Sigma}{1 + 16\pi^2(\nu - \nu_0)^2/A_\Sigma^2}.$$

10.8 Pressure broadening associated with the linear Stark effect produces profiles in the line wings varying with frequency from line centre as $\delta\omega^{-5/2}$. When convoluted with another broadening effect, the line profile $f_H(\omega)$ can vary proportionally to

$$\frac{1}{1 + (2|\omega - \omega_0|/\Delta\omega)^{5/2}}$$

where $\Delta\omega$ is the full width at half maximum of the line profile. Show that the normalised line profile with $\int_0^\infty f_H(\omega)d\omega = 1$ can be written as

$$f_H(\omega) = \left(\frac{5}{2\pi}\right) \sin\left(\frac{2\pi}{5}\right) \frac{1}{\Delta\omega} \frac{1}{1 + (2|\omega - \omega_0|/\Delta\omega)^{5/2}}.$$

Such a profile is sometimes known as a Holtsmarkian profile.

11
Collisions

To calculate quantum state densities when a plasma is not in thermodynamic equilibrium, it is necessary to examine the individual processes populating and de-populating the quantum states. We have already considered the radiative processes (spontaneous radiative decay, photo-excitation and stimulated emission in Section 4.2 and free-bound radiative recombination in Section 5.4). Typical radiative reactions involved are listed in Table 11.1.

Other processes which need to be taken into account to evaluate quantum state population densities are collision induced. In colliding with an ion, electrons (and to a lesser extent other ions) can cause transitions between quantum states and between discrete quantum states and free electrons. A list of collisional reactions affecting quantum state populations is given in Table 11.2. Models calculating plasma quantum state densities and consequent radiation emission and absorption properties using rates of radiative and collisional processes are known as collisional radiative models [89].

11.1 Collisions in Plasmas

When the electrons, ions and atoms in a plasma undergo collisions they share energy and momentum so that a thermal distribution (Equation 1.22) characterised by a temperature and number density is produced. We considered the idea of a collision of an electron with an ion in Section 5.2 where the acceleration of the electron produces bremsstrahlung emission. Only a small loss of energy to radiation occurs with each collision.

In collisions between similar mass particles such as electron–electron and ion–ion collisions, the charged particle momentum and energy can be readily exchanged between the colliding particles. Collision between electrons and ions also transfer energy, but at a rate which is slower by the relative mass of the ions and electrons.

Table 11.1. *Radiative processes in plasma.*

Process	Reaction
Spontaneous radiative decay from a quantum state q to a lower quantum state p with the emission of a photon $h\nu_{bb}$	$A_q^{Z_i+} \rightarrow A_p^{Z_i+} + h\nu_{bb}$
Radiative recombination of a free electron to quantum state p with the emission of a photon $h\nu_{fb}$	$A^{(Z_i+1)+} + e \rightarrow A_p^{Z_i+} + h\nu_{fb}$
Stimulated decay from a quantum state q to a lower quantum state p with the emission of a photon $h\nu_{bb}$ in phase with and identical to the original stimulating photon	$A_q^{Z_i+} + h\nu_{bb} \rightarrow A_p^{Z_i+} + h\nu_{bb} + h\nu_{bb}$
Photopumping by the absorption of a photon from quantum state p to higher quantum state q	$A_p^{Z_i+} + h\nu_{bb} \rightarrow A_q^{Z_i+}$

Table 11.2. *Inelastic collisional processes in plasma.*

Process	Reaction
Collisional excitation by a free electron from a quantum state p to a quantum state p' and the inverse process of de-excitation	$A_p^{Z_i+} + e \rightleftharpoons A_{p'}^{Z_i+} + e$
Collisional ionisation by a free electron of an ion and the inverse process of three-body recombination	$A_p^{Z_i+} + e \rightleftharpoons A^{(Z_i+1)+} + e + e$
Collisional excitation by an ion from a quantum state p to a quantum state p' and the inverse process of de-excitation	$A_p^{Z_i+} + H^+ \rightleftharpoons A_{p'}^{Z_i+} + H^+$
Charge exchange with a neutral atom resulting in the population of quantum state p and the ionisation of the neutral atom	$A^{(Z_i+1)+} + H \rightarrow A_p^{Z_i+} + H^+$
Dissociative recombination electron recombination with an ionised molecule with subsequent dissociation into atoms	$AB^+ + e \rightarrow A + B$

Where the kinetic energy of colliding particles is conserved (apart from the small bremsstrahlung emission), the collision is referred to as being elastic.

Inelastic collisions involve changes in the bound quantum state of an ion or atom as a result of the collision. Energy is transferred from an electron kinetic energy to the potential energy of a bound electron (or vice versa).

An important measure with all collisions is the value of the cross-section for the collision. The cross-section has the dimensions of area and can be thought of as an area around the ion or atom associated with the necessary proximity of

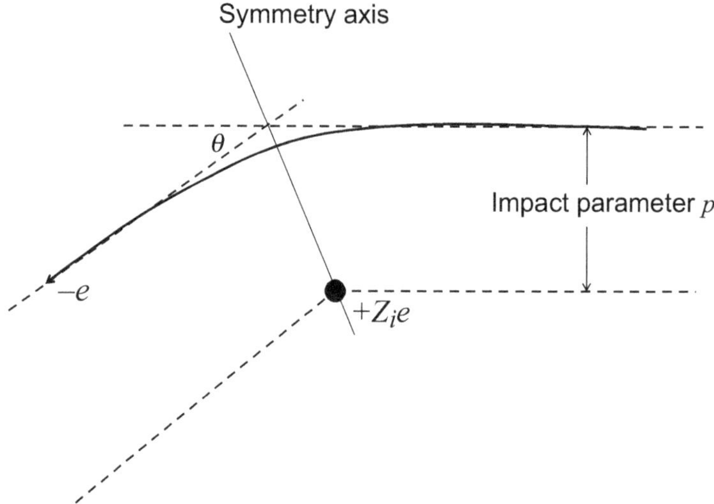

Figure 11.1 A schematic plot of the path of an electron in a collision with an ion. The impact parameter p is labelled. In an elastic collision, the path of the electron is symmetric around the labelled symmetry axis.

approach of the other colliding particle sufficient to cause the particular change of the ion quantum state (for inelastic collisions) or a particular deviation of the colliding particles (for elastic collisions). An impact factor p is the closest distance of approach of the two colliding particles if they continue undeviated by Coulomb (or other) forces as the collision ensues (see Figure 11.1). The cross-section $d\sigma$ for a particular impact parameter range from p to $p + dp$ is given by

$$d\sigma = 2\pi p \, dp.$$

The maximum impact parameter p_{max} still resulting in the appropriate collisional change is related to the total cross-section σ by $\sigma = \pi p_{max}^2$.

11.2 A Consequence of the Conservation of Angular Momentum in Collisions

In an elastic collision the angular momentum of two interacting charges remains constant. Consider an electron of initial velocity v in a collision with a much more massive ion. In the frame of reference of the ion, the angular momentum ($J = m_0 v p$) of the electron is conserved in an elastic collision, while in an inelastic collision, any change in angular momentum of the electron must be compensated by a change in the angular momentum of the ion. A small range of angular momentum dJ is related to a small range of impact parametery by $dJ = m_0 v \, dp$, so we can write for the cross-section with angular momentum from J to $J + dJ$ that

$$d\sigma = 2\pi p\, dp = 2\pi \frac{J}{m_0 v} \frac{dJ}{m_0 v} = \frac{2\pi J\, dJ}{(m_0 v)^2}.$$

In an inelastic collision, the maximum change of angular momentum of the colliding electron will be from a value J_{max} before the collision to zero after the collision. This implies a maximum value of the cross-section σ_{max} for the collision such that

$$\sigma_{max} = \frac{2\pi}{(m_0 v)^2} \int_0^{J_{max}} J dJ = \frac{\pi}{(m_0 v)^2} J_{max}^2. \tag{11.1}$$

By considering the change in angular momentum of the quantum states of the ion before and after the collision, we can evaluate J_{max} and hence obtain a maximum value for the cross-section σ_{max}. For example, if the quantum state of an ion changes from total angular momentum quantum number J_1 to J_2 in collisional excitation or de-excitation, then the maximum angular momentum change for the ion and hence also the colliding electron is given by

$$J_{max}^2 = \hbar^2 \left[J_1(J_1 + 1) - J_2(J_2 + 1) \right] \tag{11.2}$$

upon using the quantum mechanics eigenvalue for the square of angular momentum (see Equation 8.4).

Interestingly, we also see immediately from Equation 11.1 that the maximum cross-section for an inelastic collision where the free electron kinetic energy is not conserved has a dependence inversely proportional to v^2 or proportional to $1/E$, where E is the initial energy of the colliding electron. The value of cross-section given by Equations 11.1 and 11.2 is a maximum possible value that can be used for a cross-section upper bound estimate. The actual cross-sections can be considerably smaller.

11.3 The Evaluation of Collisional Cross-Sections

To evaluate collisional cross-sections between electrons and ions, it is necessary to consider the incoming electron wavefunction and its interaction with the central potential due to the charge at the nucleus and the bound electrons. An approach to evaluating the inelastic collisional excitation and ionisation cross-sections where the quantum state of the ion changes could proceed with the time-dependent Schrodinger equation in a similar way as outlined in Section 10.1.1 (where the rate of radiative absorption was evaluated). However, the interaction between an electron and an ion in a collision is dominated by the spatial variation of the distance of the electron from the ion. Consequently, the interaction is best treated by considering the spatial variation of the wavefunction as it is incident and scattered by the ion using the time-independent Schrodinger equation. The free electron

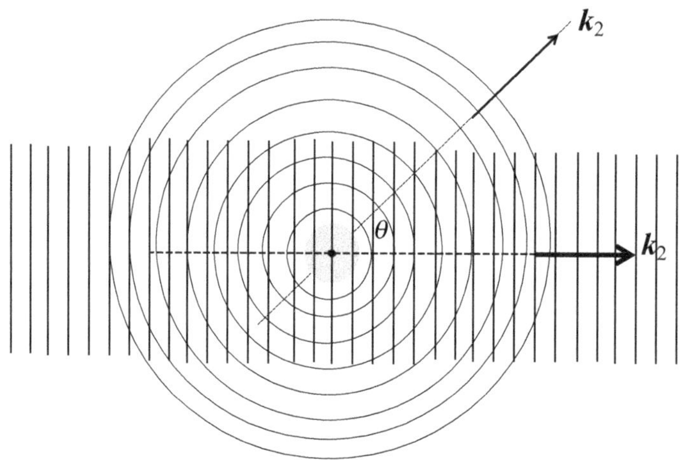

Figure 11.2 Schematic electron scatter from a central potential due to an atomic nucleus (dark center circle) and surrounding electron distribution (lighter circle). The incident $\mathbf{k}_1$ and scattered $\mathbf{k}_2$ wavefronts are illustrated for a scattering angle of θ.

can be regarded as a long wave-train which establishes a steady-state interference pattern due to the presence of the ion (see Figure 11.2). We follow a treatment originally developed by Mott and Massey [75] and Massey and Burhop [72].

The time-independent Schrodinger equation was obtained from the time-dependent Schrodinger equation in Chapter 7 (see Equation 7.3). We have

$$\left[-\frac{\hbar^2}{2m_0}\nabla^2 + V(\mathbf{r})\right]\psi(\mathbf{r}) = E\psi(\mathbf{r}). \tag{11.3}$$

Putting the potential $V(\mathbf{r})$ to zero gives a solution for the wavefunction of a free electron travelling in the z-direction of $\psi = A_{NM}\exp(ikz)$, where the wavenumber $k = \sqrt{2m_0E/\hbar^2}$. In our treatment here, the wave-train of the wavefunction is infinite in length. The normalisation constant A_{NM} is usually set with $A_{NM} = 1$ meaning that $\psi^*\psi$ is the probability density per unit length for the electron.

Where the free electron interacts with the central potential and bound electrons of an ion, the resulting wavefunction is scattered, much as an electromagnetic wave scatters at a change of refractive index. We can write for the resulting wavefunction

$$\psi = \exp(i\mathbf{k}_1 \cdot \mathbf{r}) + f(\theta,\phi)A'_{NM}\left(\frac{\exp(i\mathbf{k}_2 \cdot \mathbf{r})}{r}\right) \tag{11.4}$$

where the incident electron has wavevector $\mathbf{k}_1$ and the scattered electron has wavevector $\mathbf{k}_2$ and A'_{NM} is a normalisation constant for the scattered wavefunction. Here, θ and ϕ represent angles in spherical polar co-ordinates relative to the incident electron direction (taken to be in the z-direction; see Figure 7.1) and r

is the distance from the scattering centre (the ion nucleus) with $k_1 = \sqrt{2m_0E_1/\hbar^2}$ the incident electron wavenumber and $k_2 = \sqrt{2m_0E_2/\hbar^2}$ the scattered electron wavenumber (with E_1 and E_2 representing the electron energy 'before' and 'after' the collision). The variation of $f(\theta,\phi)/r$ represents the scattered amplitude of the electron wavefunction and is written this way as, at large distance r, the scattered electron wavefunction appears to arise from a point source at the scattering centre and hence the amplitude of the wavefunction decreases as $1/r$.

Just as the area of an opaque object placed in a light beam can be determined from the relative change in the intensity of the transmitted light, the cross-section of an ion to an electron can be determined from the relative probability density of the scattered and incident electron wavefunctions. The differential cross-section $d\sigma/d\Omega$ for the electron scattering at angles θ and ϕ is given by

$$\frac{d\sigma}{d\Omega} = \frac{k_2}{k_1}|f(\theta,\phi)|^2 \tag{11.5}$$

where the velocity ratio of the scattered and incident electrons is given by the k_2/k_1 ratio. It is necessary to allow for the different electron velocities before and after scattering as the 'counting' of the scattered electrons in a cross-section measurement will record a flux of electrons per unit time. The total cross-section for scattering into all angles is given by

$$\sigma = \int \frac{d\sigma}{d\Omega} d\Omega = \int_{\theta=0}^{\pi}\int_{\phi=0}^{2\pi} \frac{d\sigma}{d\Omega} \sin\theta d\theta d\phi = \int_{\theta=0}^{\pi}\int_{\phi=0}^{2\pi} |f(\theta,\phi)|^2 \frac{k_2}{k_1} \sin\theta d\theta d\phi. \tag{11.6}$$

The normalisation constant A'_{NM} for the scattered wavefunction in Equation 11.4 needs to be set carefully as we are assuming that the scattered wave-train extends to infinity. We are interested in the scattered wavefunction variation with angle so normalise to have the electron probability density in units of per solid angle (steradian^{-1}). This requires

$$\int \frac{A'^2_{NM}}{r^2} dV = 4\pi \tag{11.7}$$

where the integration is over unit radius and all angles. The integral is trivial and gives that $A'_{NM} = 1$. With the incident electron probability distribution measured in units of per length (e.g. m^{-1}) and the scattered electron probability measured in per unit solid angle (e.g. steradian^{-1}), the differential cross-section (Equation 11.5) represents the ratio of the scattered and incident electron probability distributions and has dimensions of area per unit solid angle. The total cross-section (Equation 11.6) has the expected dimensions of area.

Substituting Equation 11.4 into the Schrodinger equation, we can write

$$\left[-\nabla^2+\frac{2m_0}{\hbar^2}V(\mathbf{r})\right]\left(\exp(i\mathbf{k}_1\cdot\mathbf{r})+f(\theta,\phi)\frac{\exp(i\mathbf{k}_2\cdot\mathbf{r})}{r}\right)$$
$$=k_1^2\exp(i\mathbf{k}_1\cdot\mathbf{r})+k_2^2f(\theta,\phi)\frac{\exp(i\mathbf{k}_2\cdot\mathbf{r})}{r}. \tag{11.8}$$

Expanding ∇^2 in spherical polar co-ordinates it is possible to separate the radial and angular contribution:

$$\nabla^2=\nabla_r^2+\nabla_{\theta,\phi}^2.$$

Using the wavefunction (Equation 11.4), it is relatively straightforward to show that

$$\nabla_r^2\psi=\frac{1}{r^2}\frac{\partial}{\partial r}\left(r^2\frac{\partial\psi}{\partial r}\right)=-k_1^2\exp(i\mathbf{k}_1\cdot\mathbf{r})-k_2^2f(\theta,\phi)\frac{\exp(i\mathbf{k}_2\cdot\mathbf{r})}{r}.$$

Substituting in Equation 11.8 this result shows that at large r, where $V(r)=0$ and $f(\theta,\phi)$ is constant, that Equation 11.4 is a solution of the Schrodinger equation for a free electron wavefunction.

Where $V(r)\neq 0$, the value of the scattering amplitude $f(\theta,\phi)$ changes in space and we can write an equation for the angular changes of the electron wavefunction

$$\nabla_{\theta,\phi}^2\left(f(\theta,\phi)\frac{e^{i\mathbf{k}_2\cdot\mathbf{r}}}{r}\right)+\frac{2m_0}{\hbar^2}V(\mathbf{r})\left(e^{i\mathbf{k}_1\cdot\mathbf{r}}+f(\theta,\phi)\frac{e^{i\mathbf{k}_2\cdot\mathbf{r}}}{r}\right)=0. \tag{11.9}$$

Solving this equation for $f(\theta,\phi)$ at different r and integrating can give a value for $f(\theta,\phi)$ at large r. Assuming that the scattering amplitude is small suggests that $f(\theta,\phi)e^{i\mathbf{k}_2\cdot\mathbf{r}}/r$ on the right in the second term is small compared to $e^{i\mathbf{k}_1\cdot\mathbf{r}}$ and can be approximated to zero. Multiplying by $\exp(-i\mathbf{k}_2\cdot\mathbf{r})$ throughout and then integrating over all solid angles up to unit radius (to be consistent with the $f(\theta,\phi)$ normalisation of Equation 11.7), we have that

$$\int\nabla_{\theta,\phi}^2\left(f(\theta,\phi)\right)d\Omega\approx-\frac{2m_0}{\hbar^2}\int V(\mathbf{r})\exp(i(\mathbf{k}_1-\mathbf{k}_2)\cdot\mathbf{r})dV.$$

For scattering predominantly in the forward direction with θ small and only a small azimuthal asymmetry so that $\partial^2f(\theta,\phi))/\partial\phi^2\approx 0$, we have $\int\nabla_{\theta,\phi}^2f(\theta,\phi)d\Omega\approx 4\pi f(\theta,\phi)$ (see Exercise 11.4). A relatively simple approximation for the scattering amplitude follows:

$$f(\theta,\phi)\approx-\frac{m_0}{2\pi\hbar^2}\int V(\mathbf{r})\exp(i(\mathbf{k}_1-\mathbf{k}_2)\cdot\mathbf{r})dV \tag{11.10}$$

with the integration over all space. Equation 11.10 is known as the Born approximation for the scattering wavefunction amplitude. Interestingly, the Born approximation has the form of a Fourier transform of the scattering potential.

Most derivations of the Born approximation involve solutions of the Schrodinger equation which can be expanded as series to give increasingly accurate results. The Born approximation is the first term in the expansion. Comparisons with expanded solutions show that Equation 11.10 can be remarkably accurate. As an example, which we treat in Section 11.3.1, the Born approximation gives the classical Rutherford equation for the differential cross-section for elastic scattering in a Coulomb potential.

11.3.1 Elastic Collisions and the Rutherford Scattering Formula

When considering elastic collisions in plasmas, we are often concerned with effectively point charges (electrons or the central potential of ions), so that the incident particle sees a Coulomb potential associated with the particles with which it is colliding. We treat the example of an electron colliding with an ion of charge Ze with spherically symmetric central potential given by

$$V(r) = -\frac{Ze^2}{4\pi\epsilon_0}\frac{1}{r}. \tag{11.11}$$

In elastic scattering, the wavenumbers of the incident and scattered particle (k_1 and k_2) are equal so that $k_1 = k_2 = k$. We can evaluate an expression for the difference in wavevectors ($\mathbf{k}_1 - \mathbf{k}_2$) in terms of the angle θ into which the electron is scattered (see Figure 11.3). Simple geometry gives an amplitude

$$|\mathbf{k}_1 - \mathbf{k}_2| = 2k\sin\left(\frac{\theta}{2}\right). \tag{11.12}$$

This relationship between incident and scattered wavevectors was obtained for light scattering in Section 3 (and see Figure 3.3). For the purpose of integration over volume using the Born approximation (Equation 11.10) we write that

$$(\mathbf{k}_1 - \mathbf{k}_2)\cdot\mathbf{r} = 2kr\sin\left(\frac{\theta}{2}\right)\cos\theta'$$

where θ' is the angle between $\mathbf{k}_1 - \mathbf{k}_2$ and the radial vector $\mathbf{r}$. We use θ' in integrating over all space as required for the Born approximation. The integration needs to be performed for each required scattered angle θ. As the potential is uniform in the azimuthal ϕ direction, we can write

$$f(\theta) = \frac{m_0}{2\pi\hbar^2}\frac{Ze^2}{4\pi\epsilon_0}\int_{r=0}^{\infty}\int_{\theta'=0}^{\pi}\frac{1}{r}\exp\left(i2kr\sin\left(\frac{\theta}{2}\right)\cos\theta'\right)2\pi\,\sin\theta'\,d\theta' r^2 dr. \tag{11.13}$$

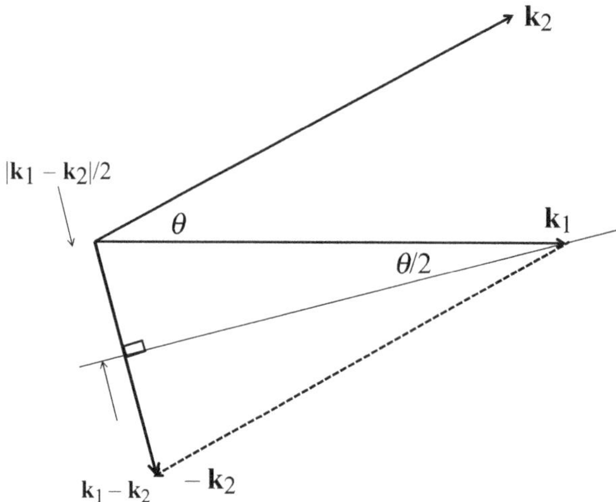

Figure 11.3 The alignment of wavevectors for the incident electron $\mathbf{k}_1$ and scattered electron $\mathbf{k}_2$ showing how the magnitude of $\mathbf{k}_1 - \mathbf{k}_2$ is related to the angle θ between the incident and scattered electron wavevectors in elastic scatter where the vector magnitudes $k_1 = k_2$. The figure illustrates that $|\mathbf{k}_1 - \mathbf{k}_2| = 2k\sin(\theta/2)$.

Integrating in angle θ' gives

$$f(\theta) = -2\frac{m_0}{\hbar^2}\frac{Ze^2}{4\pi\epsilon_0}\int_{r=0}^{\infty}\frac{\sin(2kr\sin(\theta/2))}{2kr\sin(\theta/2)}rdr \tag{11.14}$$

and then again in terms of r:

$$f(\theta) = -2\frac{m_0}{\hbar^2}\frac{Ze^2}{4\pi\epsilon_0}\frac{1}{4k^2\sin^2(\theta/2)} \tag{11.15}$$

assuming that the limit at $r = \infty$ averages to zero. The $r = \infty$ limit causes the $|f(\theta)|$ solution of Equation 11.15 to oscillate rapidly with varying θ from zero to twice the value given. In practice, there is a distribution of wavenumbers k and a small range of k at the particular r approaching ∞ assumed for the integration produces the average values of Equation 11.15.

The differential cross-section is found using Equation 11.5. Converting the wavenumber k to the energy of the electron ($E = \hbar^2 k^2/2m_0$), we have

$$\frac{d\sigma}{d\Omega} = |f(\theta)|^2 = \frac{Ze^2}{4\pi\epsilon_0}\frac{1}{16E^2\sin^4(\theta/2)}. \tag{11.16}$$

This is known as the Rutherford scattering formula. We can convert Equation 11.16 to be relevant to any colliding 'point' charges q_1 and q_2 by substituting $q_1 = Ze$ and say $q_2 = -e$. Ernest Rutherford (1871–1937) used this formula to interpret the scattering behaviour of electrons and alpha-particles (He nuclei) in the famous scattering experiments which confirmed a central point charge (the nucleus) in atoms. Rutherford published the formula in 1911 [92]. He calculated the classical Coulomb force on a point charge in the field of another point charge. The finding that scattering using the Born approximation is in exact agreement with the classically derived expression gives confidence that the Born approximation can be applied generally to particle collision calculations.

11.4 The Evaluation of Inelastic Collisional Cross-Sections

Inelastic collisions occur because bound electrons change quantum state during a collision. For ions which are not fully stripped of electrons and for atoms, the atomic or ionic potential $V(\mathbf{r})$ seen by a free electron in a collision comprises a contribution due to the central potential (the nucleus) and a contribution from the bound electrons in the ion. We considered the effect on electron scattering of the central potential in Section 11.3.1 and obtained the Rutherford scattering formula. If the density of electrons $\rho_e(\mathbf{r}')$ in the ion is considered, the potential becomes

$$V(\mathbf{r}) = \frac{1}{4\pi\epsilon_0}\left(-\frac{Ze^2}{r} + e^2\int\frac{\rho_e(\mathbf{r}')}{|\mathbf{r}-\mathbf{r}'|}dV'\right) \tag{11.17}$$

where $\mathbf{r}'$ is the vector from the nucleus used in the integral of the bound electron density over all space and $\mathbf{r}$ is a vector from the nucleus to positions at the incident electron wavefunction. The distance of separation between the incident electron wavefunction and the bound electron wavefunctions is $|\mathbf{r}-\mathbf{r}'|$. This potential can be used in the Born approximation (Equation 11.10) to evaluate the collisional cross-section taking account of the bound electrons.

The second term in Equation 11.17 for the bound electron density will produce inelastic scattering. The probability density of the bound electrons is given by $\psi_b^*\psi_b$, where ψ_b is an appropriate wavefunction for the bound electrons. In an inelastic collision, the bound wavefunction will be a linear combination of the wavefunctions ψ_1 and ψ_2 of respectively the initial and final bound quantum states. We can write that

$$\psi_b = C(\psi_1 + \psi_2)$$

so that the bound electron probability density is given by

$$\psi_b^*\psi_b = C^2(\psi_1+\psi_2)^*(\psi_1+\psi_2) = C^2\left(\psi_1^*\psi_1 + \psi_2^*\psi_2 + 2\psi_1^*\psi_2\right). \tag{11.18}$$

For ψ_b to be normalised, we require the integration of $\psi_b^*\psi_b$ over all space to be equal to unity:

$$\int \psi_b^*\psi_b dV = C^2\left(2 + 2\int \psi_1^*\psi_2 dV\right) = 1. \tag{11.19}$$

The normalisation constant $C^2 \approx 1/2$ as the integral over the two wavefunctions is small (for even parity functions) or identically zero (for odd-parity wavefunctions). The bound electron density is then given by

$$\rho_e(\mathbf{r}') = \psi_b^* \psi_b = \frac{1}{2}\left(\psi_1^* \psi_1 + \psi_2^* \psi_2\right) + \psi_1^* \psi_2. \tag{11.20}$$

Designating that $\mathbf{K} = \mathbf{k}_1 - \mathbf{k}_2$ and changing the order of the integrations, we can write that

$$\int e^{i\mathbf{K}\cdot\mathbf{r}} \left[\int \frac{\rho_e(\mathbf{r}')}{|\mathbf{r}-\mathbf{r}'|} dV'\right] dV = \int e^{i(\mathbf{K}\cdot(\mathbf{r}-\mathbf{r}')} \left[\int e^{i\mathbf{K}\cdot\mathbf{r}'} \frac{\rho_e(\mathbf{r}')}{|\mathbf{r}-\mathbf{r}'|} dV'\right] dV$$
$$= \int e^{i\mathbf{K}\cdot\mathbf{r}'} \rho_e(\mathbf{r}') \left[\int e^{i\mathbf{K}\cdot(\mathbf{r}-\mathbf{r}')} \frac{1}{|\mathbf{r}-\mathbf{r}'|} dV\right] dV'.$$

The integral in the square bracket of the last expression can be evaluated independently of the bound electron density. We have that

$$\int e^{i\mathbf{K}\cdot(\mathbf{r}-\mathbf{r}')} \frac{1}{|\mathbf{r}-\mathbf{r}'|} dV$$
$$= 2\pi \int_{\theta"=0}^{\pi} \int_{|\mathbf{r}-\mathbf{r}'|=0}^{\infty} e^{iK|\mathbf{r}-\mathbf{r}'|\cos\theta"} \frac{1}{|\mathbf{r}-\mathbf{r}'|} |\mathbf{r}-\mathbf{r}'|^2 \sin\theta" d|\mathbf{r}-\mathbf{r}'|\, d\theta"$$

where $\theta"$ is the angle between the vectors $\mathbf{K}$ and $\mathbf{r}-\mathbf{r}'$. In a similar evaluation to Equation 11.15, we find that this integral is equal to $4\pi/K^2$.

The bound electron density $\rho_e(\mathbf{r}')$ can be represented by the modulus square of the wavefunction (that is $\psi_b^*\psi_b$). Dropping the terms involving a single wavefunction in Equation 11.20 as these terms produce elastic scattering amplitudes, we have

$$\int e^{i\mathbf{K}\cdot\mathbf{r}'} \rho_e(\mathbf{r}') \left[\int e^{i\mathbf{K}\cdot(\mathbf{r}-\mathbf{r}')} \frac{1}{|\mathbf{r}-\mathbf{r}'|} dV\right] dV' = \frac{4\pi}{K^2} \int \psi_1^* \psi_2 e^{i\mathbf{K}\cdot\mathbf{r}'} dV'$$

and the inelastic scattering amplitude becomes

$$f(\theta, \phi) = \frac{m_0}{2\pi\hbar^2} \frac{e^2}{4\pi\epsilon_0} \frac{4\pi}{K^2} \int \psi_1^* \psi_2 e^{i\mathbf{K}\cdot\mathbf{r}'} dV' \tag{11.21}$$

where the integration is over the volume in the atom (with the nucleus as the origin).

The Bohr radius a_0 for an electron in the ground state of the hydrogen atom is given by

$$a_0 = \frac{4\pi\epsilon_0}{e^2} \frac{\hbar^2}{m_0}. \tag{11.22}$$

Substituting the Bohr radius for the physical constants in Equation 11.21 gives a simpler expression for the inelastic scattering cross-section:

$$f(\theta, \phi) = \frac{2}{a_0 K^2} \int \psi_1^* \psi_2 e^{i\mathbf{K}\cdot\mathbf{r}'} dV'. \tag{11.23}$$

Using Equation 11.5 the inelastic differential scattering cross-section is consequently given by

$$\frac{d\sigma}{d\Omega} = \frac{k_2}{k_1}\frac{4}{a_0^2K^4}\left|\int \psi_1^*\psi_2 e^{i\mathbf{K}\cdot\mathbf{r}'}dV'\right|^2. \tag{11.24}$$

The absolute value of the vector $\mathbf{K} = \mathbf{k}_1 - \mathbf{k}_2$ is related to the angle θ of the scattered electron to the incident electron. As kinetic energy is lost in the inelastic collision $k_1 \neq k_2$, a vector diagram of $\mathbf{k}_1$ and $\mathbf{k}_2$ at angle θ to each other shows that

$$K^2 = k_1^2 + k_2^2 - 2k_1k_2\cos\theta. \tag{11.25}$$

Differentiating both sides of this expression, we find that

$$KdK = k_1k_2\sin\theta d\theta, \tag{11.26}$$

and then the total inelastic scattering cross-section for a collisional transition can be written following Equation 11.6 as

$$\sigma = \int_{\theta=0}^{\pi}\int_{\phi=0}^{2\pi} |f(\theta,\phi)|^2\frac{k_2}{k_1}\sin\theta d\theta d\phi = \int_{K=K_{min}}^{K_{max}}\int_{\phi=0}^{2\pi} |f(\theta,\phi)|^2\frac{1}{k_1^2}KdKd\phi \tag{11.27}$$

where $K_{min} = k_1 - k_2$ and $K_{max} = k_1 + k_2$ and the scattering amplitude $f(\theta,\phi)$ is given by Equation 11.21.

11.5 Scaling of Inelastic Cross-Sections

We can approximate Equation 11.27 for the total inelastic cross-section for a transition from a lower-state wavefunction ψ_1 to an upper-state ψ_2 by expanding the exponential $e^{i\mathbf{K}\cdot\mathbf{r}'} \approx 1 + i\mathbf{K}\cdot\mathbf{r}'$ in the integral giving the inelastic scattering amplitude $f(\theta,\phi)$ (Equation 11.21). Such an approximation is reasonable as the total scattering cross-section (Equation 11.27) is largely determined by values close to K_{min} as high values of K approach an average of zero in the integral of Equation 11.21. Consequently, rather than using the maximum upper value $K_{max} = k_1 + k_2$ in the integral of Equation 11.21, we need to define an appropriate upper value where the integration is still significant. If there are many wavelengths across the dimensions of the ion, the integration values average to zero, so an appropriate upper-limit K_0 can be set as the appropriate value corresponding to the ionisation energy E_{ion} of the ion:

$$K_0^2 = \frac{2m_0E_{ion}}{\hbar^2}. \tag{11.28}$$

There is some discussion in the literature as to the optimum value of K_0, [14] but we shall see that this only affects a logarithmic term in the final expression for the collisional excitation cross-section, so does not have a large numerical effect on the cross-section values. The minimum value $K_{min} = k_1 - k_2$ can be approximated

$$K_{min} = k_1 - k_2 = \frac{k_1^2 - k_2^2}{k_1 + k_2} \approx \frac{k_1^2 - k_2^2}{2k_1} = \frac{2m_0(E_1 - E_2)}{\hbar^2 k_1} = \sqrt{\frac{2m_0}{\hbar^2}} \frac{E_1 - E_2}{\sqrt{E_1}}.$$

For smaller Kr, upon expanding the exponential in Equation 11.21, we have

$$\int \psi_1^* \psi_2 e^{i\mathbf{K}\cdot\mathbf{r}'} dV = \int \psi_1^* \psi_2 dV + iK \int \psi_1^* \psi_2 r' \cos\theta' dV$$
$$= iK \int \psi_1^* \psi_2 r' \cos\theta' dV'$$

where θ' is the angle between $\mathbf{K}$ and $\mathbf{r}$. If the wavefunctions of the lower ψ_1 and upper ψ_2 states have a different parity (see Section 10.1), then the integral over $\psi_1^* \psi_2$ is zero.

Assuming azimuthal (ϕ) symmetry in the scattering amplitude, the total cross-section becomes

$$\sigma = \frac{8\pi}{a_0^2 k_1^2} \int_{K=K_{min}}^{K_0} \left| \int \psi_1^* \psi_2 r' \cos\theta' dV' \right|^2 \frac{1}{K} dK. \tag{11.29}$$

Assuming that the scattering is mainly forward with $\theta' \approx 0$ so that $\cos\theta'$ is approximately one, the total inelastic cross-section is given by

$$\sigma = \left| \int \psi_1^* \psi_2 r' dV' \right|^2 \frac{8\pi}{a_0^2 k_1^2} \ln\left[\frac{\sqrt{E_{ion} E_1}}{E_1 - E_2} \right]. \tag{11.30}$$

We noted earlier that the upper K_0 value is uncertain. If the upper K_0 limit in Equation 11.28 replaces the ionisation energy by the energy difference between the quantum states (which is equal to the energy difference between the incoming and outgoing electron $E_1 - E_2$), the logarithmic term in Equation 11.30 becomes $(1/2)\ln[E_1/(E_1 - E_2)]$ which produces numerically similar cross-section results to the form of Equation 11.30. With both possible K_0 values, the inelastic cross–section σ varies proportionally to $\ln E_1/k_1^2$. As $k_1^2 = 2m_0 E_1/\hbar^2$, we have a cross-section dependence which is proportional to $\ln E_1/E_1$, where E_1 is the incident electron energy.

The value of the cross-section is also proportional to a similar integral as found for the evaluation of the radiative transition probability A_{21} (Section 10.1.1). Similar arguments can be made regarding the value of the integral and the parity of the lower and upper wavefunctions. If the radiative transition from state 2 to 1 is allowed (with the wavefunctions having different parity), then the collisional cross-section from state 1 to 2 will have a high value proportional to the radiative-transition probability.

11.6 Collisional Excitation for Forbidden Transitions

If the parity of the wavefunctions 1 and 2 is the same, the radiative transition between 2 and 1 is forbidden and the electric dipole transition probability between the quantum states is zero. The inelastic cross-section between the quantum states, however, does not drop to zero. The scattering amplitude can be approximated by expanding $e^{i\mathbf{K}\cdot\mathbf{r}'} \approx 1 + iKr'\cos\theta' - (1/2)K^2 r'^2 \cos^2\theta'$, where now the second term produces a zero contribution. The value of the cross-section is dominated by the third term and we have

$$\sigma = \frac{8\pi}{a_0^2 k_1^2} \int_{K=K_{min}}^{K_0} \left| \int \psi_1^* \psi_2 r'^2 \cos\theta' dV' \right|^2 \frac{1}{2} K dK \tag{11.31}$$

which gives

$$\sigma = \left| \int \psi_1^* \psi_2 r'^2 dV' \right|^2 \frac{2\pi}{a_0^2 k_1^2} \left[E_{ion} - \frac{(E_1 - E_2)^2}{E_1} \right] \frac{2m_0}{\hbar^2}. \tag{11.32}$$

The cross-section is again now proportional to $1/k_1^2 \propto 1/E_1$, but will generally be smaller than the cross-section between quantum states with allowed transitions. The integral in Equation 11.32 involves a r'^2 term with distance from the nucleus (rather than a r' term).

11.7 Inelastic Atomic and Ionic Collisions

We have explicitly considered electron collisions with ions in the evaluation of the collisional cross-section in this section, but with a change of mass m_0, the results can be generalised to atom and ion inelastic collisions with each other. The wavenumber $k_1^2 = 2m_0 E_1/\hbar$, so the significantly larger mass m_0 of an atom or ion causes the integral of the scattering amplitude (Equation 11.21) to tend to be small as there will be many wavelengths in the integral and the positive and negative values of $e^{i\mathbf{K}\cdot\mathbf{r}'}$ average to a small value. The cross-section for atom/ion inelastic collisions is usually negligible compared to the electron–ion cross-section unless k_1 is small, which implies a small energy E_1 of the impinging particle and a small energy gap between bound quantum states. Inelastic atom and ion collisions are most effective at inducing quantum states changes in an ion where the energy gap between the states is small.

11.8 Collisional Ionisation

Equation 11.27 applies for collisional ionisation with an appropriate change of the integration (Equation 11.21) for the scattering amplitude. The upper wavefunction becomes that of an unbound free electron with, say, $\psi_2 = e^{i\mathbf{k}_3\cdot\mathbf{r}'}$, where $\mathbf{k}_3$ is the wavenumber of the electron removed from the ion in the ionising collision. Equation 11.27 shows that the cross-section will have a $1/k_1^2$ or equivalently $1/E_1$ dependence.

The cross-section for collisional ionisation needs to be averaged over the possible range of energies (0 to $E_1 - E_{ion}$) of the ejected electron and over the range of K values associated with the incident and scattered electron. The total cross-section peaks when the ejected electron just obtains sufficient energy to be ionised, so we put $E_1 - E_2 \approx E_{ion}$ in the logarithmic term of Equation 11.30 to obtain $(1/2)\ln[E_1/E_{ion}]$. Similarly, the integral for the scattering amplitude varies approximately proportionally to $1/\sqrt{E_{ion}}$ which means the cross-section varies as $1/E_{ion}$. The collisional ionisation cross-section has a dependence

$$\sigma_{pC} \propto \frac{\ln(E_1/E_{ion})}{E_1 E_{ion}} \tag{11.33}$$

where E_{ion} is the ionisation energy. Expressions for collisional ionisation cross-sections with more terms, but which are accurate without fitting parameters, are available [60, 63]. We show in Chapter 12 how the rate for the inverse process to collisional ionisation, namely three-body recombination, can be calculated using detailed balance. Due to detailed balance, three-body recombination cross-sections have a similar variation as given by Equation 11.33.

11.9 Charge Exchange Recombination

Neutral atoms can undergo collisions with ions resulting in the ionisation of the atom and a drop in charge for the original ion. Writing the collision as a reaction gives

$$A^{q+} + H \rightarrow A^{(q-1)+} + H^+ \tag{11.34}$$

where A is an ion species of initial charge $+q$ in the plasma and H is an atom. Atoms of hydrogen injected into tokamak plasmas for heating and diagnostic purposes are important (as illustrated in this reaction), but charge exchange can occur between any species where there is at least one bound electron to share between the exchanging atoms or ions.

The collision in charge exchange can be regarded as coinciding with the formation of a ‘quasi-molecule’ where there is initially a weak attraction between the ion and atom arising from the electron wavefunction cloud attracting both nuclei. As the distance of approach of the nuclei decreases, the Coulomb repulsion of the nuclei increases as the ion penetrates the electron cloud. Finally, the nuclei repel each other with the electron cloud remaining in one of the quantum states of the ion.

The electronic states of the quasi-molecule are usually not bonding (there is no minimum in the energy as a function of internuclear distance; see Figure 9.1). However, the electronic states of the quasi-molecule increase in energy with the principal quantum number n for different electronic states just as occurs with stable bonding molecules. When the energy of the quasi-molecule electron states are approximately equal to the initial energy of the atom plus molecule, there is a greater probability of the electron remaining with the ion after the collision, so typically charge exchange occurs into high n quantum states. As the excited electron states later radiatively decay, the emitted photons can be used as a diagnostic of plasma conditions. The advantage of charge exchange for the diagnosis of plasma conditions is that the charge exchange only occurs where neutral atoms have been injected. Without charge exchange the plasma temperature is typically too high for the ion to have significant electron recombination and spectral line emission.

Charge exchange is important in tokamaks where energetic neutral beams of hydrogen (or deuterium) are injected into the plasma in order to create additional plasma heating or hydrogen beams are injected as a diagnostic to produce charge exchange leading to measurable radiation emission. We consider the interaction between neutral hydrogen and an ion as an example of charge exchange. There are two conditions for a charge exchange reaction between hydrogen and an ion $A^{(q-1)+}$ of charge $(q-1)+$ to take place:

1. The $A^{(q-1)+}$ ionisation energy needs to have similar energy to the H atom ionisation energy of typically $\approx$ 1 Rd (where 1 Rd = 13.6 eV is the Rydberg energy).

$$\text{The ionisation energy of } A^{(q-1)+} \approx \frac{(q-1)^2}{n^2}\text{Rd} \tag{11.35}$$

   which implies that we can expect $n \approx q-1$ to be the optimum principal quantum number for the charge exchange electron.

2. The A^{q+} and H wavefunctions need to overlap. The A^{q+} ion has a size $\sim n^2/qa_0$, whereas H has a size a_0 (where a_0 is the Bohr radius). We consequently require that

$$a_0 \approx \frac{n^2}{q}a_0 \tag{11.36}$$

   which in turn requires $n \approx \sqrt{q}$.

To obtain the most likely q and n values we can compromise and say that the final principal quantum number $n \approx q^{\frac{3}{4}}$ for the ion of initial charge q. If q is large as can occur with an impurity in a tokamak plasma, the electron will enter a high n quantum state and can decay in several principal quantum number steps, emitting a photon at each step. Each Δn step for high n often has an energy gap corresponding to visible photons which can be readily detected.

11.10 Dissociative Recombination

At moderate to high electron densities, the process of three-body recombination (the inverse of electron collisional ionisation) has a high cross-section as it is possible for both momentum and energy to be conserved during the collision. At lower electron densities, the probability of two electrons engaging in the three-body recombination collision with an ion drops, and there is no 'third' (non-recombining electron) to balance momentum exchange during the collision. At the low electron densities found in the ionosphere and laboratory 'low temperature' plasmas, electron–ion recombination is often dominated by the dissociative recombination of molecular ions.

In an electron-molecular ion collision, an incoming electron is repelled by bound electrons in the ion and attracted by the nuclear charge. The incoming electron can enter into an electronic state of the ion and excite a bound electron (thus conserving energy). This doubly excited state of neutral molecule is, however, almost always non-bonding and dissociation of the molecule into the individual atoms occurs. Alternatively, the incoming electron can enter into a highly excited electronic state of the molecular ion, forming a highly excited state of the neutral molecule. This highly excited neutral molecule state has a high probability of dissociation as the vibrational energy is high. (An alternative route is auto-ionisation back to a molecular ion and free electron.) The two possible dissociated recombination mechanisms are respectively illustrated by

$$\mathrm{AB}^{+} + \mathrm{e} \rightarrow \mathrm{AB}^{**} \rightarrow \mathrm{A} + \mathrm{B} \tag{11.37}$$

$$\mathrm{AB}^{+} + \mathrm{e} \rightarrow \mathrm{AB}^{*} \rightarrow \mathrm{A} + \mathrm{B} \tag{11.38}$$

where a single asterisk represents a singly excited electronic state and two asterisks represent a doubly excited electronic state. The detail of the mechanisms for calculating dissociative recombination cross-sections and the modelling of the different paths during the electron collision can be complicated [33].

Exercises

11.1 Evaluate the de Broglie wavelength for a free electron with a kinetic energy of 1 eV. [1.23 nm]

11.2 Use the expression for the wavefunctions of an incident and scattered electron

$$\psi = e^{i\mathbf{k}_1\cdot\mathbf{r}} + f(\theta,\phi)\frac{e^{i\mathbf{k}_2\cdot\mathbf{r}}}{r}$$

to show that

$$\nabla_r^2\psi = \frac{1}{r^2}\frac{\partial}{\partial r}\left(r^2\frac{\partial\psi}{\partial r}\right) = -k_1^2\exp(i\mathbf{k}_1\cdot\mathbf{r}) - k_2^2 f(\theta,\phi)\frac{\exp(i\mathbf{k}_2\cdot\mathbf{r})}{r}.$$

11.3 In an inelastic collision, the wavenumber of the incident and scattered electron changes from $\mathbf{k}_1$ to $\mathbf{k}_2$. Draw an appropriate example vector diagram of $\mathbf{k}_1$, $\mathbf{k}_2$, the angle θ of the electron scatter from the original electron direction and $\mathbf{k}_1 - \mathbf{k}_2$. Show that the magnitude of $|\mathbf{k}_1 - \mathbf{k}_2|$ is given by

$$|\mathbf{k}_1 - \mathbf{k}_2|^2 = k_1^2 + k_2^2 - 2k_1k_2\cos\theta.$$

11.4 In finding the Born approximation (Equation 11.10), we need to show that

$$\int (1/r)\nabla_{\theta,\phi}^2(f(\theta,\phi))dV = 4\pi f(\theta,\phi).$$

Assuming azimuthal symmetry in $f(\theta,\phi)$ (i.e. no ϕ variation), verify for spherical polar co-ordinates that the following mathematical equalities are valid:

$$\begin{aligned}\int \nabla_\theta^2 f(\theta)\, d\Omega &= \int \frac{1}{\sin\theta}\frac{\partial}{\partial\theta}\left(\sin\theta\frac{\partial f(\theta)}{\partial\theta}\right) d\Omega \\ &= \int_{\theta=0}^{\pi} \frac{1}{\sin\theta}\frac{\partial}{\partial\theta}\left(\sin\theta\frac{\partial f(\theta)}{\partial\theta}\right) 2\pi\sin\theta\, d\theta \\ &\approx 4\pi f(\theta).\end{aligned}$$

The last approximate equality requires that $f(\theta)$ is strongly peaked in the forward ($\theta \approx 0$) direction.

11.5 In elastic collisions of electrons with the central potential of an ion, the value of the electron wavefunction amplitude ($|f(\theta)|$) exhibits a solution to the Born approximation which oscillates spatially from zero to twice the value given by the Rutherford scattering formula. At a large distance from the scattering ion, show that the oscillation is radially symmetric around the initial direction of the electron with a spatial period of length equal to the de Broglie wavelength of the electron.

12
Collisional-Radiative Models

In a plasma in complete thermodynamic equilibrium, the radiation field is given by the Planck black-body expression (see Section 4.1), the ionisation is determined by the Saha-Boltzmann equation (see Section 1.4.1) and the population ratios of bound quantum states are determined by the Boltzmann ratio. In local thermodynamic equilibrium (LTE), quantum state populations are given by the Saha-Boltzmann equation and Boltzmann ratio, but the radiation field is not in equilibrium with the particles. We discuss the plasma conditions needed to establish equilibrium later in this chapter, but it is worthwhile to note that LTE often occurs when collisional processes dominate the populating and de-populating of the quantum state populations and radiative processes are not significant.

Radiative rates of decays for bound quantum states were determined in Section 10.1 and between free and bound states in Section 5.4. The cross-sections for collisional processes were discussed in Chapter 11. The cross-sections depend on the energy of the incident colliding electron, but in a plasma we have a Maxwellian distribution of the energies of the free electrons. The cross-section values need to be averaged over the Maxwellian distribution to produce a rate coefficient which when multiplied by the density of free electrons and the initial quantum state density yields the rate of change of the quantum state. The radiative reactions involved were listed in Table 11.1. A list of collisional reactions affecting quantum state populations has been given in Table 11.2. Models calculating plasma quantum state densities and consequent radiation emission and absorption properties using rates of radiative and collisional processes are known as collisional radiative models [89].

12.1 Collisional Excitation and De-Excitation

Our investigation of cross-sections for excitation by inelastic electron collisions has shown a variation with the energy E of the incident electron approximately proportional to $1/E$ (see Section 11.4). The cross-section for collisional excitation can be written in terms of a collision strength $\Omega_{pq}(E)$ such that

$$\sigma_{pq}(E) = \frac{\Omega_{pq}(E)\,\pi a_0^2}{g_p\,E} \tag{12.1}$$

where πa_0^2 is a cross-section for the ground state of the hydrogen atom (taken as the area associated with the Bohr radius a_0) and g_p is the degeneracy of the initial quantum state. 'Effective collision strengths' γ_{pq} are tabulated for different temperatures where the collision strength has been averaged over the electron distribution.

The dependence $\propto 1/E$ of cross-sections with the energy E of incident electrons enables the averaging of cross-sections when there is a distribution of electron energies to proceed more readily. When calculating quantum state populations, it is useful to consider the net transition rate, taking into account all the different energies of the colliding particles. For an electron collision-induced transition between bound states p and q with cross-section σ_{pq} we have a rate coefficient K_{pq} given by

$$\begin{aligned} K_{pq} &= \int_{\sqrt{2(E_p-E_q)/m_0}}^{\infty} \sigma_{pq}(v) v \hat{f}_v(v)\,dv \\ &= \sqrt{\frac{2}{m_0}} \int_{E_p-E_q}^{\infty} \sigma_{pq}(E)\,E^{1/2} \hat{f}_E(E) dE \end{aligned} \tag{12.2}$$

where $\hat{f}_v(v)$ and $\hat{f}_E(E)$ are the Maxwellian distribution of speeds and energy respectively normalised so that $\int_0^\infty \hat{f}_v(v) = \int_0^\infty \hat{f}_E(E) = 1$. We have previously used $f_v(v)$ or $f_E(E)$, where $f_v(v) = n_e \hat{f}_v(v)$ and $f_E(E) = n_e \hat{f}_E(E)$ for the Maxwellian distribution (see Section 1.2). The rate of dN_p/dt electron collision-induced transitions from the p to q state is $n_e N_p K_{pq}$ where N_p is the population density of the initial quantum state. The value of K_{pq} varies with the temperature of the particles colliding with the ion.

In equilibrium, the principle of detailed balance requires the inverse processes listed in Table 12.1 to occur at equal rates. The equilibrium populations are known, so knowing the rate coefficient for one process it is possible to find the rate coefficient for the inverse process. As we discussed in Section 4.2 for the radiative Einstein coefficients, the collisional-rate coefficients are simply atomic parameters and so the rate coefficients found in this way are independent of the state of the plasma and apply to plasmas not in equilibrium (provided the electron energy distribution is Maxwellian).

Electron collisional excitation from quantum state p to q is balanced in equilibrium by electron collisional de-excitation from quantum state q to p. We have

$$n_e N_p K_{pq} = n_e N_q K_{qp} \tag{12.3}$$

where the populations of the quantum states N_p and N_q are the equilibrium populations and the electrons of total density n_e have a Maxwellian distribution. Using the Boltzmann relation for the quantum state population ratio N_p/N_q, we can relate the rate coefficient for collisional de-excitation K_{qp} to that for collisional excitation K_{pq} by re-arranging Equation 12.3:

$$K_{qp} = K_{pq}\frac{g_p}{g_q}exp\left[\frac{E_p - E_q}{k_B T_e}\right] \tag{12.4}$$

where g_p, g_q are the statistical weights and E_p, E_q the ionisation energies of the two quantum states p and q.

The Maxwellian distribution $\hat{f}_E(E)$ of electron energies are normalised so that $\int_0^\infty \hat{f}_E(E) = 1$ can be obtained from Equation 1.22. We have

$$\hat{f}_E(E) = \frac{2}{\sqrt{\pi}}\left(\frac{1}{k_B T}\right)^{3/2} E^{1/2}\exp\left(-\frac{E}{k_B T}\right). \tag{12.5}$$

We can expand Equation 12.2 by substituting for the Maxwellian distribution of electron energies and introduce Equation 12.1 for the collision strength $\Omega_{pq}(E)$. We obtain

$$\begin{aligned}
K_{pq} &= \sqrt{\frac{2}{m_0}}\int_{E_p-E_q}^{\infty} \sigma_{pq}(E)\, E^{1/2}\hat{f}_E(E)dE \\
&= 2\sqrt{\frac{2}{\pi m_0}}\left(\frac{1}{k_B T}\right)^{\frac{3}{2}}\int_{E_p-E_q}^{\infty} \sigma_{pq}(E)\, E\exp\left(-\frac{E}{k_B T}\right) dE \\
&= 2\sqrt{\frac{2}{\pi m_0}}\left(\frac{1}{k_B T}\right)^{\frac{3}{2}}\frac{\pi a_0^2}{g_p}\int_{E_p-E_q}^{\infty} \Omega_{pq}(E)\, \exp\left(-\frac{E}{k_B T}\right) dE.
\end{aligned} \tag{12.6}$$

An effective collision strength γ_{pq} is obtained by averaging the collision strength over the Maxwellian distribution of electron energies. Defining

$$\gamma_{pq} = \frac{\int_{\Delta E/k_B T}^{\infty} \Omega_{pq}(E)\exp\left(-E/k_B T\right) d\left(E/k_B T\right)}{\int_{\Delta E/k_B T}^{\infty} \exp\left(-E/k_B T\right) d\left(E/k_B T\right)} \tag{12.7}$$

the rate coefficient for collisional excitation is written as

$$\begin{aligned}
K_{pq} &= \frac{2}{\sqrt{\pi}}\left(\frac{2}{m_0}\right)^{1/2}\frac{\gamma_{pq}}{g_p(k_B T)^{1/2}}\exp\left(-\frac{\Delta E}{k_B T}\right)\pi a_0^2 \\
&= 3.7\times 10^{11}\frac{\gamma_{pq}}{g_p(k_B T)^{1/2}}\exp\left(-\frac{\Delta E}{k_B T}\right).
\end{aligned} \tag{12.8}$$

after substituting numerical values so that K_{pq} is given in units of cm^3 s^{-1} with temperature $k_B T$ measured in electron volts. The energy difference between the ionisation energies of the p and q states is $\Delta E = E_p - E_q$, also measured in electron volts.

The detailed balance relationship of Equation 12.4 can be used to find the rate coefficient K_{qp} for collision de-excitation in terms of the averaged collision strength γ_{pq} for collisional excitation. We obtain

$$K_{qp} = \frac{2}{\sqrt{\pi}} \left(\frac{2}{m_0} \right)^{1/2} \frac{\gamma_{pq}}{g_q (k_B T)^{1/2}} \pi a_0^2. \tag{12.9}$$

Values of the effective collision strength γ_{pq} derived from accurate cross-section calculations or measurements have been tabulated by many authors [96, 115]. The collision strengths tend not to vary greatly with temperature whereas the raw collisional excitation rate coefficient K_{pq} values vary approximately as $\exp(-\Delta E/k_B T)/(k_B T)^{1/2}$ (see Equation 12.8) and the de-excitation rate coefficients K_{qp} vary as $1/(k_B T)^{1/2}$ (see Equation 12.9). It is, consequently, easier and usually more accurate to interpolate a table of effective collision strengths for more accurate calculations or simply use a typical collision strength value for the temperature and transition under consideration.

The most widely used approximation for the collisional excitation rate coefficient is a formula developed by van Regemorter [115] (see Exercise 12.3). In units of cm^3 s^{-1}, this approximation gives

$$K_{pq} = 3.2 \times 10^{-7} \frac{R_d^{3/2}}{(k_B T)^{1/2} \Delta E} \exp\left(-\frac{\Delta E}{k_B T} \right) f_{pq}\, G_{pq} \tag{12.10}$$

where $R_d = 13.6\,\text{eV}$ is the Rydberg energy, f_{pq} is the oscillator strength for the transition and G_{pq} is the Gaunt factor. Van Regemorter showed that $G_{pq} \approx 0.2$ for ions with $\Delta E > k_B T$. The oscillator strength f_{pq} approaches one for a strong radiative transition. Some caution has been recommended in the literature if using the van Regemorter formula where high accuracy is required [96]. The accurate alternative involves using values of averaged collision strength tabulated in the literature (for example, [24, 41]) or the use of a computer code to calculate cross-sections and collision strengths (see, for example, Gu [39], Bar-Shalom et al. [7]).

12.2 Collisional Ionisation and Three-Body Recombination

The cross-section for collisional ionisation by free electrons varies in a similar way to collisional excitation, exhibiting a $1/E$ dependence on the energy of the incident electron. We can write for the cross-section for collisional ionisation from the level p with ionisation energy E_{ion} (see Equation 11.33) that

$$\sigma_{pC}(E) = C_{ion}\, g_p \frac{\ln(E/E_{ion})}{E E_{ion}} \tag{12.11}$$

where C_{ion} is a proportionality constant and g_p is the degeneracy of the level p. For electron energy $E \leq E_{ion}$, $\sigma_{pC}(E) = 0$. Experimentally, C_{ion} is a constant between 2.6 and 4.5×10^{-18} m^2 (eV)^2 [68]. The rate coefficient for collisional ionisation can be written as

$$K_{pC} = \sqrt{\frac{2}{m_0}} \int_{E_{ion}}^{\infty} \sigma_{pC}(E)\, E^{1/2} \hat{f}_E(E)\, dE$$

$$= 2\sqrt{\frac{2}{\pi m_0}} \left(\frac{1}{k_B T}\right)^{\frac{3}{2}} \int_{E_{ion}}^{\infty} \sigma_{pC}(E)\, E \exp\left(-\frac{E}{k_B T}\right) dE$$

$$= 2\sqrt{\frac{2}{\pi m_0}} \left(\frac{1}{k_B T}\right)^{\frac{3}{2}} \frac{C_{ion}\, g_p}{E_{ion}} \int_{E_{ion}}^{\infty} \ln(E/E_{ion}) \exp\left(-\frac{E}{k_B T}\right) dE$$

$$= 2\sqrt{\frac{2}{\pi m_0}} \left(\frac{1}{k_B T}\right)^{\frac{3}{2}} C_{ion}\, g_p \int_{1}^{\infty} \ln(x) \exp(-xy)\, dx \qquad (12.12)$$

where $y = E_{ion}/k_B T$. The last integral here can be written in terms of the exponential integral $E_i(y)$ (see Figure 12.1). We have that

$$\int_{1}^{\infty} \ln(x) \exp(-xy)\, dx = \left(\frac{1}{y}\right) \int_{y}^{\infty} \frac{e^{-x}}{x}\, dx = \left(\frac{1}{y}\right) E_i(y).$$

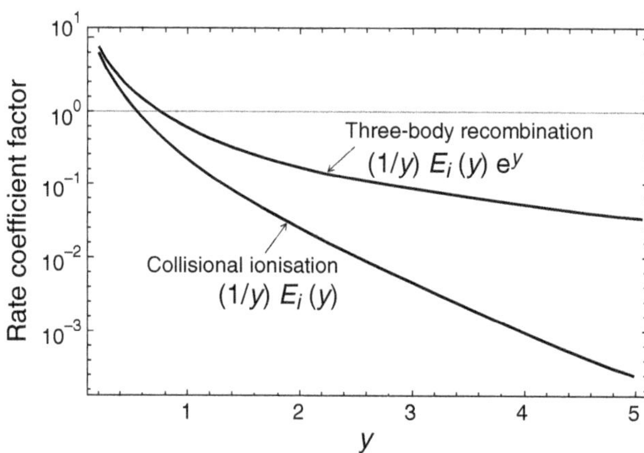

Figure 12.1 The value of $(1/y)E_i(y)$ and $(1/y)E_i(y)e^y$ as a function of y, where $E_i(y)$ is the exponential integral. The values of these integrations are proportional to the rate coefficient for collisional ionisation K_{pC} (see Equation 12.12) or three-body recombination (see Equation 12.16) with y defined as the ionisation energy/electron temperature, i.e. $y = E_{ion}/k_B T$.

The rate coefficient for collisional ionisation becomes

$$K_{pC} = 2\sqrt{\frac{2}{\pi m_0}}\left(\frac{1}{k_BT}\right)^{3/2} C_{ion}\, g_p \left(\frac{k_BT}{E_{ion}}\right) E_i\!\left(\frac{E_{ion}}{k_BT}\right). \tag{12.13}$$

This approximation for the collisional ionisation rate coefficient was first proposed by Lotz [68] and has been found to be reasonably accurate. A useful approximation [95] in units of $\text{cm}^3\ \text{s}^{-1}$ can be written

$$K_{pC} = 3 \times 10^{-6} g_p \frac{1}{(k_BT)^{3/2}} \frac{E_i(y)}{y} \tag{12.14}$$

where the temperature k_BT is measured in units of electron volts.

Detailed balance can be used to determine the inverse rate coefficient K_{Cp} for three-body recombination. In equilibrium the rate of collisional ionisation will be equal to the rate of three-body recombination. We have

$$n_e n_p K_{pC} = n_e n_{Z_i+1} K_{Cp}$$

where n_{Z_i+1} is the population density of the $Z_i + 1$ charged ion, n_p is the population density of the p state in the Z_i-charged ion and n_e is the electron density, which is assumed to have a Maxwellian distribution. Re-arranging and using the Saha-Boltzmann equation (Equation 1.29) for the equilibrium ratio of n_p/n_{Z_i+1} gives

$$K_{Cp} = \frac{g_p}{g_{Z_i+1}} \frac{n_e}{2} \left(\frac{h}{m_0}\right)^3 \left(\frac{m_0}{2\pi k_BT}\right)^{3/2} \exp\left(\frac{E_p}{k_BT}\right) K_{pC} \tag{12.15}$$

which, using Equation 12.12, can be written as

$$K_{Cp} = \frac{g_p}{g_{Z_i+1}} \frac{n_e}{2} \left(\frac{h}{m_0}\right)^3 \frac{m_0}{\pi^2} \frac{1}{(k_BT)^3} C_{ion} \left[\frac{E_i(y)}{y} e^y\right]. \tag{12.16}$$

The variation in square brackets is shown in Figure 12.1. We see that the three-body recombination rate coefficent varies $\propto n_e$ so that the rate of three-body recombination will vary as $\propto n_e^3$ for constant ionisation (as $n_{Z+1} \propto n_e$). The temperature dependence is also high. After allowing for the integration in the square bracket of Equation 12.16, the three-body recombination rate coefficient varies approximately $\propto 1/T^{1.5}$.

12.3 Collisional and Radiative Processes

The rate coefficients for collisional transitions can be combined with transition rates for radiative transitions to form rate equations for all the populating and de-populating processes affecting the population density of each quantum state. We have listed typical collisional transitions in Table 11.2 and determined appropriate rate coefficients for the collisional transitions in Section 12.1 (collisional excitation and de-excitation) and Section 12.2 (collisional ionisation and three-body recombination). Radiative transitions are listed in Table 11.1 and we determined transition probabilities and cross-sections for photo-excitation in Chapter 10 and photo-ionisation in Chapter 5.

We make some observations on the relative magnitude of the collisionally induced transition rates listed in Table 11.2. Electron collisionally induced transitions occur at a much greater rate than ion collisionally induced transitions, unless the gap between the energy levels is much smaller than the ion temperature $k_B T_i$. Similarly, the process of charge exchange is only important if the density of neutral atoms in the plasma is large. This sometimes occurs in tokamaks, for example.

At high densities, collisional processes (Table 11.2) tend to dominate over radiative processes (Table 11.1) as the collisional rates from a quantum state p increase as $n_e N_p$, while the radiative decay from a quantum state q only increases as N_q. For this reason, at a sufficiently high density, plasmas are in local thermal equilibrium. The principle of detailed balancing requires the inverse collisional rates to balance and if collisional processes dominate the population and de-population of the quantum states, the quantum states must have thermal populations.

At lower densities, it is, in general, necessary to consider the collisional and radiative processes populating and de-populating each quantum state. If we ignore the usually less important ion-collisional transitions and charge-exchange collisions, we can write a differential equation of the following form for each bound quantum state, here labelled p in an ion of charge Z_i. We have a rate equation for each quantum state of form

$$\begin{aligned}\frac{dN_p}{dt} = \sum_{q \neq p} \Bigg[&-N_p A_{pq} + N_q A_{qp} + \int \left(-N_p B_{pq} f_{pq}(\nu) + N_q B_{qp} f_{qp}(\nu) \right) W(\nu) d\nu \\ &+ n_e \left(-N_p K_{pq} + N_q K_{qp} \right) \Bigg] - n_e N_p K_{pC} + n_e n_{Z_i+1} K_{Cp} \\ &- N_p \int_{E_p/h}^{\infty} \sigma_{pf}(\nu) c W(\nu) d\nu + n_e n_{Z_i+1} A_{Cp} \end{aligned} \tag{12.17}$$

where the sum is over all bound states q (but not the state p). The collisional rate coefficients to and from the free-electron states are summed here by a single rate coefficient K_{pC} or K_{Cp}, where C represents all the continuum states. The Einstein coefficient for radiative transitions from the free-electron states (labelled C here again) is represented by a single coefficient A_{Cp}. The radiation energy between frequency ν and $\nu + d\nu$ is given by $W(\nu)$ and we use $f_{qp}(\nu)$ for the lineshape function for the q to p transition. The population of the next-higher ionisation stage is labelled n_{Z_i+1}. The cross-section for photo-ionisation from the p state is written as $\sigma_{pf}(\nu)$ and for the complete photo-ionisation rate is multiplied by the radiation intensity $cW(\nu)$ integrated over all frequencies corresponding to photon energies greater than the ionisation energy of the p state.

For an isolated atom or ion the number of discrete bound quantum states p extends to infinity with the degeneracy increasing as p increases. For atoms or ions in a plasma, the number of bound states is, however, finite as continuum lowering of the ionisation potential due to the presence of other ions within the spatial reach of the high p state wavefunction causes high-lying p states to become effectively free-electron states (see Section 13.7 for a detailed treatment of continuum lowering). In denser plasmas, continuum lowering of the ionisation potential can reduce the number of discrete quantum states considerably so that only the small number of remaining discrete quantum states needs to be explicitly considered for each ionisation stage.

In lower-density plasmas, the bound p states can extend to large principal quantum numbers so that some simplification of the system of Equation 12.17 is necessary. Discrete, excited quantum states with energies close to the ionisation energy of an ion have equilibrium populations as the states are close (in energy) to the thermalised free electrons and the density of states is high so that the populations of these highly excited quantum states are given by the Saha-Boltzmann equation. It is possible to truncate the number of equations which need to be solved. Only the lower quantum state densities need to be evaluated using rate equations. The populating contributions to the lower quantum states from high quantum states in Saha-Boltzmann equilibrium can be summed to a required accuracy by considering a finite number of the high quantum states as their contribution converges with increasing quantum number.

Collisional rates between quantum states are approximately constant with the energy difference ΔE between the states (see Equation 12.9), while radiative decay rates increase rapidly with energy differences between states ($\propto \Delta E^3$; see Equation 10.28 noting that $\Delta E = \hbar\omega$). In addition, both collisional and radiative rates are proportional to the overlap of the lower- and upper-state wavefunctions, which means that transitions where the principal quantum number n changes by one ($\Delta n = 1$) have both the largest collisional and radiative rates. Low-lying (small n) quantum states decay by $\Delta n = 1$ radiative transitions and high-lying (large n) states decay by $\Delta n = 1$ collisional transitions. There is an intermediate quantum state, say principal quantum number n^* where the collisional decay rate n^* to $n^* - 1$ equals the radiative decay rate. Due to the rapid collisional rate of transitions between states $p > n^*$, the quantum state n^* is a convenient boundary between higher states with populations given by the Saha-Boltzmann equation and lower states where it is necessary to solve rate equations such as Equation 12.17.

The state n^* where radiative and collisional rates are equal is referred to as the 'bottleneck' state as electron recombination from n^* to $n^* - 1$ implies that recombination will most likely continue down to the ground state of the ionisation stage due to the large radiative decay rate for states $p < n^*$. Recombination through n^* controls the rate of recombination to the ground state: the state n^* forms the 'bottleneck' controlling the maximum 'flow' of electrons from high-lying quantum states [83]. The state n^* is also referred to as the 'collision limit'.

Sample solutions of rate equations of the form of Equation 12.17 show that when a plasma initially in a steady state ($dN_p/dt = 0$ for all quantum states p) is perturbed, dN_p/dt for the excited quantum states return to zero much more quickly than dN_1/dt for the ground state. To calculate the excited quantum state densities, we can set $dN_p/dt = 0$ ($p > 1$). The excited quantum states are then said to be quasi-steady state and their populations can be found by solving linear equations (for the non-thermalised excited quantum states) and one differential equation (for the ground state) of each ionisation stage.

However, the ground state population: (1) may not have time to adjust to changing temperature and density conditions in the plasma; (2) may also be affected by macroscopic particle transport processes, such as diffusion.

Rapidly varying plasma parameters (condition 1) occur in, for example, laser-produced plasmas. Particle transport (condition 2) adds inflow and outflow terms to Equation 12.17 for the ground state of each ionisation stage and is particularly important in long-lasting, non-uniform plasmas such as in tokamaks. If plasma conditions remain stable for sufficient time for the ground state to approach equilibrium and the inflow minus outflow of ground state particles of the particular ionisation stage is small, the time derivative dN_1/dt for the ground state can be set equal to zero and the ionisation stage is steady state ($dN_p/dt = 0$ for all quantum states p).

Comprehensive computer codes solving the collisional-radiative equations represented by Equation 12.17 are available. Examples include the Atomic Data and Analysis Structure (ADAS), (see, for example, [41]) the FLYCHK code [18], the Astrophysical Plasma Emission Code (APEC) [101] and the Cloudy code [31].

12.4 The Escape Factor Approximation for the Effects of Radiation

The radiation field is sometimes taken into account by setting $W(\nu) = 0$ in Equation 12.17 and then by artificially modifying the radiative transition probabilities A_{qp} to $T_{qp}A_{qp}$, where, for example, the factor $T_{qp} = 1$ when absorption and stimulated emission for the particular transition are zero (or perhaps balanced) and $T_{qp} = 0$ when all radiation for the particular transition is absorbed. This factor T_{qp} is known variously as the trapping factor, the escape factor and in the astrophysical literature the net radiative bracket. For laboratory plasmas, the radiation field $W(\nu)$ is usually much smaller than that for a black body of temperature $\sim$ electron temperature and so the contributions to the quantum state populations caused by absorption and stimulated emission are small. T_{qp} is then close to unity for all transitions. The absorption of continuum radiation affecting radiative recombination has a smaller effect than the absorption of resonance line radiation and is usually neglected in the escape factor approximation.

Equation 12.17 can now be re-written to incorporate the modifications and simplifications we discussed. The excited states are quasi-steady state, but we still need to solve (in general) a differential equation for the ground state. We have

$$
\begin{aligned}
0 = & -\sum_{q<p} N_q T_{pq} A_{pq} + \sum_{p<q} N_q T_{qp} A_{qp} \\
& + n_e n_{Z_i+1} A_{Cp} - n_e N_p \left[\sum_{q \neq p} K_{pq} + K_{pC} \right] \\
& + n_e \left[\sum_{q \neq p} N_q K_{qp} + n_{Z_i+1} K_{Cp} \right] \qquad \text{for } p \neq 1 \\
\frac{dN_1}{dt} = & \sum_{q>1} N_q T_{q1} A_{q1} + n_e n_{Z_i+1} A_{C1} \\
& - n_e N_1 \left[\sum_{q \neq 1} K_{1q} + K_{1C} \right] \\
& + n_e \left[\sum_{q \neq 1} N_q K_{q1} + n_{Z_i+1} K_{C1} \right] + \Gamma_i \qquad \text{for } p = 1
\end{aligned} \tag{12.18}
$$

where n_{Z_i+1} is the population density of the next-higher ionisation state to that containing the p and q quantum states and Γ_i is the net flux of ground state ions caused by, for example, diffusion into the plasma volume under consideration. Tables of solutions of these equations for hydrogen-like and helium-like ions (i.e. ions with one or two captured electrons) have been published (see Bates et al. [8] for the first published table), but in many cases the equations are solved as part of large computer codes for particular plasma conditions [18].

There are several schemes to calculate the escape factors T [54]. Many involve an approximation of the form

$$
T = \frac{1 - e^{-\tau}}{\tau} \tag{12.19}
$$

where τ is an optical depth obtained by an integration of the absorption coefficient ($\tau = \int K dz$) over an appropriate path. Equation 12.19 has the correct asymptotic behaviour, tending to one for small optical depth τ and dropping to zero at large optical depth.

We have seen that the intensity of radiation (see Equation 6.4) varies with a constant source function $S_0 = \epsilon/K$ as

$$I = S_0(1 - e^{-\tau}).$$

Expanding the source function in terms of the emissivity ϵ and absorption coefficient K and multiplying by a length Δz over which these quantities are approximately constant, we can write for the intensity from a section of plasma of thickness Δz:

$$I = \frac{\epsilon \Delta z}{K \Delta z}(1 - e^{-\tau}) = \frac{I_0}{\tau}(1 - e^{-\tau})$$

where I_0 is a measure of the intensity arising from the plasma thickness of Δz without absorption. With the escape factor T defined as in Equation 12.19, the intensity decrease due to absorption is given by

$$I = I_0 T.$$

The escape factor T represents the radiation fraction escaping a local plasma element. To a first approximation, the absorption of the radiation not escaping causes an effective drop in the rate of emission, so multiplying the radiative transition probability by T in collisional-radiative calculations is appropriate.

12.5 Coronal Equilibrium

At sufficiently low densities, there is a regime of ionisation and quantum state population equilibrium known as coronal equilibrium or collisional ionisation equilibrium (sometimes referred to as CIE). The 'coronal equilibrium' title arises as plasmas with this type of equilibrium are observed emitting from the solar corona. At low densities radiative processes dominate collisional processes in de-populating quantum states and so collisional-radiative equations such as Equation 12.18 can be greatly simplified.

Radiative de-populating processes dominate collisional de-populating processes at sufficiently low densities as the collisional rates from a quantum state q increase as $n_e N_q$, while the radiative decay rates only increase as N_q. At sufficiently low density, the 'bottleneck' state discussed in Section 12.3 effectively extends to the continuum states, three-body recombination becomes negligible and radiative recombination to the ground state of an ion is the dominant recombination process. The ground state populations are given by a balance between collisional ionisation and radiative free-bound recombination. Using the notation for the population densities and rate coefficients introduced in this chapter, we can write for two ionization stages Z_i and $Z_i + 1$ in steady-state coronal equilibrium:

$$n_e n_{Z_i} K_{1C} = n_e n_{Z_i+1} A_{C1} \tag{12.20}$$

so that the ratio of the populations of the ground states of the $Z_i + 1$ and Z_i charged ions is

$$\frac{n_{Z_i+1}}{n_{Z_i}} = \frac{K_{1C}}{A_{C1}}. \tag{12.21}$$

In coronal equilibrium, the de-exciting processes are dominated by radiative processes which are not balanced by equivalent exciting processes, so the ground state densities of the ions are much greater than excited quantum state densities. As the populations of excited states are small compared to the ground state, we can neglect contributions to the ground state density due to transitions from excited quantum states.

Values for the coronal ionisation balance can be obtained using expressions previously derived for K_{1C} (Equation 12.12) and A_{C1} (Equation 5.22). We have

$$\frac{n_{Z_{ie+1}}}{n_{Z_i}} = \frac{K_{1C}}{A_{C1}} = \frac{\sqrt{3}}{2\pi} m_0 C_{ion} g_{Z_i} \left(\frac{4\pi\epsilon_0}{e^2}\right)^3 \frac{3c^3}{4} \frac{\hbar n^3}{2R_d Z_i^4} \Theta_C(T) \tag{12.22}$$

where the temperature-dependent components are grouped together and given by

$$\Theta_C(T) = \frac{1}{\tilde{G}_{fb}} \frac{k_B T}{E_{ion}} \exp\left(-\frac{E_{ion}}{k_B T}\right). \tag{12.23}$$

Here g_{Z_i} is the degeneracy of the ground state of principal quantum number n of the Z_i charged ion and C_{ion} is a constant introduced in our discussion of collisional ionisation with value between 2.6 and 4.5×10^{-18} m^2 (eV)^2 [68]. The frequency-averaged Gaunt factor $\tilde{G}_{fb} \approx 1$ for free-bound recombination has a small variation with temperature.

The excited quantum state populations in coronal equilibrium are small, but still produce radiation emission. The excited state populations are determined by a balance between collisional excitation from the ground state and radiative decay. We have:

$$n_e N_1 K_{1q} = N_q \sum_{p'<q} A_{qp'} \tag{12.24}$$

where the summation is over all energy levels $p' < q$. The emission coefficient for the transition q to p into 4π steradians is

$$\begin{aligned}\varepsilon &= N_q A_{qp} h\nu_{qp} \\ &= n_e N_1 K_{1q} h\nu_{qp} \frac{A_{qp}}{\sum\limits_{p'<q} A_{qp'}}\end{aligned} \tag{12.25}$$

where $h\nu_{qp}$ is the photon energy for the transition q to p.

When the radiative decay of the spectral line $q \rightarrow p$ is much more likely than from q to any other levels, we have that $A_{qp} \approx \sum\limits_{p'<q} A_{qp'}$. Equation 12.25 reduces to the much simpler form

$$\varepsilon = n_e N_1 K_{1q} h\nu_{qp}. \tag{12.26}$$

The intensity of the spectral line is then directly proportional to the rate coefficient for collisional excitation.

Excited quantum states with small transition probabilities can exist in coronal equilibrium with the ground state provided collisional de-excitation is sufficiently small. Such states are referred to as metastable if their radiative lifetimes are much longer than other excited states. The population of an excited metastable state increases in coronal equilibrium above the populations in states where fast radiative decay can occur. Typically Equation 12.26 is then valid with the collisional excitation rate coefficient determining the spectral emission coefficient.

The emission coefficient for a metastable state can be generalised further from Equation 12.26 to consider any populating process, not just collisional excitation from the ground state. If the rate of population per unit time of a metastable state q is R and there is only the slow radiative de-populating process (with transition probability A_{qp}), there is a rate balance with $R = N_q A_{qp}$. The emission coefficient for radiative decay from the metastable state is then given by

$$\epsilon = N_q A_{qp} h\nu_{qp} = R h\nu_{qp}. \tag{12.27}$$

Populating processes contributing to R may include allowed radiative recombination from a higher quantum state to the metastable state, photo-excitation or, as discussed above, collisional excitation.

The upper state of the 21 cm transition between the ground state hyperfine states of hydrogen discussed in Section 7.6 is an example of a metastable state. The upper state has the electron and nuclear spins aligned, and in the lower state they are anti-parallel to each other. Due to the small energy difference between the two states (5.9×10^{-6} eV) and because the transition is forbidden and only occurs due to a magnetic dipole interaction (see Section 10.1.4), the transition probability between the two hyperfine states is small ($2.85 \times 10^{-15}\ \mathrm{s}^{-1}$). When collisional processes are significant, the population of the upper hyperfine state decays by collisions to the lower hyperfine state and 21 cm radiation emission is small due to the small 21 cm radiative transition probability. However, in hydrogen clouds within a galaxy, for example, the average electron density is sufficiently small (typically 0.2–0.5 cm^{-3}) that electron collisional de-excitation between the hyperfine states is even smaller. The upper hyperfine state population of hydrogen can increase significantly with the emission coefficient given by Equation 12.27. The populating pathways for the hydrogen hyperfine upper state include radiative emission from the higher quantum states of hydrogen and absorption of 21 cm radiation [47]. Emission and absorption at 21 cm is used by radio astronomers to map hydrogen density. Radio astronomy at 21 cm established the evidence for our galaxy being a spiral [10], enables measurements of the early universe during the formation of the first galaxies [74] and established the existence of dark matter in rotating galaxies [119].

12.6 Dielectronic Recombination and Auto-Ionisation

An interesting type of spectral line known as a dielectronic satellite line is emitted from a plasma when there are two electrons in excited quantum states. One of the electrons undergoes a transition resulting in the emission of a photon, while the other (known as the spectator electron) remains in its original quantum state. The energy levels of the quantum states are affected by the presence of the spectator electron, but much less than if the spectator electron was in the ground state. The doubly excited or dielectronic quantum state energies are shifted by only a small amount (to lower energy) from the energies of the ion with the spectator electron completely removed, that is, the next-higher ionisation stage. Consequently, the dielectronic lines are satellites of the lines of the next-higher ionisation stage appearing at slightly higher wavelength.

The doubly excited satellite lines of hydrogen-like ions, with two excited electrons, and helium-like ions where there is a single electron in the $n = 1$ shell and two excited electrons have been particularly studied. Due to the large number of quantum states possible for a spectator electron, there are a large number of dielectronic satellite lines. A widely used lettering system $(a, b, c, \ldots)$ to designate different satellite lines to the helium-like 1s2p$\rightarrow$1s^2 line was introduced in the first comprehensive treatment of dielectronic satellite line intensities [37]. The initially singly excited emission line is often referred to as the resonance line and in the lettering description designated as w.

Doubly excited ions are affected by additional collisional and radiative processes to those listed in Tables 11.1 and 11.2. The most important reactions of doubly excited ions are listed in Table 12.1 and schematically illustrated in Figure 12.2. Dielectronic recombination occurs when a free electron recombines to an excited bound quantum state and a bound electron (usually in the ground state) is excited to an excited state. The energy released in the dielectronic recombination of the free

Table 12.1. *Collisional and radiative processes affecting doubly excited ions.*

Auto-ionisation of a double excited state to produce an ion in the next ionisation stage and the reverse process of dielectronic recombination	$A^{Z_i+}_{pp''} \rightleftharpoons A^{Z_i+1} + e$
Inner shell excitation of an electron from a complete shell by a free electron to produce a doubly excited ion	$A^{Z_i+}_{l} + e \rightarrow A^{Z_i+}_{pp''} + e$
Radiative decay of a doubly excited ion to form a singly excited ion	$A^{Z_i+}_{pp''} \rightarrow A^{Z_i+}_{p''} + h\nu_{bb}$

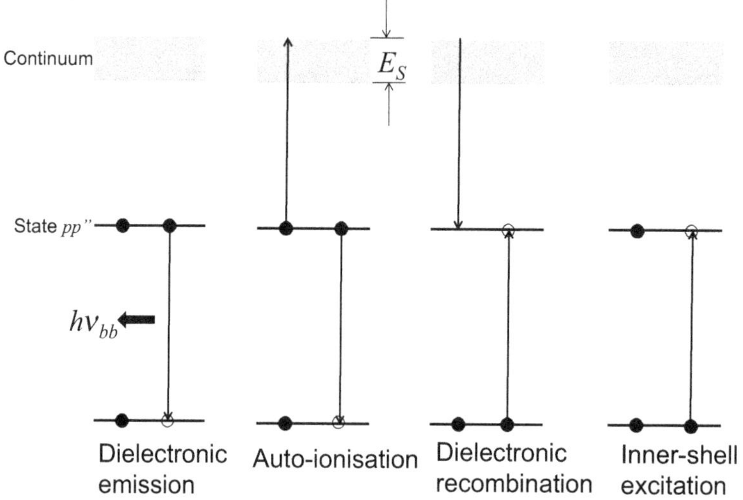

Figure 12.2 Schematic energy-level diagram showing the labelled transitions associated with doubly excited states. Filled states are indicated by solid dots, while vacant quantum states are shown as hollow dots.

electron is equal to the energy gained in exciting the initially bound electron. Due to the large number of doubly excited quantum states, the process of dielectronic recombination can have a significant effect on the degree of ionisation in plasmas.

Auto-ionisation is the inverse process to dielectronic recombination. In auto-ionisation, an ion with two electrons in excited states undergoes transitions where one excited electron decays to a lower level and the other excited electron is ejected from the ion to the continuum of free electrons. The ejected electron has a kinetic energy (which we designate as E_S) such that there is conservation of energy in the auto-ionisation process. Auto-ionisation is also known as Auger decay, where the term is usually applied to the non-radiative decay as illustrated in Table 12.1 of excited neutral atoms.

As they are inverse processes, there is a detailed balance relationship between the rates of dielectronic recombination C_{DR} and auto-ionisation A_A. We can write for the rate of change of the population $n_{pp''}$ of a doubly excited state due to auto-ionisation and dielectronic recombination that

$$\frac{dn_{pp''}}{dt} = -n_{pp''}A_A + n_{Z_i+1}C_{DR} \tag{12.28}$$

assuming that the double excited state is in a Z_i-charged ion and n_{Z_i+1} is the population density of the ground state of the $Z_i + 1$ charged ion. The rate of dielectronic recombination C_{DR} can be expressed in terms of the auto-ionisation rate A_A by assuming a steady-state balance in Equation 12.28 (so that $dn_{pp''}/dt = 0$) and using the population ratio $n_{pp''}/n_{Z_i+1}$ as found for thermal equilibrium. Using the Saha-Boltzmann equation for the population ratio in thermal equilibrium (Equation 1.29), we have

$$C_{DR} = \frac{n_{pp''}}{n_{Z_i+1}}A_A = \frac{g_{pp''}}{g_{Z_i+1}}\frac{n_e}{2}\left(\frac{h}{m_0}\right)^3\left(\frac{m_0}{2\pi k_B T}\right)^{3/2}\exp\left(-\frac{E_S}{k_B T}\right)A_A. \tag{12.29}$$

Here the potential energy of the doubly excited state in the Z_i-charged ion is higher than the potential energy of the Z_i+1 ion ground state by the energy E_S corresponding to the kinetic energy of the free electron created in the auto-ionising process.

At low densities far from thermal equilibrium, the three processes, dielectronic emission, auto-ionisation and dielectronic recombination, can be in balance and the inner-shell excitation rate and other processes populating and de-populating the double excited states are small. This is a particular example of coronal equilibrium discussed in Section 12.5. We can write that the total rate of dielectronic satellite emission (coefficient $\sum A_R$) and auto-ionisation (coefficient $\sum A_A$) from the double excited $n_{pp''}$ state to the ionic ground state is balanced by the rate of dielectronic recombination C_{DR} of electrons with the $Z_i + 1$-charged ion. We have

$$n_{pp''}\left(\sum A_R + \sum A_A\right) = n_{Z_i+1}C_{DR}.$$

The population density of the double excited state in this non-equilibrium situation is consequently

$$n_{pp''} = \frac{n_{Z_i+1}C_{DR}}{\sum A_R + \sum A_A}$$

and the emission coefficient of the dielectronic satellite line is given by

$$\epsilon_R = n_{pp''}A_R h\nu_{bb} = \frac{n_{Z_i+1}A_R C_{DR}}{\sum A_R + \sum A_A}h\nu_{bb} \tag{12.30}$$

where A_R is the rate coefficient for dielectronic recombination for the particular dielectronic transition to the ground state under investigation. Substituting Equation 12.29 for C_{DR}, we obtain for the dielectronic satellite line emission coefficient

$$\epsilon_R = N_{Z_i+1} n_e F_1(T) F_2 \tag{12.31}$$

where the temperature-dependent parameters are grouped in the function $F_1(T)$ given by

$$F_1(T) = \left[\frac{1}{2} \left(\frac{h}{m_0} \right)^3 \left(\frac{m_0}{2\pi k_B T} \right)^{3/2} \exp\left(-\frac{E_S}{k_B T} \right) \right] \tag{12.32}$$

and the atomic physics parameters for the particular doubly excited transition to the ground state are grouped in the function F_2 with

$$F_2 = \left[\frac{g_{pp''}}{g_{Z_i+1}} \frac{A_R A_A}{\sum A_R + \sum A_A} h\nu_{bb} \right]. \tag{12.33}$$

The emission coefficient arising from dielectronic satellite lines is seen to be composed of components varying with the electron density n_e and density n_{Z_i+1} of ions in the next ionisation stage and temperature. For ease of use in deducing satellite-line intensities, the parameters F_2 (and E_S) are often tabulated (see, for example, Bombarda et al. [13]).

The summations in the denominator of Equation 12.33 are over all transitions where dielectronic satellite emission ($\sum A_R$) and auto-ionisation ($\sum A_A$) can occur, but often the decay to the ground state is dominated by the rate coefficients (A_R and A_A) for the particular dielectronic satellite line under consideration. Auto-ionisation rate coefficients (A_A) are approximately independent of atomic number in ions of the same iso-electronic series (e.g. hydrogen-like or helium-like), while radiative decay rate coefficients (A_R) increase rapidly with the ion charge as Z_i^4 (see Equation 5.22). In low atomic number ions of the same iso-electronic series, the auto-ionisation rate coefficient (A_A) is much greater than the dielectronic rate coefficient (A_R) and so the A_A rates in the numerator and denominator of Equation 12.33 cancel and $F_2 \propto A_R$.

The form of Equation 12.32 shows that all dielectronic satellite line intensities when there is a balance between dielectronic emission, auto-ionisation and dielectronic recombination are similarly affected by temperature changes, largely varying $\propto 1/T^{3/2}$. Typically the value of the free-electron energy created in auto-ionisation $E_S < k_B T$, so that $\exp(-E_S/k_B T) \approx 1$. We have seen that collisional excitation of a singly excited electron varies with temperature as $\propto 1/T^{1/2}$ (see Equation 12.8). In coronal equilibrium, where collisional excitation is immediately followed by radiative decay (see Section 12.5), the singly ionised state consequently radiatively emits at a rate $\propto 1/T^{1/2}$. With the emission of the dielectronic satellite lines occuring at a rate $\propto 1/T^{3/2}$ (Equation 12.32), the relative intensity of the satellite lines to the singly excited state emission (the resonance line) has a variation $\propto 1/T$. The satellite line to resonance line intensity ratio is, consequently, a useful diagnostic of electron temperature for plasmas in coronal equilibrium (as found, for example, in solar flares [85] and in tokamak plasmas [13]).

12.7 Criteria for LTE

Complete thermodynamic equilibrium occurs when the populations of ion quantum states are in equilibrium and the radiation field is in equilibrium with them. With many interactions, the radiation field is then Planckian, the ionisation balance is given by the Saha-Boltzmann equation and population-density ratios are given by the Boltzmann ratio.

In LTE, the populations of quantum states are given by the Boltzmann ratio and the degree of ionisation is determined by the Saha-Boltzmann equation. The radiation field is usually not in equilibrium (i.e. it is not a Planckian black-body distribution). In LTE the plasma can be optically thin to emitted radiation. In LTE, the equilibrium populations are achieved because collisional processes (excitation, ionisation and their inverses) dominate the radiative processes. Inverse collisional processes have a detailed balance relationship to each other (see, for example, Section 12.1), so if no other process occurs apart from the detailed balance process, the populations have the equilibrium values for the particular density and temperature of the plasma.

The transition probability A_{qp} for radiative decay scales with the energy gap ΔE between the quantum states proportionally to ΔE^3 (see Equation 10.28 noting that $\Delta E = \hbar\omega_0$), while rate coefficients for collisional processes are approximately constant with the energy gap ΔE (see Equation 12.9). Both radiative and collisional processes are larger when the initial and final quantum state wavefunctions overlap and hence are large when the principal quantum number n changes by $\Delta n = 1$. As discussed in Section 12.3, the two very different collisional and radiative scalings cause larger energy gaps to have radiative transitions between bound quantum states dominating collisional de-excitation. Larger energy quantum steps occur typically in the lower-energy quantum states, so the higher states (closer to the continuum), which are more closely spaced, have populations determined by collisional excitation and de-excitation processes. Due to the balanced collisional processes with little radiative decay, the higher energy, more closely spaced states are in equilibrium with each other with populations determined by the Boltzmann ratio.

High-energy states are also in equilibrium with the ground state of the next-higher ionisation stage. Exercise 12.6 shows that collisional three-body recombination dominates over radiative recombination at high principal quantum number n (there is n^7 scaling for the ratio of collisional to radiative recombination to hydrogen-like ions). The balance between collisional ionisation and three-body recombination ensures that high n quantum states are also in Saha-Boltzmann equilibrium with the ground state of the next ionisation stage.

As high principal quantum number states have populations given by the Saha-Boltzmann equation, it is only necessary to calculate the populations using rate equations (for example, Equation 12.17) for the lower-energy states below the 'bottleneck' state (see Section 12.3), where the rates of radiative decay and collisional de-excitation between quantum states are equal. LTE between the ground states of different ionisation stages can be said to occur at densities sufficient for the ground state to be the 'bottleneck' state.

Radiative transition probabilities A_{qp} for dipole-allowed transitions scale proportionally to the square of the matrix element integral $|\int \psi_p^* r \psi_q dV|^2$, where ψ_p and ψ_q are the respective lower and upper quantum states involved in the transition (see Equation 10.28). As discussed above, the radiative transition probabilities also scale proportionally to ΔE^3. The rate coefficients for collisional de-excitation have a similar proportionality to the matrix element integral. Taking the ratio of the rate of collisional de-excitation $N_q n_e \int \sigma v \hat{f}_v(v) dv$ to the radiative decay rate $N_q A_{qp}$, we have that the matrix element integral and the upper-state density N_q cancel. Here $\hat{f}_v(v)$ is the Maxwellian speed distribution and σ is the cross-section for collisional de-excitation. The collisional rate/radiative decay rate ratio becomes independent of the detailed atomic physics of the particular transition and varies proportionally to $n_e/(\sqrt{T_e}\,\Delta E^3)$.

McWhirter [88] noticed the scaling for the collisional rate/radiative decay rate and proposed that it should be more than ten in order to ensure LTE between quantum states. Once numerical factors are included, this gives a criterion for LTE of a minimum electron density. We have as the condition for LTE to exist between two quantum states of energy separation ΔE that

$$n_e > 1.7 \times 10^{14} \sqrt{T_e}\,\Delta E^3 \tag{12.34}$$

where the energy gap ΔE and electron temperature T_e are measured in electron volts and the electron density n_e is measured in cm^{-3}.

A further criterion for LTE is that there is sufficient time for the populations to equilibrate. The necessary times for LTE to be achieved with respect to the ground state, however, are typically short (<200 fs) at the necessary densities for the collisional processes to dominate radiative decay [3].

12.8 Spectral Line Intensity Ratios

As the volume of plasma associated with emission and absorption from both laboratory and astrophysical plasmas is not often known accurately and spectrometers are often not accurately calibrated to measure absolute intensities, plasma parameters are usually best measured using the relative intensity of emission or absorption. In the exercises at the end of this chapter (Exercises 12.4 and 12.5), we show that, in both coronal equilibrium and in LTE, the ratio of the spectral line intensities of strongly emitting allowed transitions can be used to determine the electron temperature as there is an $\exp(-\Delta E/k_B T)$ dependence on temperature. Here, ΔE is the energy difference between the energies of the upper quantum states for the spectral lines.

In Section 12.6, the intensity ratio of dielectronic satellite lines to the resonance line is shown to have a $1/T$ dependence and hence the potential for the measurement of temperature. A technique for measuring temperature using the emitted intensity ratio of a spectral line in the same iso-electronic series from two nearby atomic number elements has been utilised [71]. This temperature measurement depends on the accuracy of code calculations to evaluate the degree of ionisation (as discussed in Section 12.9).

Plasma density measurement from spectral line emission ratios is possible where there is an effect due to electron density on one spectral line, which is different to the effect on another. For a plasma in coronal equilibrium, quantum states which are metastable to radiative decay increase in population so that the emission is still determined by the rate of collisional excitation and given by, for example, Equation 12.26. The collisional excitation rate can readily be significant (see Section 11.6), but the radiative decay rate is small. With increasing density, collisional transitions to other excited quantum states and the ground state start to de-populate the excited metastable state. At some density, the population of the metastable state will be given by the Saha-Boltzmann equation, with a much-reduced relative population and a much-reduced spectral line emission (relative to the emission from a non-metastable quantum state). The change from the coronal metastable emission to the reduced relative emission in LTE is density dependent so that comparing the metastable line intensity to a reference strongly emitting allowed transition gives a strong electron density variation.

In the emission spectra of moderate to high atomic number ions isoelectronic with helium (helium-like ions), there are metastable quantum states, such as 1s2s $^3S_1$, 1s2p $^3P_1$ and 1s2p $^3P_2$ (see Section 8.4). Radiative decay to the ground state $1s^2$ $^1S_0$ is metastable as angular momentum can only be conserved if the electron spin changes during the transition, which considerably reduces the radiative decay rate (see the discussion in Section 10.2). Transition to the ground state from these quantum states in helium-like ions is 'forbidden', but not 'strictly forbidden'. The 1s2s $^3S_1$ – $1s^2$ $^1S_0$ transition is usually referred to as the forbidden line (line z in Gabriel [37]), while transitions 1s2p ^3P – $1s^2$ $^1S_0$ are referred to as intercombination lines (lines x and y in Gabriel [37]). Both the intercombination and forbidden lines are satellites (along with the dielectronic satellites; see Section 12.6) at slightly higher wavelengths to the resonance 1s2p $^1P_1$ – $1s^2$ $^1S_0$ line (line *w* following Gabriel [37]). The ratio of these intercombination and forbidden lines to the resonance line of helium-like ions has been shown to be an indicator of electron density in the range where the upper quantum state populations change from the coronal value to LTE values [11, 12].

The intensity ratio of the two fine-structure components of the Lyman alpha line ($2p_{3/2}$ – 1s and $2p_{1/2}$ – 1s) of hydrogen-like ions have been utilised as a density diagnostic. In the coronal regime at low density, the rate of collisional excitation from the ground state into both $j = 3/2$ and $1/2$ is proportional to the degeneracy of the upper quantum state, so following Equation 12.26, the intensity ratio $\epsilon_{j=3/2}/\epsilon_{j=1/2} = 2$. In LTE, the populations are proportional to the degeneracy of the upper quantum state, so again $\epsilon_{j=3/2}/\epsilon_{j=1/2} = 2$. However, at densities intermediate between coronal and LTE populations, electron and ion collision-induced transitions between the fine-structure sub-levels result in the ratio $\epsilon_{j=3/2}/\epsilon_{j=1/2}$ varying in the range one to two depending on the electron density [11, 107].

12.9 The Average Ionisation

In Chapter 1 we discussed how the average degree of ionisation $Z_{av} = n_e/n_i$ allows (i) fluid codes to relate the material density ρ and pressure P, and (ii) plays a role in determining the sound speed in a plasma which in turn determines the propagation speed of shock waves and the free-expansion velocity of a plasma. It is worthwhile to return to consider how collisional-radiative calculations determine Z_{av}.

The material density ρ is related to the total ion density n_i for an ion of atomic number A by

$$\rho = Am_p n_i$$

where m_p is the mass of a proton. The calculation of the average ionisation typically involves an evaluation of the electron density given a value of material density and the temperature, though a diagnostic measurement of the electron density of a plasma may require the calculation of the degree of ionisation and hence material density.

The electron and total ion densities can be determined from the individual quantum state populations by adding all the populations. We have

$$n_e = \sum_{Z_i=1}^{Z} Z_i \sum_p N_p(Z_i) \tag{12.35}$$

$$n_i = \sum_{Z_i=0}^{Z} \sum_p N_p(Z_i) \tag{12.36}$$

where $N_p(Z_i)$ represents the population density in the pth quantum state of the $+Z_i$ charged ion. The atomic number of the material and hence maximum ion charge is Z. The Saha-Boltzmann equation gives the ratio of the population of two quantum states in adjacent ionisation stages if equilibrium or LTE conditions are valid (see Section 1.29). In this chapter (see Equation 12.21) we have also seen that coronal equilibrium calculations similarly enable the determination of a relative population ratio. However, we need absolute values of the ion populations (not relative populations) to evaluate, for example, Equation 12.36.

The summation of populations over the individual quantum states within an ionisation stage can be represented by a sum over the ratio relative to the ground state. We have

$$\sum_p N_p(Z_i) = N_1(Z_i) \sum_p \frac{N_p(Z_i)}{N_1(Z_i)}. \tag{12.37}$$

In coronal equilibrium the excited state populations are small compared to the ground state populations and can usually be neglected so that

$$\sum_p N_p(Z_i) = N_1(Z_i).$$

In LTE, the quantum state populations are given by summing the Boltzmann ratios (see Equation 1.27):

$$\sum_p N_p(Z_i) = N_1(Z_i) \sum_p \frac{g_p(Z_i)}{g_1(Z_i)} \exp\left(-\frac{E_p(Z_i) - E_1(Z_i)}{k_B T}\right) \tag{12.38}$$

where $g_p(Z_i)$ is the degeneracy of the p state in the Z_i charged ion and $E_p(Z_i)$ is the ionisation energy of the p state in the Z_i charged ion.

The relative ground state populations can all be related to, say, the singly ionised or neutral populations so that

$$\frac{N_1(Z_i)}{N_1(1)} = \frac{N_1(Z_i)}{N_1(Z_i-1)} \frac{N_1(Z_i-1)}{N_1(Z_i-2)} \cdots \frac{N_1(2)}{N_1(1)} = \prod_{Z_i'=1}^{Z_i-1} \frac{N_1(Z_i'+1)}{N_1(Z_i')},$$

$$\frac{N_1(Z_i)}{N_1(0)} = \frac{N_1(Z_i)}{N_1(Z_i-1)} \frac{N_1(Z_i-1)}{N_1(Z_i-2)} \cdots \frac{N_1(1)}{N_1(0)} = \prod_{Z_i'=0}^{Z_i-1} \frac{N_1(Z_i'+1)}{N_1(Z_i')}.$$

We can write for the ground state ion density of a Z_i charged ion

$$N_1(Z_i) = N_1(0) \left(\prod_{Z_i'=0}^{Z_i-1} \frac{N_1(Z_i'+1)}{N_1(Z_i')} \right). \tag{12.39}$$

Equations 12.39 and 12.36 can be solved iteratively by assuming a value of electron density n_e and $N_1(1)$ or $N_0(1)$ for the first evaluation. and then multiplying by the appropriate factor to get the known electron density n_e or, if the material density ρ is known, by the known value of $n_i/(Am_p)$.

In an equilibrium plasma or where LTE applies to all quantum states, the ratios $N_1(Z_i'+1)/N_1(Z_i')$ of the ground state populations are given by the Saha-Boltzmann expression (Equation 1.29). The Saha-Boltzmann equation derived in Section 1.4.1 represents the high-density limit of the ratio of the ionisation stages. In the present notation

$$\frac{N_{Z_i'+1}}{n_{Z_i'}} = \frac{g_{Z_i'+1}}{g_{Z_i'}}\frac{2}{n_e}\left(\frac{m_0}{h}\right)^3\left(\frac{2\pi k_B T}{m_0}\right)^{3/2}\exp\left(-\frac{E_{ion}(Z_i')}{k_B T}\right) \tag{12.40}$$

where $E_{ion}(Z_i')$ is the ionisation energy of the Z' charged ion. We can write for the Saha-Boltzmann equation that

$$\frac{N_{Z_i'+1}}{n_{Z_i'}} = \frac{g_{Z_i'+1}}{g_{Z_i'}}\frac{C_{SB}(k_B T)^{3/2}}{n_e}\exp\left(-\frac{E_{ion}(Z_i')}{k_B T}\right) \tag{12.41}$$

where C_{SB} is a constant of proportionality which is independent of the ionisation.

The ionisation ratio in the low-density coronal equilibrium situation has been considered for Equation 12.21. We have been using Equation 12.22 that

$$\frac{n_{Z_i'+1}}{n_{Z_i'}} = \frac{K_{1C}}{A_{C1}} = C_C\frac{g_{Z_i'}\, k_B T\ (n(Z_i'))^3}{Z_i'^4\, E_{ion}(Z_i')}\exp\left(-\frac{E_{ion}(Z_i')}{k_B T}\right) \tag{12.42}$$

where C_C is a constant of proportionality which is independent of the ionisation, and $g_{Z_i'}$ is the degeneracy of the ground state of principal quantum number $n(Z_i')$ of the Z_i' charged ion. In the coronal limit, the ionisation ratios are independent of the electron density n_e, but have a similar exponential dependence on the temperature $\propto \exp\left(-E_{ion}(Z_i')/k_B T)\right)$ as the LTE limit.

For both coronal and equilibrium evaluations of the product in Equation 12.39, the exponential dependence causes the product to rapidly approach zero for $E_{ion}(Z_i') \gg k_B T$. With the dependence of the ionisation ratios for LTE given by Equations 12.41 and coronal equilibrium given by Equation 12.42, the product in Equation 12.39 can be expanded. For LTE conditions the ground state density of the Z_i charged ion is given by

$$N_1(Z_i) = N_1(1)\frac{g_{Z_i}}{g_1}\left(\frac{C_{SB}(k_B T)^{3/2}}{n_e}\right)^{Z_i-1}\exp\left(-\sum_{Z_i'=1}^{Z_i'=Z_i-1}\frac{E_{ion}(Z_i')}{k_B T}\right), \tag{12.43}$$

while in the coronal equilibrium limit, the ground state density of the Z_i charged ion is given by

$$N_1(Z_i) = N_1(1)\, G_C\, (C_C\, k_B T)^{Z_i-1} \exp\left(- \sum_{Z_i'=1}^{Z_i'=Z_i-1} \frac{E_{ion}(Z_i')}{k_B T} \right) \tag{12.44}$$

where

$$G_C = \left(\prod_{Z_i'=1}^{Z_i'=Z_i-1} g_{Z_i'} \frac{(n(Z_i'))^3}{Z_i'^4 E_{ion}(Z_i')} \right).$$

For both LTE and coronal equilibrium, these evaluation of $N_1(Z_i)$ have a sharply peaked maximum at a particular charge Z_i. For coronal equilibrium the peak ionisation Z_i corresponds approximately to where the ionisation energy $E_{ion}(Z_i-1) \approx k_B T$ (see Figure 12.3). For LTE, the peak ionisation is dependent on the electron density with typical charges Z_i of maximum population with $E_{ion}(Z_i) \gg k_B T$. An example of the variation of the ionisation of a plasma with temperature for LTE conditions is shown in Figure 12.4. Typically only one to three ionisation stages have significant populations at a particular temperature.

Ionisation stages with full, closed-shell electron configurations, such as helium-like and neon-like ions, have ionisation energies which are much greater than slightly less ionised ions with, say, an additional electron, but have comparable ionisation energy to slightly more ionised ions with, say, one less bound electron.

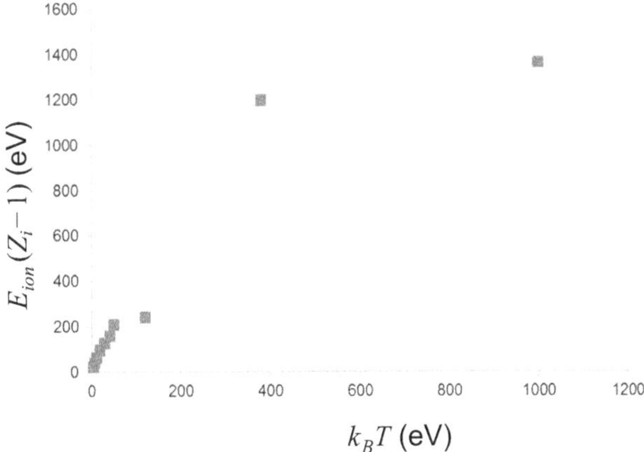

Figure 12.3 The ionisation energy $E_{ion}(Z_i - 1)$ for neon ions corresponding to the ionisation one below the peak charge Z_i as a function of the electron temperature $k_B T$ where the charge Z_i is the most abundant. The neon ion abundances are calculated for coronal equilibrium (see Abels-van Maanen [114]) with the ionisation energies given by Kramida [63].

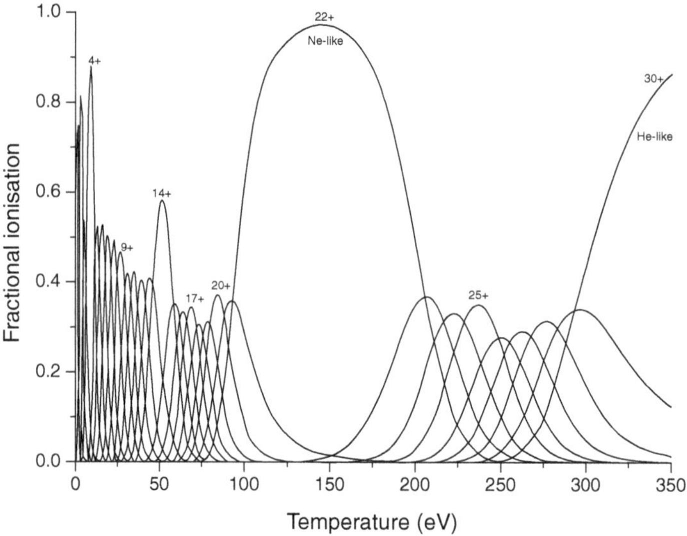

Figure 12.4 The fraction of different ionisation stages as a function of temperatures for a germanium plasma in LTE at an electron density of 10^{21} cm^{-3}. (Taken from A. K. Rossall, PhD thesis, University of York, 2011).

Closed-shell electron configurations, consequently, represent the dominant ionisation stage for a broad temperature range. For example, a broad temperature range for the high abundance of neon-like germanium is apparent in Figure 12.4.

Exercises

12.1 Following Equation 11.30, the cross-section for collisional excitation by an electron of energy E from quantum states p to q can be written as

$$\sigma_{pq} = \left| \int \psi_p^* \psi_q r' dV' \right|^2 \frac{8\pi}{a_0^2 k_1^2} \frac{1}{2} \ln \Lambda$$

where $k_1^2 = 2m_0 E/\hbar^2$. The rate coefficient for this collisional excitation is given by

$$K_{pq} = \sqrt{\frac{2}{m_0}} \int_{\Delta E}^{\infty} \sigma_{pq}(E)\, E^{1/2} \hat{f}_E(E) dE$$

where ΔE is the energy gap between the p and q states. Show that if the $\ln \Lambda$ term is constant with electron energy E that the rate coefficient can be written as

$$K_{pq} = 4\sqrt{\frac{2\pi}{m_0} \frac{\hbar^2}{m_0 a_0^2}} \left| \int \psi_p^* \psi_q r' dV' \right|^2 \left(\frac{1}{k_B T} \right)^{1/2} \exp\left(-\frac{\Delta E}{k_B T} \right) \ln \Lambda.$$

12.2 Show that the rate coefficient K_{pq} obtained in the previous question can be written in terms of the absorption oscillator strength f_{pq} as

$$K_{pq} = 12\sqrt{\frac{2\pi}{m_0}\frac{\hbar^4}{m_0^2 a_0^2}}\frac{1}{\Delta E}\left(\frac{1}{k_B T}\right)^{1/2}\exp\left(-\frac{\Delta E}{k_B T}\right) f_{pq}\ln\Lambda.$$

12.3 The formula for the collisional excitation rate coefficient published by van Regemorter [115] is written in units of cm^3 s^{-1} with the energy gap between the p and q quantum states ΔE and the electron temperature $k_B T$ measured in electron volts. Use the answer to Exercise 12.2 to show that in these units we have

$$K_{pq} = 3 \times 10^{-7}\frac{R_d^{3/2}}{(k_B T)^{1/2}\Delta E}\exp\left(-\frac{\Delta E}{k_B T}\right) f_{pq}\ \ln\Lambda$$

where $R_d = 13.6\,\text{eV}$ is the Rydberg energy. This equation is identical to Equation 12.10 taken from the van Regemorter paper if we exchange the Gaunt factor G_{pq} by $\ln\Lambda$ (where $\ln\Lambda$ is known as the Coulomb logarithm).

12.4 Consider the ratio of emission coefficients from two excited states (1 and 2) in LTE for radiative transitions back to the ground state. If the degeneracies of the two states are g_1 and g_2 and the states are separated in energy such that state 2 is ΔE higher in energy than state 1, show that the emission coefficient ratio $\epsilon_{20}/\epsilon_{10}$ for decay to the ground state (labelled 0) is given by

$$\frac{\epsilon_{20}}{\epsilon_{10}} = \frac{g_2 A_{20}\,\nu_{20}}{g_1 A_{10}\,\nu_{10}}\exp\left(-\frac{\Delta E}{k_B T}\right)$$

where the radiative transition probabilities from state 2 and 1 to the ground state are respectively A_{20} and A_{10} and the frequencies of emission of these transitions are respectively ν_{20} and ν_{10}.

12.5 Consider when the emission for the states of Exercise 12.4 is determined by coronal equilibrium such that collisional excitation to states 1 and 2 is balanced by radiative decay back to the ground state 0. Using the result of Exercise 12.3 and the expression found earlier for the oscillator strength (Equation 4.29), show that the ratio of the emission coefficient becomes

$$\frac{\epsilon_{20}}{\epsilon_{10}} = \frac{g_2 A_{20}\,\nu_{10}^2}{g_1 A_{10}\,\nu_{20}^2}\exp\left(-\frac{\Delta E}{k_B T}\right)$$

provided $\ln\Lambda$ is approximately the same for collisional excitation to both states 1 and 2. [The similarity of the emission coefficient ratios here to the result for LTE equilibrium (Exercise 12.4) enables the assumption of a Boltzmann distribution of populations to be used widely to determine the electron temperature k_BT from the relative intensity of closely spaced $\nu_{20} \approx \nu_{10}$ spectral lines. However, for good accuracy the technique needs $\Delta E \gg k_BT$.]

12.6 Consider the relative rates of three-body recombination given by Equation 12.16 and radiative recombination given by Equation 5.22. Show that the ratio of the rate of collisional to radiative recombination from a fully stripped ion to principal quantum number n in a hydrogen-like ion is given by

$$\frac{K_{Cn}}{A_{Cn}} = C_0 n^7 \frac{n_e}{(k_BT)^{1/2}} \frac{1}{Z^4}$$

where C_0 is a constant, n_e and k_BT are the electron density and temperature, and Z is the atomic number of the ion.

12.7 Consider the ground state (population density $N_1(t)$) and an excited quantum state (population $N_2(t)$) in an atom or ion. If collisional excitation and de-excitation and radiative decay between the two quantum states are the main de-populating processes out of the two states, we can write for the de-population rates of the states 1 and 2:

$$\frac{N_1(t)}{dt} = -K_{12} n_e N_1,$$

$$\frac{N_2(t)}{dt} = -K_{21} n_e N_1 - A_{21}$$

where K_{12} and K_{21} are the rate coefficients for collisional excitation and de-excitation, A_{21} is the radiative transition probability between the two states and n_e is the electron density. Show that the populations relax with time t to equilibrium values from high-value perturbed initial values $N_1(0)$ and $N_2(0)$ such that

$$N_1(t) \approx N_1(0) \exp(-t/\tau_1)$$

$$N_2(t) \approx N_2(0) \exp(-t/\tau_2)$$

where

$$\tau_1 = \frac{1}{K_{21} n_e + A_{21}},$$

$$\tau_1 = \frac{1}{K_{12} n_e}.$$

12.8 From the result of Exercise 12.7, show that the maximum of the ratio τ_2/τ_1 of the relaxation times of the excited and ground states is given by

$$\frac{\tau_2}{\tau_1} = \frac{g_2}{g_1} \exp\left(-\frac{\Delta E}{k_B T}\right)$$

where g_2 and g_1 are the degeneracies of the two states, ΔE is the energy difference between the states and $k_B T$ is the electron temperature. [For $\Delta E \gg k_B T$, we have that $\tau_2 \ll \tau_1$, which illustrates the comment made in Section 12.3 that the excited states relax much more quickly to quasi-steady-state values than the ground state.]

12.9 Lasers use the excited states of atoms or ions to amplify radiation by stimulated emission. A schematic energy-level diagram for a laser is shown in Figure 12.5, where net stimulated emission can occur between states 2 and 1. Collisional excitation from the ground state 0 to level 3 leads to radiative decay to level 2 (the upper laser level), with subsequent radiative decay to levels 1 (the lower laser level) and 0 (the ground state). A population inversion between levels 1 and 2 can be produced such that $N_2 - (g_2/g_1)N_1 > 0$, where N_2 and N_1 are the number densities of levels 2 and 1 and g_2 and g_1 are their degeneracies. We assume a so-called small signal condition where the stimulated emission between levels 2 and 1 is negligible. With the rates of the collisional excitation

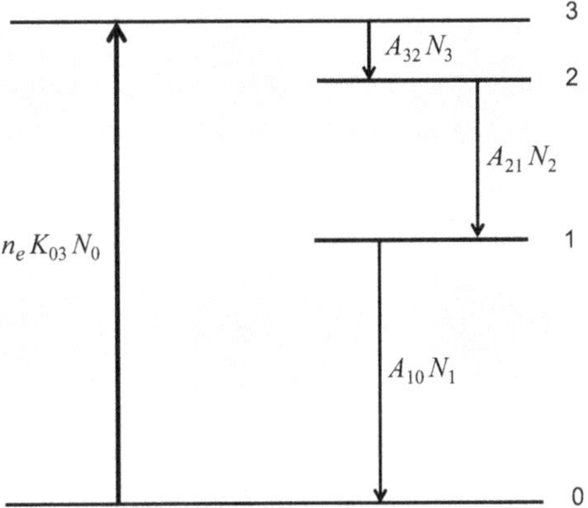

Figure 12.5 The energy levels of a four-level laser to be considered for Exercises 12.9 and 12.10. The rates of population change (e.g. $dN_3/dt = n_e K_{03} N_0$) are appended beside the arrow indicating a collisional or radiative transition.

and radiative decay indicated on Figure 12.5, show that for steady-state conditions

$$N_2 - \frac{g_2}{g_1}N_1 = \frac{N_0\, n_e\, K_{03}}{A_{21}}\left(1 - \frac{g_2 A_{21}}{g_1 A_{10}}\right).$$

12.10 The rate of change dN_2/dt of the population of a quantum state due to the absorption of radiation of intensity $I(\omega)$ associated with a lineshape function $f(\omega)$ was illustrated by Equation 4.26:

$$\frac{dN_2}{dt} = \frac{\pi^2 c^3}{\hbar\omega_{21}^3} A_{21}\left(\frac{g_2}{g_1}N_1 - N_2\right)\int_0^\infty \left(\frac{I(\omega)}{c}\right) f(\omega)d\omega.$$

Consider the rates of collisional excitation, radiative decay, radiative absorption and stimulated emission as illustrated in Figure 12.5 when stimulated emission and absorption between levels 2 and 1 is significant. Show that the steady-state rate of net stimulated emission (minus absorption) associated with lasing between levels 2 and 1 as illustrated in Figure 12.5 is given by

$$\frac{dN_2}{dt} = W_{ef}\frac{N_0\, n_e\, K_{03}\,(1 - (g_2 A_{21}/(g_1 A_{10}))}{1 + W_{ef}}$$

where

$$W_{ef} = \frac{\pi^2 c^3}{\hbar\omega_{21}^3}\int_0^\infty \left(\frac{I(\omega)}{c}\right) f(\omega)d\omega.$$

Here $I(\omega)$ is the intensity of laser light of central frequency ω_{21} and $f(\omega)$ is the lineshape function for the 2 to 1 transition.

13
High-Density Plasmas

The definition of the necessary density to have a high-density plasma probably depends on the research area of the person seeking the definition. Many regard departure from coronal equilbrium (see Section 12.5) a suitable definition of high density. However, we will consider plasma material at sufficiently high densities to be 'high density' if LTE or near-LTE occurs between the ground states of different ionisation stages (see Section 12.7). With LTE conditions, new concepts (not, for example, considered in Chapter 12) often need to be considered. Even with comparatively low temperatures (e.g. a few eV), plasmas formed at high density have a high energy content per unit volume and form a subset of the research field of high-energy density physics [20, 28]. A high-energy density is defined as an energy exceeding 10^{11} J m^{-3} by most authors working in the field (e.g. [20]). Such energy densities are found in materials at solid density and above when temperatures exceed a few eV.

The physics of plasmas at high density requires an understanding of equilibrium relationships. Equilibrium relations are valid at sufficiently high densities which often makes for simpler calculations of plasma ionisation and radiation emission. In plasmas at high density, many interactions can lead to statistical distributions, so that equilibrium ionisation populations and equilibrium radiation distributions as introduced in Sections 1.4.1 and 4.1, respectively, are present. The equilibrium relationships for ionisation and radiation distribution were used to deduce rates of inverse processes by invoking detailed balance (see Chapter 4 for radiative processes and Chapter 12 for collisional processes).

At very high densities and low temperatures, we need to modify the equilibrium-ionisation relation (the Saha-Boltzmann equation) as it is necessary to allow for free-electron quantum states becoming fully occupied so that Fermi-Dirac rather than Maxwellian electron energy distribution is required (see Section 13.4). At high density, photons are more likely to interact with particles. However, the Planck black-body radiation distribution does not require modification at high-photon or particle density as photons are bosons and any number can occupy an excitation state.

If the density is high and the temperature of the plasma is not high, the chemical potential and Fermi energy associated with the near-full occupancy of free-electron quantum states are important. We use the concept of the chemical potential in two different ways: (i) the chemical potential of the free electrons – the energy required to add another electron to the free electrons, and (ii) in the Thomas-Fermi model (see Section 13.5) to treat the chemical potential of an ion – the energy required to add another electron to the bound states of an ion. In both cases, the chemical potential is found, in practise, by integrating the Fermi-Dirac distribution containing the relevant chemical potential, so that the total number of electrons in either unit volume (for the free electrons) or an ion (the Thomas-Fermi model) is correct.

The energy associated with free electrons having near-full occupancy of the free-electron states creates a degeneracy pressure as electrons cannot occupy low-energy filled states and require extra energy to be in unoccupied higher-energy states. We discussed in Chapter 1 how the free-electron energy per unit volume is equivalent to a pressure (see Equation 1.10). Pressure has the same dimensions as energy density per unit volume and even at high density a plasma is often close to being a perfect gas, where numerically the gas pressure is equal to 2/3 of the energy density per volume.

At the extremes of high density, the assumption that ions have quantum states dominated by the central nuclear potential starts to break down. Some methods to allow for high-density atomic physics effects are discussed in this chapter. We present high-density corrections to the Saha-Boltzmann equation and to collisional and radiative rates, introduce the idea of continuum lowering (also known as ionisation potential depression) and present the Thomas-Fermi and average atom models, which can account for some of this loss of ionic individuality. For more discussion of the physics at the extremes of high density, the reader is referred to the books by Larsen [65], Colvin and Larsen [20], Drake [28] and Zel'dovich and Raizer [124].

13.1 Examples of High-Density Plasmas

Atomic and radiation physics often dominates the movement of energy and the physical state of plasma material at the high densities where LTE or near-LTE applies. This section gives some background to examples of high-density plasmas.

High-density plasma is formed when a solid is heated by a laser. Laser-produced plasma densities range from above solid density to vanishingly small densities as the plasma expands into vacuum with densities approaching those achieved by the vacuum pumping system. The physics of the interaction of the laser with expanding plasma has been considered in Section 2.4. Lasers operating in the infra-red to ultra-violet interact with high-density plasma expanding from a solid surface. The first free electrons are created by multi-photon absorption associated with photons in the leading edge of the laser pulse. Once a substantial number of free electrons is present, the laser absorption proceeds by inverse bremsstrahlung (see Section 5.3) and other processes such as resonance absorption (see Section 2.4.6). Energy is conducted from the maximum density for the laser light penetration (the critical density) into the solid material where further high-density and low-temperature plasma is formed. All laser-produced plasma material created from a solid passes through a high-density/low-temperature state as material ablates.

The plasmas formed by focusing X-ray and extreme ultra-violet lasers onto targets can create plasma with greater high-density thickness and uniformity than achieved with focused infra-red to ultra-violet lasers. X-ray and extreme ultra-violet lasers have been developed using plasma media [91, 108] and with free electron lasers [1, 30, 55] where linear particle accelerators create lasing action using the oscillations of electrons in alternating magnetic fields (known as 'undulators'). This last technology has allowed a new regime of high-density plasma physics to be investigated [76, 116]. X-ray laser radiation frequencies are much greater than the plasma frequency formed at solid density, so X-ray lasers are not reflected or absorbed due to collective plasma effects (such as resonance absorption; see Section 2.4.6) and heat a large thickness of, for example, a solid target. The plasma produced by an X-ray laser typically absorbs the laser radiation by photo-ionisation (see Section 5.5) initially, and subsequently by photo-ionisation and inverse bremsstrahlung (see Section 5.3). Much of the laser energy is accounted for in direct photo-ionisation, so that electron temperatures can be small ($\ll 100$ eV) and as the pulse durations are typically <100 fs, the ion temperatures remain cold close to room temperature during the laser interaction [76].

The compressed plasmas investigated for inertial fusion represent an extreme of high-density plasma [42, 66]. Inertial fusion or inertial confinement fusion (ICF) seeks to achieve energy production by compressing a fuel target of deuterium and tritium to high density so that fusion reactions occur. Compression is achieved either by using X-rays created in a cylindrical black-body (*hohlraum*) cavity by laser heating (known as indirect drive; see Section 4.1) or by direct laser irradiation of a pellet containing the fuel (known as direct drive). The deuterium-tritium fuel is typically frozen on the inside of a spherical plastic pellet. X-ray or direct laser ablation of the shell of the fuel pellet produces the fuel compression as a consequence of momentum conservation to the outward-directed momentum of ablated material. In inertial fusion the plasma is not confined by magnetic fields (or other means) so that an ideal scenario involves all the deuterium and tritium fuel reacting before plasma expansion causes the fusion reactions to stop. At the centre of a compressed spherical pellet, a small 'hot spot' of material undergoes a 10^3 increase in density enabling the first fusion reactions producing helium nuclei (α particles) with kinetic energy of 3.5 MeV and neutrons of energy 14.1 MeV. The α particles heat an inner layer of the fuel, producing further fusion and a small expanding spherical shell of material undergoing fusion (a so-called burn-wave). The burn-wave needs to propagate outwards at a sufficiently fast rate so that all the fuel undergoes fusion before it starts to significantly expand. In inertial fusion, the deuterium-tritium fuel is ideally kept relatively cool during compression, so that compression can be optimised. Free-electron degeneracy effects on the plasma pressure are important for such cold, dense plasmas.

Inertial fusion research has benefitted in some ways by an overlap of some physics with processes important in nuclear weapons, but in other ways has been held back due to nuclear proliferation concerns. The NIF in the US, based at the Lawrence Livermore National Laboratory, has led research in indirect drive inertial fusion [42]. Other laser systems have been constructed in order to produce experimental results which can validate simulation codes used to certify the performance and safety of weapons [90].

Astrophysical objects such as dwarf stars have high densities where the degeneracy of the free electrons is a significant component of the total pressure. The first investigations of electron degeneracy effects in plasmas were undertaken in order to understand dwarf stars [16]. The interiors of giant planets are also at high densities, and in plasma terms, relatively low temperatures ($\approx$1,000 K), so electron degeneracy affects the plasma ionisation and other parameters [40].

13.2 The Ion–Ion Plasma Coupling Constant

A measure of a high-density plasma is the ratio of the potential energy required to assemble the plasma relative to the thermal plasma energy. The ion–ion plasma coupling constant reflects the ratio of the potential energy between adjacent ions and the average kinetic energy. We can write for the ion–ion coupling constant

$$\Gamma_{ii} = \frac{Z_{av}^2 e^2}{4\pi \epsilon_0 r_0 k_B T_i}$$

where Z_{av} is the average degree of ionisation, r_0 is the average spacing between ions and T_i is the ion temperature. We are using SI units so that the parameters e, ϵ_0 and k_B are the electron charge, electric dielectric constant and Boltzmann constant, respectively. The volume occupied by an ion is on average $4/3\pi r_0^3 = 1/n_i$, where n_i is the ion number density, so we can write that

$$r_0 = \left(\frac{4\pi}{3} n_i\right)^{-1/3}.$$

We then have

$$\Gamma_{ii} = \frac{1}{4\pi\epsilon_0}\left(\frac{4\pi n_i}{3}\right)^{1/3}\frac{Z_{av}^2 e^2}{k_B T_i}.$$

The average degree of ionisation Z_{av} is calculated from the ratio of the number density of electrons n_e to ions (i.e. $Z_{av} = n_e/n_i$). If $\Gamma_{ii} \geq 10$, plasmas are referred to as strongly coupled and if $\Gamma_{ii} \leq 0.1$, as weakly coupled [52]. Inside an evolved star and, for example, in the plasmas formed by free-electron laser irradiation of solid targets, plasmas with $\Gamma_{ii} \geq 10$ are produced. An interesting state of matter with large Γ_{ii} resulting from photo-ionisation of either dense [110] or laser-cooled material [59] by lasers with photon energy just above the ionisation threshold can be produced.

The ion coupling factor gives a guide to the likelihood of ions forming ordered structures. At high values of Γ_{ii}, ions in a plasma can arrange themselves in regular, crystal-like structures similar to solid-state crystals. Light and X-ray scattering from high Γ_{ii} plasmas can show evidence of the formation of regular structures [52].

Interestingly, plasmas containing charged microparticles (known as 'dusty' plasmas) can be regarded as having a high Γ_{ii} if the electrostatic energy between microparticles is used to evaluate Γ_{ii}. The charged particle size can range from millimetre to nanometre size with either positive or negative charges of typically 1,000 e. Just as in the solid state, crystal-like arrangements of the microparticles can occur. The rings of the planet Saturn are an example of a dusty plasma. Light scatter from the dust particles is the dominant radiation interaction with a dusty plasma.

13.3 The Fermi Energy and Pressure of Free Electrons

Fermi-Dirac statistics govern the occupancy of electronic states in high-density plasmas as the Pauli exclusion principle does not allow more than one electron in each quantum state. In high-density plasmas, the probability of occupancy of a free-electron quantum state becomes high, so the Pauli exclusion principle needs to be invoked in order to determine energy density and pressure calculations.

To gain an estimate of the Fermi energy and pressure of the free electrons, we can assume a temperature T approaching zero. The distribution of electron energies (discussed in Section 1.3) becomes straightforward as there is then full occupancy of the free-electron quantum states for electron energies less than the Fermi energy and zero occupancy for electron energies greater than the Fermi energy. For electron energies less than the Fermi energy, we have seen (Equation 1.26) that the number of electrons (at zero temperature) with energy in the range E to $E + dE$ per unit volume is given by

$$g(E)dE = 4\pi \left(\frac{2m_0}{h^2}\right)^{3/2} E^{1/2} dE.$$

The electron density associated with the Fermi energy E_F can be found by integrating this expression over energy. We only need to integrate up to the Fermi energy as at zero temperature there is zero occupancy of electrons at higher energy. The electron density for a particular Fermi energy is given by

$$\begin{aligned} n_e &= 4\pi \left(\frac{2m_0}{h^2}\right)^{3/2} \int_0^{E_F} E^{1/2} dE \\ &= 4\pi \left(\frac{2m_0}{h^2}\right)^{3/2} \frac{2}{3} E_F^{3/2}. \end{aligned} \tag{13.1}$$

It is more common for the electron density to be known from estimation or a measurement rather than the Fermi energy, so we manipulate this expression to give an expression for the free-electron Fermi energy as a function of the electron density. We obtain

$$E_F = \left(\frac{3}{8\pi}\right)^{2/3} \frac{h^2}{2m_0} n_e^{2/3}. \tag{13.2}$$

In practical units, if we measure the electron density in units of cm^{-3}, the Fermi energy in electron volts (eV) is given by

$$E_F(eV) = 0.3646 \left(\frac{n_e}{10^{21}}\right)^{2/3}.$$

The average energy U of Fermi degenerate electrons per unit volume is calculated by integrating the electron energy multiplied by the number density of electrons with that energy. We have

$$\begin{aligned} U_F &= \int_0^{E_F} E f_E(E) dE \\ &= 4\pi \left(\frac{2m_0}{h^2}\right)^{3/2} \int_0^{E_F} E^{3/2} dE \\ &= 4\pi \left(\frac{2m_0}{h^2}\right)^{3/2} \frac{2}{5} E_F^{5/2}. \end{aligned} \tag{13.3}$$

Using Equation 13.1, we see that $U = (3/5) n_e E_F$.

Equation 1.10 relates energy density per unit volume $U = \rho\epsilon$ to pressure P. Even for Fermi degenerate electrons, the assumption that $\gamma = 5/3$ (see Equation 1.10) for an ideal gas is a good approximation, so we find that the pressure is related to the energy density by $P = (2/3)U$. This means that our calculation of the average energy U of the Fermi degenerate electrons corresponds to a Fermi pressure where

$$P_F = (2/5)n_e E_F = \frac{2}{5}\left(\frac{3}{8\pi}\right)^{2/3} \frac{h^2}{2m_0} n_e^{5/3}.$$

In practical units, the Fermi pressure in Mbar ($= 10^{11}\,\mathrm{Nm}^{-2}$) is related to the electron density in units of cm^{-3} by

$$P_F(Mbar) = 0.5\left(\frac{n_e}{10^{23}}\right)^{5/3}.$$

Partially Degenerate Free Electrons

The above treatment of electron degeneracy is valid at zero temperature where the Fermi energy equals the chemical potential. Considering an effectively zero temperature is useful for evaluating the Fermi pressure, but is not often found. At finite temperatures, we can evaluate the internal electron energy by an integral of the energy E over the Fermi-Dirac distribution function $f_{FD}(E)$. We obtain

$$U = \int_0^\infty E f_{FD}(E) dE = \frac{4}{\sqrt{\pi}}\left(\frac{2\pi m_0 k_B T}{h^2}\right)^{3/2} k_B T\, I_{3/2}(\mu/k_B T) \tag{13.4}$$

where $I_m(\eta_C)$ is the Fermi-Dirac integral of order m. We introduce a reduced chemical potential $\eta_C = \mu/k_B T$ and write for the Fermi-Dirac integrals

$$I_m(\eta_C) = \int_0^\infty \frac{x^m dx}{\exp(x - \eta_C) + 1}. \tag{13.5}$$

The chemical potential μ is the energy needed to add one more electrons to the free-electron population at constant entropy and volume. It is related to the electron density by the requirement that an integration of the electron distribution function over all energies gives the total free-electron density n_e. We can write

$$n_e = \int_0^\infty f_{FD}(E) dE = \frac{4}{\sqrt{\pi}}\left(\frac{2\pi m_0 k_B T}{h^2}\right)^{3/2} I_{1/2}(\mu/k_B T) \tag{13.6}$$

where $I_m(\eta_C)$ is the Fermi-Dirac integral of order m.

The reduced chemical potential $\eta_C = \mu/k_B T$ can be determined using Equation 13.6 (see Figure 13.1 where we plot a scaling of electron density and temperature as a function of the reduced chemical potential η_C). The vertical axis of Figure 13.1 represents values of different energy content (see Section 1.1.1). The ratio of the internal energy U evaluated using Equation 13.4 to the Fermi internal energy U_F

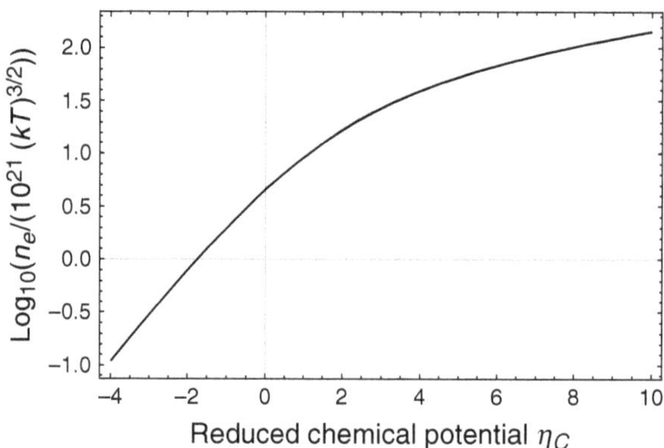

Figure 13.1 The scaling of electron density n_e in units of cm^{-3} and electron temperature k_BT in units of eV as a function of the reduced chemical potential $\eta_C = \mu/k_BT$. At negative chemical potential, the free-electron degeneracy is not significant and there is a log linear variation.

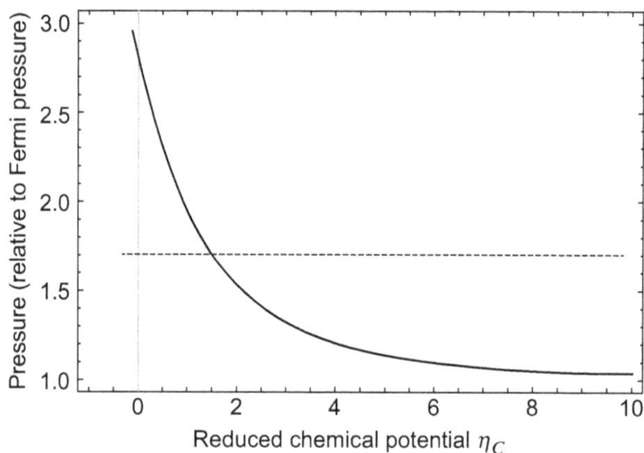

Figure 13.2 The total pressure of an electron gas including the degeneracy pressure as a function of reduced chemical potential. The horizontal line shows the maximum pressure for compression in inertial fusion design studies [42].

using Equation 13.3 gives the ratio of the pressure to the Fermi pressure (see Figure 13.2).

Figure 13.2 illustrates the increasing contribution of electron degeneracy to the total internal energy and pressure with increasing positive values of the chemical potential. This 'degeneracy pressure' arises because near-full occupancy of lower-energy free-electron quantum states requires additional electrons to be placed in higher-energy states, thus requiring more energy (and more pressure) to increase the electron density. The Fermi pressure is the degeneracy pressure as temperatures approach zero.

Degeneracy pressure prevents the gravitational collapse of dwarf and neutron stars. The degeneracy pressure is similarly the minimum pressure that can be achieved in inertial fusion compression. Typical inertial fusion scenarios require the total pressure (including degeneracy pressure) to be less than 1.7× the Fermi pressure (see Figure 13.2).

In the low-density limit where $\eta_C = \mu/k_BT \ll -1$, the Fermi-Dirac integral has an analytic solution with

$$I_{1/2}(\mu/k_BT) \approx \frac{\sqrt{\pi}}{2}\exp(\mu/k_BT).$$

Consequently, at large negative chemical potential, electron degeneracy effects become negligible and the chemical potential and electron density are related through

$$\exp\left(\frac{\mu}{k_BT}\right) = \frac{n_e}{2}\left(\frac{h^2}{2\pi m_0 k_BT}\right)^{3/2}. \tag{13.7}$$

This exponential variation of the reduced chemical potential $\eta_C = \mu/k_BT$ with $n_e/(k_BT)^{3/2}$ at large negative values of chemical potential is clearly seen in Figure 13.1.

13.4 The Saha–Boltzmann Equation at High Density

The Saha-Boltzmann equation for the equilibrium population-density ratio between adjacent ionisation stages for a low-density plasma was determined in Section 1.4.1. At high density, there is an equivalent equation which takes account of the Pauli exclusion principle by utilising Fermi-Dirac statistics for the free-electron distribution. At high density, free-electron quantum states can be full to their degeneracy limit and the available states for a free electron to occupy on ionisation is reduced.

In the ionization process of converting a Z_i charged ion to a $Z_i + 1$ charged ion, a free electron with energy (say E) is created and a $Z_i + 1$ ion quantum state is populated. We assume that a particular $Z_i + 1$ charge quantum state has a probability P of existing. In a low-density plasma with low occupancy of quantum states by the electrons, P approaches unity, but we shall see later that in a plasma with near-full occupancy of the free-electron quantum states, the probability that the state can exist is less than unity.

The population ratio of the free-electron number density $f(E)dE$ and the population N_{Z_i} of bound electrons of charge Z_i per unit volume can be found using Equation 1.19 after allowing for the degeneracy of the free electrons (given by Equation 1.26) and the degeneracy g_{Z_i} of the bound quantum state. For each $Z_i + 1$ ion, there are $g(E)dE/N_{Z_i+1}$ free-electron states, where N_{Z_i+1} is the number density of $Z_i + 1$ ionization states. The bound state of the $Z_i + 1$ ion may also be degenerate with degeneracy g_{Z_i+1}, so that the total degeneracy of the 'upper state' created by ionization is $g_{Z_i+1}g(E)dE/N_{Z_i+1}$. The ratio of the Fermi-Dirac populations for the free electrons $f_{FD}(E)dE$ and bound electrons N_Z can be written as a ratio of the 'upper' and 'lower' state populations using Equation 1.19. We have

$$\frac{f_{FD}(E)dE}{N_{Z_i}} = \frac{g_{Z_i+1}g(E)dE/N_{Z_i+1}}{g_{Z_i}} \frac{\exp[-(\mu + E_{ion})/k_BT] + 1}{\exp[-(\mu - E))/k_BT] + 1} P. \tag{13.8}$$

Here E_{ion} is the ionization energy of the Z charged ion quantum state being considered. Ionization energy is negative on our free-electron energy scale, hence our assumed positive E_{ion} has the opposite sign to the free-electron kinetic energy E in Equation 13.8.

If we integrate both sides of Equation 13.8, we obtain

$$n_e = \frac{N_{Z_i}}{N_{Z_i+1}} \frac{g_{Z_i+1}}{g_{Z_i}} \frac{4}{\sqrt{\pi}} \left(\frac{2\pi m_0 k_BT}{h^2}\right)^{3/2} \left[\exp\left(-\frac{\mu + E_{ion}}{k_BT}\right) + 1\right] I_{1/2}(\mu/k_BT)P \tag{13.9}$$

if P is independent of energy E. Using the definition of the chemical potential (Equation 13.16), Equation 13.9 reduces to

$$\frac{N_{Z_i+1}}{N_{Z_i}} = \frac{g_{Z_i+1}}{g_{Z_i}} \left[\exp\left(-\frac{\mu + E_{ion}}{k_BT}\right) + 1\right] P. \tag{13.10}$$

The existence of a Z_i+1 charged quantum state depends on the number of 'holes' in the Z_i charged ion, i.e. the number of states not fully occupied. For example, if all the quantum states of the Z_i charged ion are occupied, no electrons have been removed to create the $Z_i + 1$ charge ion, so the probability P that the $Z_i + 1$ state exists is zero. More generally, considering ionization from a quantum state with ionization potential E_{ion} we can write

$$P = 1 - \frac{1}{\exp\left(-\frac{\mu+E_{ion}}{k_BT}\right) + 1}. \tag{13.11}$$

P has the form of a 'blocking factor' representing the probability that the bound state of ionization energy E_{ion} has a 'hole' or quantum state which is unfilled. The number of holes determines the probability that an electron has been removed to produce a $Z_i + 1$ quantum state. Substituting this expression for P into Equation 13.10 gives

$$\frac{N_{Z_i+1}}{N_{Z_i}} = \frac{g_{Z_i+1}}{g_{Z_i}} \exp\left(-\frac{E_{ion}}{k_BT}\right) \exp\left(-\frac{\mu}{k_BT}\right). \tag{13.12}$$

This expression for the Saha ionization balance is consistent with a thermodynamic understanding of ionization where the energy of the free electrons changes by the ionization energy plus the chemical potential ($E_{ion} + \mu$) upon ionization. The chemical potential is defined as the energy required to add one more electron to the free-electron population. If we substitute the low-density limit (Equation 13.7) for the chemical potential, we obtain our original Equation 1.4.1, i.e.

$$\frac{n_e N_{Z_i+1}}{N_{Z_i}} = \frac{g_{Z_i+1}}{g_{Z_i}} 2 \left(\frac{2\pi m_0 k_B T}{h^2} \right)^{3/2} \exp\left(-\frac{E_{ion}}{k_B T} \right). \tag{13.13}$$

If we assume that the probability P for the existence of the $Z_i + 1$ quantum state is as given by Equation 13.11, then the ratio of the populations N_2 and N_1 of two bound states with ionization energies E_2 and E_1 and degeneracies g_2 and g_1 from Equation 13.12 is given by

$$\frac{N_2}{N_1} = \frac{g_2}{g_1} \exp\left(-\frac{E_2 - E_1}{k_B T} \right) \tag{13.14}$$

as the term involving the chemical potential cancels. This Boltzmann ratio (Equation 13.14) is the same as found at low densities where free-electron degeneracy is not important.

13.5 The Thomas–Fermi Model

The Thomas-Fermi model allows an approximate method of calculating the degree of ionisation Z_{av} of a plasma and a method to obtain the charge distribution $\rho_e(\mathbf{r}')$ of electrons in ions. The last here can be used to obtain the total ionic potential $V(\mathbf{r})$ around the nuclei, which enables collision cross-sections using, for example, the Born approximation (Equation 11.10) to be determined.

In the Thomas-Fermi model, the quantum nature of the wavefunctions is largely ignored and a classical calculation of the charge density of electrons around the nucleus is evaluated using Poisson's equation. Electrons are allowed to fill the available quantum states according to the Fermi-Dirac distribution (Equation 1.20) with the quantum state density as for free electrons [77]. The essential features of the Thomas-Fermi model were devised to explain atomic structure before quantum mechanics was fully developed, but it is still useful in atomic physics calculations relevant to plasmas.

The dropping of the effects of the quantum nature of bound electron distributions in the Thomas-Fermi model is partly justified by considering that the ensemble average of the ions with random orientation in a plasma will have an average electron distribution which is spherically symmetric around the nucleus. The model is more accurate for calculations with high atomic number material which are only partly ionised as there are then a significant number of bound electrons so that the orbital and discrete nature of the individual electron probability distributions can be averaged. It is also more accurate at high density where the internuclear distance is a good guide to the 'size' of the atom.

The number of particles following a Fermi-Dirac distribution with energies between E and $E + dE$ is given by multiplying the Fermi-Dirac occupancy $n(E)$ of quantum states (see Equation 1.20) by the density $g(E)$ of quantum states (see Equation 1.26). We have

$$f_{FD}(E) = g(E)n(E) = \frac{4}{\sqrt{\pi}}\left(\frac{2\pi m_0}{h^2}\right)^{3/2} E^{1/2}\frac{1}{\exp\left(-\frac{\mu-E}{k_BT}\right)+1}. \tag{13.15}$$

The chemical potential μ is the energy needed to add one more electron to the electron population. The energy E of the electrons is determined by the potential energy $V_P(r')$ in the field of the nucleus plus their kinetic energy. The chemical potential is determined by integrating over the range of electron kinetic energies (from zero to ∞) or (equivalently) by integrating the total electron energy from $V_P(r')$ (the local potential energy) to ∞. The integration over all possible electron energies must give the total density of electrons $\rho_e(r')$. We have

$$\rho_e(r') = \int_{V_P(r')}^{\infty} f_{FD}(E)dE = \frac{4}{\sqrt{\pi}}\left(\frac{2\pi m_0 k_B T}{h^2}\right)^{3/2} I_{1/2}\left(\frac{(\mu - V_P(r'))}{k_BT}\right) \tag{13.16}$$

where $I_m(\eta)$ is the Fermi-Dirac integral of order m and r_0 is the radius of the atom. We introduce the reduced chemical potential $\eta' = (\mu - V_P(r'))/k_BT$ and write for the Fermi-Dirac integrals

$$I_m(\eta') = \int_0^{\infty} \frac{x^m dx}{\exp(x-\eta')+1}. \tag{13.17}$$

The Fermi-Dirac integrals require numerical solutions unless η' is large and negative or large and positive.

Poisson's equation describes the relationship between a potential V_P on a closed surface and the charge $-e\rho_e$ enclosed by the surface (see Appendix A.2). We have

$$\nabla^2 V_P = \frac{e\rho_e}{\epsilon_0}.$$

For a spherically symmetric charge distribution in an ion

$$\begin{aligned}\nabla^2 V_p(r') &= \frac{1}{r'}\frac{d^2(r'V_P(r'))}{dr'^2}\\ &= \frac{d^2V_P(r')}{dr'^2} + \frac{2}{r'}\frac{dV_P(r')}{dr'}\end{aligned}$$

so that

$$\frac{d^2V_P(r')}{dr'^2} + \frac{2}{r'}\frac{dV_P(r')}{dr'} = -\frac{1}{\epsilon_0}\left(-e\rho_e(r') + Ze\right). \tag{13.18}$$

The boundary conditions to solve this differential equation can be set:

$$\begin{aligned} V_P(r_0) &= 0 \\ \left[r'V_P(r')\right]_{r'=0} &= eZ \\ \left[\frac{dV_P(r')}{dr'}\right]_{r'=r_0} &= 0. \end{aligned} \tag{13.19}$$

where Z is the atomic number of the ion and hence eZ is the charge on the nucleus. Equations 13.16 and 13.18 can be solved together if the temperature T and the radius r_0 of the ion are known. When applied to high-density plasmas, the ion radius is usually defined in terms of the internuclear spacing. If the material density is ρ (in units of mass per unit volume), then the number density of the ion is $\rho/(Am_p)$, where A is the atomic number of the ion and m_p is the proton mass. We can write for the ion sphere radius r_0 that

$$\left(\frac{4}{3}\right)\pi r_0^3 = \frac{Am_p}{\rho}. \tag{13.20}$$

The volume on the left-hand side is referred to as the average atom cell. If calculations are done for low-density plasmas where there is a significant unfilled volume between nuclei, then another estimate of the volume of the atom needs to be employed. The average ionisation Z_{av} in the plasma can be found by integrating the number density of the electrons over the radius of the ion. We have

$$Z_{av} = 4\pi \int_0^{r_0} \rho_e(r')r'^2 dr'. \tag{13.21}$$

Once the bound electron density $\rho_e(\mathbf{r}')$ is determined, the electron scattering potential (Equation 11.17) can be evaluated and then Equation 11.10 can be used to calculate the scattering amplitude and cross-section.

13.6 The Average Atom Model

The average atom model is an extension of the Thomas-Fermi model of ionisation using more accurate physics. The Thomas-Fermi model assumes that electrons occupy quantum states in the vicinity of the nuclear potential with quantum state densities as for free electrons. The average atom model takes account of the energy eigenvalues of discrete energy levels, usually associated with the principal quantum number n and an assumption of hydrogen-like energy levels. The different shells (different n values) can be partially occupied by a fraction of electron charge to represent the distribution of ionisation stages. The equilibrium populations are determined using Fermi-Dirac statistics with the number density per ion of an energy level given by

$$f_{AA} = \frac{g_n}{\exp(-E_{ion} + \mu/k_B T) + 1} \tag{13.22}$$

where g_n is the degeneracy of the energy level and E_{ion} is the (negative) ionisation energy of the level. The chemical potential μ is now determined by assuming the bound electrons are in equilibrium with the free electrons so that Equation 13.16 is valid (see Figure 13.1). The equilibrium ionisation level of the average atom is given by

$$Z_{av} = Z - \sum_n f_{AA} \tag{13.23}$$

where Z is the atomic number of the ion.

The ionisation energies E_{ion} in average atom calculations are usually determined using corrected hydrogen-like energies (see Equation 1.30) for each principal quantum number n such that

$$E_{ion} = -\frac{Z_n^2 R_d}{n^2} + \Delta E_n \tag{13.24}$$

where Z_n represents the nuclear charge reduced by screening from electrons in inner ($<n$) levels and ΔE_n is a correction due to electrons in outer levels ($>n$) and the continuum electrons (see Section 13.7). Average atom calculations of ionisation are often coupled with fluid code modelling [22, 27].

The average atom model for plasma ionisation is a high-temperature version of density functional theory used in solid-state physics and chemistry [22, 61]. The Nobel Prize in Chemistry was awarded to W. Kohn and J. A. Pople in 1998 for the development of density functional theory.

13.7 Continuum Lowering

Ions in a high-density plasma have wavefunctions and energy levels which are perturbed by the plasma environment. The plasma environment lowers the ionisation potential required to remove an electron from one of the lower-bound levels into the continuum of free-electron states and causes the highest-energy discrete levels to disappear. The effect is known as ionisation potential depression and also as continuum lowering, as the continuum of free-electron quantum states subsumes some of the higher-energy ionic states. The amount of continuum lowering depends on the proximity of other perturbing ions and free electrons to the ion under examination.

We can define an appropriate radius a_C associated with an individual ion as either the distance r_0 to the next ion (known as the ion sphere radius; see Equation 13.20) or the Debye length λ_D (introduced in Chapter 1). An estimate of the ion sphere radius r_0 is obtained by associating the volume of a sphere of radius r_0 with the average volume occupied by an ion. We can write

$$\frac{4\pi}{3}r_0^3 = \frac{1}{n_i} = \frac{Z_i}{n_e} \tag{13.25}$$

where n_i is the ion density and Z_i is the charge of the ion.

If we make the assumption that electron wavefunctions enclosed within a_C are similar and the electrons have similar energies to those in an isolated ion, the effect of the plasma is to eliminate some higher principal quantum number n states occurring at higher radius, while leaving lower quantum states n largely unperturbed. This is, in fact, what is seen experimentally. Higher n quantum levels disappear so that the ionisation energy is reduced and the continuum of free-electron states subsumes some of the higher energy states that are discrete in an isolated ion.

An estimate of the ionisation potential depression ΔE_{ion} in a dense plasma can be obtained by using the energy level and wavefunction scaling applicable to hydrogen-like ions. The higher energy levels of ions with large wavefunction radius see a central potential which is close to being hydrogen-like (i.e. a nuclear charge with a few shielding electrons close to the nucleus). Our Bohr model calculations of Section 1.5 show that the radius of the wavefunction for a hydrogen-like ion of nuclear charge Z varies as $a_0 n^2/Z$, where a_0 is the Bohr radius. The energies of levels for hydrogen-like ions scale as $R_d Z^2/n^2$, where R_d is the Rydberg energy which is the ionisation energy of the hydrogen ground state. The degree of ionisation potential depression ΔE_{ion} is inversely proportional to a_C and we can fix the scale of ΔE_{ion} absolutely by taking a ratio to the ground state ionisation energy E'_{ion} and hydrogen-like ion wavefunction radius $a_0/(Z_i + 1)$. We write

$$\frac{\Delta E_{ion}}{E'_{ion}} = \frac{\Delta E_{ion}}{(Z_i + 1)^2 R_d} = \frac{2a_0/(Z_i + 1)}{a_C}.$$

Substituting the ion sphere radius (Equation 13.25) or Debye length (Equation 1.4) for a_C, we can obtain an expression for the ionisation potential depression ΔE_{ion} due to neighbouring ions or plasma:

$$\frac{\Delta E_{ion}}{E'_{ion}} = \frac{a_0}{Z_i + 1}\left(\frac{2}{a_C}\right). \tag{13.26}$$

A more detailed treatment of continuum lowering that has become a standard (see e.g. [32, 48]) is available in the literature (Stewart and Pyatt [104]), though there is some uncertainty as to the correct treatment for some dense plasma measurements where heating is rapid on a femtosecond timescale [21, 19, 117]. The paper by Stewart and Pyatt [104] uses a Thomas-Fermi model (see Section 13.5) to evaluate the potential ΔE_{ion} due to the presence of plasma. An approximate expression for continuum lowering is found such that

$$\Delta E_{ion} = \frac{(Z_i + 1)e^2}{4\pi \epsilon_0 \, a_C} \tag{13.27}$$

where the distance a_C recommended by Stewart and Pyatt is given by the larger of the Debye length λ_D (see Equation 1.4) or 2/3 of the ion sphere radius (i.e. $(2/3)r_0$; Equation 13.25). To achieve more accurate values of the effect of continuum lowering, numerical adjustments to Equation 13.27 are presented by Stewart and Pyatt with the adjustments representing multiplicative corrections of range 0.92–1.3 in the value of ΔE_{ion} (though for a large parameter range the correction is close to unity).

The standard continuum-lowering expression from Stewart and Pyatt (Equation 13.27) can be re-arranged into a form identical to Equation 13.26. Using the results of Equation 1.30, Equation 13.27 becomes

$$\frac{\Delta E_{ion}}{E'_{ion}} = \frac{a_0}{Z_i + 1}\left(\frac{2}{a_C}\right). \tag{13.28}$$

Expressions where the ionisation energy is employed need to be modified to take account of the ionisation potential depression. For example, in the Saha-Boltzmann equation (Equation 13.9), we have

$$\frac{n_{Z+1}}{n_Z} = \frac{g_{Z+1}}{g_Z}\frac{2}{n_e}\left(\frac{m_0}{h}\right)^3\left(\frac{2\pi k_B T}{m_0}\right)^{3/2} \exp\left(-\frac{E_{ion} - \Delta E_{ion}}{k_B T}\right)$$

where E_{ion} is the ionisation energy of a ground or excited state.

13.8 Collisional Rates at High Density

We now return to consider ionic bound state population evaluations using rate coefficients. At high density, the free-electron states may be ‘blocked’ by electrons occupying a quantum state and the energy distribution function is modified to the form of the Fermi-Dirac distribution. We may need to take account of both of these effects to determine high-density collisional rate coefficients applicable to collisional-radiative models.

The condition for the likely importance of ‘degeneracy’ effects on collisional rates is the value of the chemical potential of the free electrons which we have seen is determined by relating the integration of the Fermi-Dirac distribution over all energies to the total electron density (Equation 13.16). If the chemical potential is positive, the effects of degeneracy are important, while large negative values are related directly to the electron density (Equation 13.7). At large negative values of the free-electron chemical potential, a Maxwellian electron distribution function is applicable and the free-electron quantum states are not blocked. The rate coefficients outlined in Section 12.1 are then valid.

At high density, the quantum state populations are often in LTE if, for example, collisional processes dominate radiative processes. However, high-density departures from LTE can occur when, for example, radiative processes dominate ionisation and excitation, while three-body recombination and collisional de-excitation dominate recombination. Such plasmas are referred to as being radiation dominated.

13.8.1 Collisional Excitation

Including the effects of degeneracy, the transition rate per ion for collisional excitation from ionization energy E_p to E_q is given by

$$K_{pq}n_e = \int_{\Delta E}^{\infty} \left(\frac{2E}{m_0}\right)^{1/2} \sigma_{pq}(E) f_{FD}(E)\, P(E - \Delta E)\, dE \qquad (13.29)$$

where the blocking factor P is calculated for the final free-electron energy $E - \Delta E$, where E is the initial electron energy and $\Delta E = E_q - E_p$. We have

$$P(\epsilon) = 1 - \frac{1}{\exp\left(-\mu - \epsilon/k_BT\right) + 1} \qquad (13.30)$$

and the free-electron distribution $f(E)$ is given by Equation 13.15. In Equation 13.29, $\sigma_{pq}(E)$ is the cross-section for the collisional transition between bound states.

The typical dependence of inelastic cross-sections on the incident electron energy varies as $1/E$. With this dependence, Equation 13.29 can be written as

$$K_{pq}n_e \approx \frac{4}{\sqrt{\pi}} \left(\frac{2\pi m_0}{h^2}\right)^{3/2} \left(\frac{2}{m_0}\right)^{1/2} \Delta E\, \sigma_{pq}(\Delta E)\, J_{pq}(\Delta E)\, k_BT \qquad (13.31)$$

where

$$J_{pq}(\Delta E) = \int_{\Delta E/k_BT}^{\infty} \frac{1}{\exp\left(-\frac{\mu - E}{k_BT}\right) + 1} \left(1 - \frac{1}{\exp\left(-\frac{\mu - E + \Delta E}{k_BT}\right) + 1}\right) d\left(\frac{E}{k_BT}\right) \qquad (13.32)$$

$$= \frac{\exp(-\Delta E/k_BT)}{1 - \exp(-\Delta E/k_BT)} \ln\left[\frac{1 + \exp(\mu/k_BT)}{1 + \exp((\mu - \Delta E)/k_BT)}\right].$$

In the absence of degeneracy, the integral $J_{pq}(\Delta E)$ simplifies to

$$J_{pq}(\Delta E) = \int_{\Delta E/k_BT}^{\infty} \exp\left(\frac{\mu - E}{k_BT}\right) d\left(\frac{E}{k_BT}\right) = \exp\left(\frac{\mu - \Delta E}{k_BT}\right) \qquad (13.33)$$

which means that the ratio R_{pq} of the collisional excitation rate coefficient with degeneracy to the rate coefficient without degeneracy is given by

$$R_{pq} = \frac{(2/n_e)(2\pi m_0 k_B T/h^2)^{3/2}}{1 - \exp(-\Delta E/k_BT)} \ln\left[\frac{1 + \exp(\mu/k_BT)}{1 + \exp((\mu - \Delta E)/k_BT)}\right]. \qquad (13.34)$$

13.8.2 Collisional Ionisation

The collisional ionisation rate coefficient K_{ion} evaluation requires a knowledge of the differential cross-section $\sigma(E, E_1)$ where we assume, say, that the incident electron has an energy E and the ejected electrons have energy E_1 and $E_2 = E - E_1 - E_{ion}$. We can write that

$$K_{ion} n_e = \int_{E_{ion}}^{\infty} \left(\frac{2E}{m_0}\right)^{1/2} f(E) \left[\frac{\int_0^{E-E_{ion}} \sigma(E, E_1)\, P(E_1)\, P(E - E_1 - E_{ion})\, dE_1}{\int_0^{E-E_{ion}} dE_1}\right] dE \qquad (13.35)$$

where the blocking factors $P(E_1)$ and $P(E - E_1 - E_{ion})$ are appropriate for the two ejected electrons. We assume that the initial bound state has an ionization energy of E_{ion}. The integrations in the square bracket average the differential cross-section and blocking factors over the range of ejected electron energies (from zero energy to the impinging electron energy minus the ionization energy). As the two electrons in collisional ionization are indistinguishable, the differential cross-section $\sigma(E, E_1)$ is symmetric around energy $(E - E_{ion})/2$, so for enhanced computational speed, it is possible to evaluate the integrals over the reduced range 0 to $(E - E_{ion})/2$ (see e.g. [102]).

If we assume as a first treatment that the differential cross-section is constant with ejected electron energy E_1 and simply varies as $\sigma(E, E_1) = \sigma(E_{ion}, 0)E_{ion}/E$, we can proceed with a similar approximation as we made for collisional excitation. With the assumption that the differential cross-section is independent of the energy distribution between the two electrons, the rate coefficient can be written as

$$K_{ion} n_e \approx \frac{4}{\sqrt{\pi}} \left(\frac{2\pi m_0}{h^2}\right)^{3/2} \left(\frac{2}{m_0}\right)^{1/2} E_{ion}\, \sigma(E_{ion}, 0)\, J_{ion}(E_{ion})\, (k_BT) \qquad (13.36)$$

with

$$J_{ion}(E_{ion}) = \int_{E_{ion}}^{\infty} \frac{1}{\exp(-\mu - E/k_BT) + 1} \frac{1}{E - E_{ion}}$$

$$\left[\int_0^{\frac{E-E_{ion}}{k_BT}} \left(1 - \frac{1}{\exp(-\mu - E_1/k_BT) + 1}\right)\left(1 - \frac{1}{\exp(-\mu - E + E_1 + E_{ion}/k_BT) + 1}\right) d\left(\frac{E_1}{k_BT}\right)\right] dE. \quad (13.37)$$

The integral in the square brackets is over the blocking factors for the two electrons after the collision. Letting $y = (E - E_{ion})/k_BT$ and $\eta = \mu/k_BT$, the integral can be re-written and solved analytically, so that

$$J^*_{ion}(y) = \frac{1}{y}\int_0^y \left(1 - \frac{1}{\exp(-\eta + x) + 1}\right)\left(1 - \frac{1}{\exp(-\eta + y - x) + 1}\right) dx \quad (13.38)$$

$$= \frac{1}{1 - \exp(2\eta - y)}\left[1 + \frac{2}{y}\ln\left(\frac{e^{\eta - y} + 1}{e^{\eta} + 1}\right)\right].$$

The double integral is not analytically soluble. We can write for $J_{ion}(E_{ion})$ that

$$J_{ion}(E_{ion}) = \int_0^{\infty} J^*_{ion}(y) \frac{1}{e^{-\eta + y + \beta} + 1} dy \quad (13.39)$$

where $\beta = E_{ion}/k_BT$.

For non-degenerate free electrons, the solution of Equation 13.37 is

$$J_{ion}(E_{ion}) = \exp\left(\frac{\mu - E_{ion}}{k_BT}\right) \quad (13.40)$$

For non-degenerate electrons, we define a ratio R_{ion} for the value of $J_{ion}(E_{ion})$ relative to the low degeneracy value. We have

$$R_{ion} = \frac{J_{ion}(E_{ion})}{\exp(-E_{ion}/k_BT)}(2/n_e)(2\pi m_0 k_b T/h^2)^{3/2}. \quad (13.41)$$

Substituting into Equation 13.36 using the non-degenerate expression for $\exp(\mu/k_BT)$, (Equation 13.7) gives

$$K_{ion} \approx \frac{2}{\sqrt{\pi}}\left(\frac{2}{m_0}\right)^{1/2} \frac{E_{ion}}{(k_BT)^{1/2}} \sigma(E_{ion}, 0) \exp\left(-\frac{E_{ion}}{k_BT}\right). \quad (13.42)$$

The ionization rate coefficient changes by several orders of magnitude at high positive chemical potential, but the change is almost independent of the electron temperature for a constant value of the reduced chemical potential.

13.8.3 Detailed Balance of the Collisional Processes at High Density

Allowing for the free electrons to be affected by degeneracy, it is still possible to use detailed balance to determine the rates of three-body recombination and collisional de-excitation [109]. Instead of Equation 12.16 we can write generally for the relationship between the collisional ionisation and three-body recombination rate coefficients that

$$\frac{K_{ion}}{K_{rec}} = \frac{g_{Z+1}}{g_Z} \exp\left(-\frac{\mu}{k_B T}\right) \exp\left(-\frac{E_{ion}}{k_B T}\right). \tag{13.43}$$

The right-hand side of Equation 13.43 is, of course, the Saha ratio of populations N_{Z+1}/N_Z (see Equation 13.12).

For the collisional excitation and de-excitation rate coefficients, the relationship found previously when electron degeneracy effects are small (Equation 12.9) is valid.

13.9 Radiative Rates at High Density

The occupation of free-electron states at high density affects the rates of radiative absorption and emission between free electrons (free-free processes) and between free and bound quantum states (free-bound processes). We considered the radiative emission process in Chapter 5, but did not consider the possibility of degeneracy of the free electrons (for bremsstrahlung or free-free emission, see Section 5.2, and for radiative recombination or free-bound emission, see Section 5.4).

The rate of bremsstrahlung taking account of the free-electron degeneracy can be evaluated by noting the form of the integration over the free-electron energy used in Section 5.2. If we divide by the appropriate integration over the Maxwellian distribution and multiply by the appropriate integration over the Fermi-Dirac distribution, including the blocking factor for the final energy of the free electron, it is possible to obtain the emission coefficient ϵ_{ff}^* for bremsstrahlung including degeneracy from the emission coefficient ϵ_{ff} neglecting degeneracy. We have

$$\epsilon_{ff}^* = \epsilon_{ff} \frac{\int_{\hbar\omega}^{\infty} (1/E^{1/2}) f_{FD}(E)\, P(E - \hbar\omega)\, dE}{\int_{\hbar\omega}^{\infty} (1/E^{1/2}) f_M(E)\, dE} \tag{13.44}$$

where we have designated the electron energy distribution as $f_{FD}(E)$ for a Fermi-Dirac distribution and $f_M(E)$ for a Maxwellian distribution. $P(E - \hbar\omega)$ is the blocking factor for the electron final energy after the emission of a photon of energy $\hbar\omega$. The $1/E^{1/2}$ terms cancel with a $E^{1/2}$ term in both Fermi-Dirac and Maxwellian distributions functions. For the Maxwellian distribution we can replace the electron density by the non-degenerate expression for chemical potential μ (Equation 13.7). We have

$$f_M(E)dE = n_e \frac{2}{\sqrt{\pi}} \left(\frac{1}{k_B T}\right)^{\frac{3}{2}} E^{1/2} \exp\left(-\frac{E}{k_B T}\right) dE$$
$$= \frac{4}{\sqrt{\pi}} \left(\frac{2\pi m_0}{h^2}\right)^{3/2} \exp\left(\frac{\mu}{k_B T}\right) \exp\left(-\frac{E}{k_B T}\right).$$

The Fermi-Dirac distribution is given by Equation 13.15. We have

$$f_{FD}(E) = \frac{4}{\sqrt{\pi}} \left(\frac{2\pi m_0}{h^2}\right)^{3/2} E^{1/2} \frac{1}{\exp\left(-\frac{\mu - E}{k_B T}\right) + 1}$$

and for the blocking factor

$$P(E - \hbar\omega) = 1 - \frac{1}{\exp\left(-\frac{\mu - (E - \hbar\omega)}{k_B T}\right) + 1}.$$

With the above substitutions, the denominator of Equation 13.44 can be analytically integrated, leaving only the multiplication of the Fermi-Dirac and blocking factor integral requiring numerical solution. We have

$$\epsilon_{ff}^* = \epsilon_{ff} R_{ff}^*$$

with

$$R_{ff}^* = (2/n_e)(2\pi m_0 k_B T/h^2)^{3/2} \exp\left(\frac{\hbar\omega}{k_B T}\right) I_{int} \tag{13.45}$$

where

$$I_{int} = \int_{\frac{\hbar\omega}{k_B T}}^{\infty} \left(\frac{1}{\exp(-\mu - E/k_B T) + 1}\right) \left(1 - \frac{1}{\exp(-\mu - (E - \hbar\omega)/k_B T) + 1}\right) d\left(\frac{E}{k_B T}\right).$$

The ratio R_{ff}^* of the bremsstrahlung emission coefficient ϵ_{ff}^* including degeneracy to the emission coefficient ϵ_{ff} neglecting degeneracy varies with the chemical potential μ and the photon energy $\hbar\omega$. Due to detailed balance, the modification R_{ff}^* of the bremsstrahlung rate at frequency $\hbar\omega$ with and without degeneracy effects is the same as the modification of the inverse bremsstrahlung rate for frequency $\hbar\omega$.

Modifications to the rates of photo-ionisation can be evaluated by considering the blocking factor for the final energy of the ejected electron. We have for the ratio R_{bf}^* of the photo-ionisation cross-section σ_{bf}^* including degeneracy to the photo-ionisation cross-section σ_{bf} without degeneracy (see Section 5.4):

$$\sigma_{bf}^* = \sigma_{bf} R_{bf}^*$$

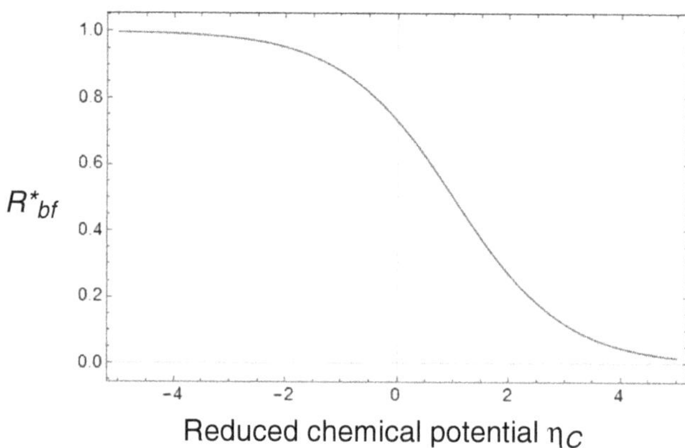

Figure 13.3 The ratio R^*_{bf} of the degenerate rates for photo-ionization to the rate calculated assuming the electrons are non-degenerate. Curves are shown for a photon energy $\hbar\omega$ equal to the electron temperature k_BT, while the ionization energy E_{ion} is assumed to have $E_{ion} = 2k_BT$.

where

$$R^*_{bf} = P(E_{ion} - \hbar\omega) = 1 - \frac{1}{\exp(-\mu - (E_{ion} - \hbar\omega)/k_BT) + 1} \tag{13.46}$$

where E_{ion} is the ionisation energy of the ion. Again, detailed balance means that the modification R^*_{bf} due to electron degeneracy of a cross-section for free-bound recombination at the same frequency $\hbar\omega$ is also R^*_{bf}. For plasmas with the same reduced chemical potential $\eta_C = \mu/k_BT$ for the free electrons, the degeneracy modifications for free-free and free-bound processes are similar (see Figure 13.3).

Exercises

13.1 In a laser-produced plasma experiment using a carbon target, emission from the hydrogen-like carbon Lyman β line ($n = 3 \rightarrow 1$) is observed, but not the Lyman γ ($n = 4 \rightarrow 1$) transition, implying that continuum lowering has affected this line. Estimate the electron density in the plasma. [5×10^{23} cm^{-3}]

13.2 In an inertial fusion experiment at the NIF (see [50]), the peak electron density of compressed deuterium-tritium fuel was measured as 10^{25} cm^{-3} with a minimum volume of the fuel of 3×10^5 μm^3. Assuming that the temperature of the fuel remains cold so that the chemical potential of the fuel remains large and positive, calculate the chemical potential energy of the ensemble of compressed electrons. [49 J]

13.3 Use Figure 13.1 to estimate the electron temperature for a plasma with an electron density of $10^{25}\,\text{cm}^{-3}$ if the reduced chemical potential $\eta = 8$. [21.5 eV]

13.4 Inertial fusion design studies require the electron pressure relative to the Fermi pressure of compressed deuterium-tritium fuel to be less than 1.7 [42]. Use Figures 13.1 and 13.2 to determine the required maximum electron temperature of the fuel if the electron density at peak compression is $10^{25}\,\text{cm}^{-3}$. [100 eV]

13.5 The high-density form of the Saha-Boltzmann equation (Equation 13.12) enables a straightforward evaluation of the equilibrium population $N_1(Z_i)$ of any ionisation stage of charge Z_i relative, say, to a singly charged ion population $N_1(1)$. Show that the ratio in thermal equilibrium of the ground state populations of Z_i charged ions to singly charged ions is given by

$$\frac{N_1(Z_i)}{N_1(1)} = \frac{g_{Z_i}}{g_1} \exp\left(-\frac{(Z_i - 1)\mu}{k_B T}\right) \exp\left(-\sum_{Z_i'=1}^{Z_i'=Z_i-1} \frac{E_{ion}(Z_i')}{k_B T}\right),$$

where μ is the chemical potential and $E_{ion}(Z_i')$ is the ionisation energy of the ions of charge Z_i'.

Appendix

Vectors, Maxwell's Equations, the Harmonic Oscillator and a Sum Rule

A.1 Vector Analysis

In the study of many aspects of physics, vector analysis can greatly simplify the writing of equations and bring to prominence the physical meaning of the equations. Vector analysis is essential in dealing with the more complex aspects of the propagation of light, the interaction of electric and magnetic fields on charged particles, in fluid and plasma dynamics, and in quantum mechanics and atomic physics. A brief review of vector analysis is given here.

A scalar quantity is completely characterised by its magnitude, while a vector quantity is characterised by its magnitude and direction. Examples of scalar quantities include the mass of an object, the density of a fluid (mass per unit volume), the energy of a particle and the volume of a space. Examples of vectors include the velocity of an object, the acceleration of an object and the position of a point relative to another point in space. In Cartesian co-ordinates, a vector $\mathbf{v}$ is specified by its components along the three Cartesian axes, namely v_x, v_y and v_z. We can write that

$$\begin{aligned} v_x &= |\mathbf{v}| \cos \alpha_1 \\ v_y &= |\mathbf{v}| \cos \alpha_2 \\ v_z &= |\mathbf{v}| \cos \alpha_3 \end{aligned} \tag{A.1}$$

where α_1, α_2 and α_3 are the angles of the vector $\mathbf{v}$ to the x, y and z axes, respectively. The quantity $|\mathbf{v}|$ is the magnitude of the vector. We adopt the standard convention of typing vectors in bold, un-italicised symbols. When handwriting vectors and in some texts, it is common to write a vector symbol with an underline or sometimes with an arrow above the symbol.

The sum of two vectors is defined as the vector whose components are the sums of the corresponding components of the original vectors. This means that vectors obey the familiar rules of addition, subtraction and association within parentheses that are familiar for scalar quantities. For example, we can write for three vectors **a**, **b** and **c** that

$$\mathbf{a} + (\mathbf{b} + \mathbf{c}) = (\mathbf{a} + \mathbf{b}) + \mathbf{c} = (\mathbf{a} + \mathbf{c}) + \mathbf{b} = \mathbf{a} + \mathbf{b} + \mathbf{c}. \tag{A.2}$$

A.1.1 Multiplication of Vectors

Multiplication of vector quantities is not as straightforward as addition. The simplest multiplication product is a scalar times a vector. This operation results in a vector with each component equal to the scalar times the corresponding component of the original vector. If c is a scalar and **a** is a vector, the product $c\mathbf{a}$ is defined by a vector **b** with components

$$\begin{aligned} b_x &= ca_x \\ b_y &= ca_y \\ b_z &= ca_z. \end{aligned} \tag{A.3}$$

If two vectors are to be multiplied, we consider two possibilities. The scalar and vector products produce, respectively, a scalar and another vector. Due to the symbols used to represent the scalar and vector products, the operations are often referred to as the dot and cross products. In the scalar or dot product, the components of the vectors along the Cartesian axes are multiplied together and added up. We write for the scalar product $\mathbf{a} \cdot \mathbf{b}$ of two vectors **a** and **b** that

$$\mathbf{a} \cdot \mathbf{b} = a_x b_x + a_y b_y + a_z b_z. \tag{A.4}$$

By considering a Cartesian co-ordinate system with one axis rotated parallel to one of the vectors, it is straightforward to show that this definition is equivalent to multiplying the magnitude of the vectors together with the cosine of the angle θ between the vectors, i.e.

$$\mathbf{a} \cdot \mathbf{b} = |\mathbf{a}|\,|\mathbf{b}| \cos\theta. \tag{A.5}$$

The vector or cross product of two vectors results in a vector. We write $\mathbf{a} \times \mathbf{b}$ for the vector product so that the resulting vector **c** is defined to have components

$$\begin{aligned} c_x &= a_y b_z - a_z b_y \\ c_y &= a_x b_z - a_z b_x \\ c_z &= a_x b_y - a_y b_x. \end{aligned} \tag{A.6}$$

The vector product is more easily remembered in terms of a determinant. If we define unit vectors $\hat{x}$, $\hat{y}$ and $\hat{z}$ along the respective x, y and z Cartesian axes, the vector product can be written in terms of a determinant, such that

$$\mathbf{a} \times \mathbf{b} = \begin{vmatrix} \hat{x} & \hat{y} & \hat{z} \\ a_x & a_y & a_z \\ b_x & b_y & b_z \end{vmatrix} = \hat{x}(a_y b_z - a_z b_y) + \hat{y}(a_x b_z - a_z b_x) + \hat{z}(a_x b_y - a_y b_x). \quad \text{(A.7)}$$

The vector resulting from the vector product can be shown to be directed perpendicular to the plane formed by the two multiplying vectors with a magnitude given by multiplying the magnitude of the two vectors with the sine of the angle θ between the vectors:

$$|\mathbf{a} \times \mathbf{b}| = |\mathbf{a}|\,|\mathbf{b}| \sin\theta. \quad \text{(A.8)}$$

The right-hand rule is used to determine the positive direction of the vector resulting from a vector product. The first vector (here **a**) is imaginarily pushed with the fingers of the right hand into the second vector (here **b**) and the direction of the thumb on the hand gives the direction of $\mathbf{a} \times \mathbf{b}$. The right-hand rule implies that $\mathbf{a} \times \mathbf{b} = -\mathbf{b} \times \mathbf{a}$.

A.1.2 Vector Calculus

The extension of vector operations to include differentiation and integration (i.e. vector calculus) will now be considered. The simplest operation to consider here is the directional derivative of a scalar quantity. Say we are interested in how the scalar quantity ψ, which may be a function of several variables such as x, y and z, varies in a particular direction $\mathbf{s}$. To find the infinitesimal change $d\psi$ in the direction $d\mathbf{s}$, we can write

$$\frac{d\psi}{ds} = \frac{\partial \psi}{\partial x}\frac{dx}{ds} + \frac{\partial \psi}{\partial y}\frac{dy}{ds} + \frac{\partial \psi}{\partial z}\frac{dz}{ds} \quad \text{(A.9)}$$

where ds represents the magnitude of $d\mathbf{s}$. A particular directional derivative is the maximum directional derivative $\mathbf{s}_{max}$. The maximum directional derivative is given a special operator symbol ∇ such that

$$\mathbf{s}_{max} = \nabla \psi.$$

The expression $\nabla\psi$ is termed the gradient of ψ. In Cartesian co-ordinates, the gradient is given by

$$\nabla \psi = \hat{x}\frac{\partial \psi}{\partial x} + \hat{y}\frac{\partial \psi}{\partial y} + \hat{z}\frac{\partial \psi}{\partial z}. \quad \text{(A.10)}$$

A.1.3 Integration

With integration involving vectors, there are three kinds of integral: line, surface and volume, and the integrand can be either a vector or a scalar. The most interesting and useful combinations are the line integral of a vector, the surface integral of a vector, and volume integrals of both vectors and scalars. If **F** is a vector, the line integral of **F** is written as a scalar product of **F** and vectors for increments of distance $d\mathbf{l}$ along the line C of interest. We have

$$\int_{aC}^{b} \mathbf{F} \cdot d\mathbf{l}$$

where C is the curve in x, y and z co-ordinates along which the integration is performed, a and b represent the initial and final points on the curve, and $d\mathbf{l}$ is an infinitesimal vector displacement along the curve C.

The surface integral of a vector **F** over an area S is written as a scalar product of **F** and vectors for the increments of area $d\mathbf{A}$. We have

$$\int_{S} \mathbf{F} \cdot d\mathbf{A}$$

where incremental areas dA are written as vectors $d\mathbf{A}$ of magnitude dA with a direction normal to the surface.

An important operator in vector calculus, which is essentially a derivative, is the divergence operator. The divergence of a vector is the limit of its surface integral per unit volume as the volume enclosed by the surface goes to zero. We use the symbol $\nabla\cdot$ to represent the divergence and write for the divergence of a vector **F**:

$$\nabla \cdot \mathbf{F} = \mathrm{Lim}_{V \to 0} \int_{S} \mathbf{F} \cdot d\mathbf{A} \tag{A.11}$$

where the integration is over the closed surface S enclosing the volume V.

The third interesting vector differential operator is the curl. The curl of a vector (written $\nabla\times$) can be defined as the limit of the surface integral of the vector product of vector **F** per unit volume with a unit vector in the direction of the surface normal as the volume enclosed by the surface goes to zero. We can write:

$$\nabla \times \mathbf{F} = \mathrm{Lim}_{V \to 0} \frac{1}{V} \int_{S} \mathbf{F} \times d\mathbf{A}. \tag{A.12}$$

The form of the curl in Cartesian co-ordinates can be evaluated using the determinant expression given by

$$\nabla \times \mathbf{F} = \begin{vmatrix} \hat{x} & \hat{y} & \hat{z} \\ \frac{\partial}{\partial x} & \frac{\partial}{\partial y} & \frac{\partial}{\partial z} \\ F_x & F_y & F_z \end{vmatrix} \tag{A.13}$$

where F_x, F_y and F_z are the components of the vector **F** in the directions x, y, and z.

A.1.4 Combinations of Operations

It is possible to combine grad, div and curl operations. An important double operation is the divergence of the gradient of a scalar field. This double operation is sufficiently important that it is given its own name (the Laplacian) and symbol ∇^2 (instead of writing $\nabla \cdot \nabla$ which would be the normal way of writing the divergence of a gradient in the notation used thus far). In Cartesian co-ordinates it is straightforward to show that the Laplacian of a scalar function ψ is given by

$$\nabla^2 \psi = \frac{\partial^2 \psi}{\partial x^2} + \frac{\partial^2 \psi}{\partial y^2} + \frac{\partial^2 \psi}{\partial z^2}. \tag{A.14}$$

Manipulating the definitions of grad, div and curl in Cartesian co-ordinates, it is straightforward to show that the curl of a gradient ($\nabla \times \nabla$) of a scalar and the divergence of a curl ($\nabla \cdot \nabla \times$) of a vector are identically equal to zero. Taking the curl of the curl ($\nabla \times \nabla \times$) of a vector $\mathbf{F}$ gives

$$\nabla \times \nabla \times \mathbf{F} = \nabla(\nabla \cdot \mathbf{F}) - \nabla^2 \mathbf{F} \tag{A.15}$$

where the Laplacian (∇^2) of a vector is defined as a vector with Cartesian components in x, y and z, which are the Laplacian values of the original vector components in x, y and z. We can write the Laplacian of a vector as

$$\nabla^2 \mathbf{F} = \left(\frac{\partial^2 F_x}{\partial x^2} + \frac{\partial^2 F_x}{\partial y^2} + \frac{\partial^2 F_x}{\partial z^2} \right) \hat{x} + \left(\frac{\partial^2 F_y}{\partial x^2} + \frac{\partial^2 F_y}{\partial y^2} + \frac{\partial^2 F_y}{\partial z^2} \right) \hat{y} + \left(\frac{\partial^2 F_z}{\partial x^2} + \frac{\partial^2 F_z}{\partial y^2} + \frac{\partial^2 F_z}{\partial z^2} \right) \hat{z}. \tag{A.16}$$

The Laplacian of the electric field vector is found in the wave equation describing propagation of electromagnetic waves (see Section 2.1).

A.1.5 Spherical Polar Co-Ordinates

Atomic physics calculations involve a central potential due to the charge of the nucleus, so a spherical polar co-ordinate system, rather than Cartesian co-ordinates (as described above), is usually employed, for example, to solve the Schrodinger equation (see Section 7.2) or the distribution of charge in an ion (see Section 13.5). The relationship of the spherical polar co-ordinates r, θ and ϕ to the Cartesian co-ordinates x, y and z is shown in Figure 7.1 in the discussion of the hydrogen atom and hydrogen-like ions. We have

$$x = r \sin\theta \, \cos\phi,$$
$$y = r \sin\theta \, \sin\phi,$$
$$z = r \cos\theta.$$

The vector operations of grad, div and curl are obtained from these relations. Using unit vectors $\hat{\mathbf{r}}$, $\hat{\theta}$ and $\hat{\phi}$ for the co-ordinates r, θ and ϕ, the grad, div, and curl operations can be shown to produce, respectively:

$$\nabla\psi = \hat{\mathbf{r}}\frac{\partial\psi}{\partial r} + \hat{\theta}\frac{1}{r}\frac{\partial\psi}{\partial\theta} + \hat{\phi}\frac{1}{r\sin\theta}\frac{\partial\psi}{\partial\phi}, \tag{A.17}$$

$$\nabla\cdot\mathbf{F} = \frac{1}{r^2}\frac{r^2\partial F_r}{\partial r} + \frac{1}{r\sin\theta}\frac{\partial F_\theta\sin\theta}{\partial\theta} + \frac{1}{r\sin\theta}\frac{\partial F_\phi}{\partial\phi}, \tag{A.18}$$

$$\begin{aligned}\nabla\times\mathbf{F} = {}& \frac{1}{r\sin\theta}\left[\frac{\partial(F_\phi\sin\theta)}{\partial\theta} - \frac{\partial F_\theta}{\partial\phi}\right]\hat{r} \\ &+ \frac{1}{r}\left[\frac{1}{\sin\theta}\frac{\partial F_r}{\partial\phi} - \frac{rF_\phi}{\partial r}\right]\hat{\theta} + \frac{1}{r}\left[\frac{\partial rF_\theta}{\partial r} - \frac{\partial F_r}{\partial\theta}\right]\hat{\phi}.\end{aligned} \tag{A.19}$$

The Laplacian of a scalar quantity in spherical polar co-ordinates is used in the solution of the Schrodinger equation for hydrogen and hydrogen-like ions (see Section 7.2). We have

$$\nabla^2 = \frac{1}{r^2}\left[\frac{\partial}{\partial r}\left(r^2\frac{\partial}{\partial r}\right) + \frac{1}{\sin\theta}\frac{\partial}{\partial\theta}\left(\sin\theta\frac{\partial}{\partial\theta}\right) + \frac{1}{\sin^2\theta}\frac{\partial^2}{\partial\phi^2}\right]. \tag{A.20}$$

A.2 Maxwell's Equations

The behaviour of electrical and magnetic fields is determined by a set of four equations known as Maxwell's equations. Maxwell's equations describe experimental observations and can be taken as fundamental equations describing experimental observations. They can, however, be written in different forms. In the differential form as first discussed here, they describe experimental observations relating current flow density **J** (charge per unit area per unit time), magnetic field strength **B**, and electric field strength **E**. Quantities called the magnetic intensity **H** and the electric displacement **D** defined by $\mathbf{H} = (\mu_0\mu_r)^{-1}\mathbf{B}$ and $\mathbf{D} = \epsilon_0\epsilon_r\mathbf{E}$ are useful in dealing with the fields inside media (other than vacuum). Here $\mu_0 = 4\pi \times 10^{-7}\,\mathrm{N\,A^{-2}}$ is the magnetic permeability; $\mu_r(\approx 1)$ is the relative magnetic permeability, which is dependent on the medium; $\epsilon_0 = 8.854 \times 10^{-12}\,\mathrm{F\,m^{-1}}$ is the permittivity or dielectronic constant; and $\epsilon_r(\approx 1)$ is the relative dielectronic constant, which is also dependent on the medium. In vacuum μ_r and ϵ_r are both in unity. We are using Systeme International (SI) units here. The units mentioned are Newtons (N), Amps (A) and Farads (F) as well as the more familiar metres (m).

Maxwell's equations can be written using vector notation as

$$\nabla\times\mathbf{H} = \mathbf{J} + \frac{\partial\mathbf{D}}{\partial t}$$

$$\nabla\times\mathbf{E} = -\frac{\partial\mathbf{B}}{\partial t}$$

$$\nabla \cdot \mathbf{D} = \rho_c$$
$$\nabla \cdot \mathbf{B} = 0 \tag{A.21}$$

where ρ_c is the charge density of the medium. The first (Ampere's law) and the second (Faraday's law) of Maxwell's equations use the 'curl' operator ($\nabla\times$) which determines the 'rotation' of a vector around a point, while the third (Gauss's law) and fourth use the divergence operator ($\nabla\cdot$) which determines the 'outflow' of a vector from a point. Ampere's law relates the production of a magnetic field by a flowing current or changing electric field. Faraday's law shows how a changing magnetic field can produce an electric field. Gauss's law shows that electrical charge is a source of electric field. There is no such source of magnetic field – 'magnetic monopoles' do not exist, so the divergence of a magnetic field is always zero.

There are a number of ways of relating the electric displacement **D** and electric field **E**. We have

$$\mathbf{D} = \epsilon_0 \mathbf{E} + \mathbf{P} = \epsilon_0(1+\chi)\mathbf{E} = \epsilon_0 n^2 \mathbf{E}$$

where **P** is the polarisation of the medium, χ is the susceptibility of the medium and n is the refractive index of the medium. The way of relating **D** and **E** is determined by the problem involved and certain conventions for different areas of study have developed. For example, linear optics uses the refractive index.

The differential form of Maxwell's equation can be converted to the integral form using Stokes' theorem. This theorem states that the line integral of a vector **F** around a closed curve is equal to the integral of the normal component of the curl of the vector over the area bounded by the curve. We can write

$$\int_C \mathbf{F} \cdot d\mathbf{l} = \int_S \nabla \times \mathbf{F} \cdot d\mathbf{A}$$

where integration around the contour is indicated by C and over the area by S. Applying Stokes' theorem, Ampere's law becomes

$$\int_C \mathbf{B} \cdot d\mathbf{l} = \mu_0 I + \mu_0 \epsilon_0 \frac{\partial}{\partial t}\left[\int_S \mathbf{E} \cdot d\mathbf{A}\right]$$

where I is the current flowing through the surface S as defined by the closed curve C. For Faraday's law

$$\int_C \mathbf{E} \cdot d\mathbf{l} = -\frac{\partial}{\partial t}\left[\int_S \mathbf{B} \cdot d\mathbf{A}\right].$$

Gauss's law converts to integral form by requiring that the integral of the electric field over a closed area is related to the charge enclosed by the area:

$$\int_S \mathbf{E} \cdot d\mathbf{A} = \frac{1}{\epsilon_0} \int_V \rho_c dV.$$

The equivalent equation for the magnetic field is

$$\int_S \mathbf{B} \cdot d\mathbf{A} = 0.$$

Instead of directly using the vectors for the electric field **E** and magnetic field **B**, it is possible to treat electromagnetic fields using two quantities known as the vector potential **A** and the scalar potential V_P. The vector potential is related to the magnetic field by

$$\mathbf{B} = \nabla \times \mathbf{A}.$$

The scalar potential represents a contribution to the electric field due to electrostatic effects associated with electric charges. The total electric field is the electrostatic contribution plus the electric field created by a changing magnetic field, so that

$$\mathbf{E} = -\nabla V_p - \frac{\partial \mathbf{A}}{\partial t}.$$

A purely electrostatic electric field **E** is related to an electric potential V_P by $\mathbf{E} = -\nabla V_P$. Using Gauss's law, we have that the electric potential is related to the charge density ρ_c, so that

$$\nabla^2 V_P = -\frac{\rho_c}{\epsilon_0}.$$

This expression is known as Poisson's equation.

A.3 The Harmonic Oscillator

The one-dimensional quantum mechanical harmonic oscillator is relevant to our derivation of the Planck black-body radiation energy density (Chapter 4) and to molecular vibrational energies (Chapter 9). A classical harmonic oscillator has a mass m_{red} oscillating in position r with a restoring force F proportional to the displacement r. We can write that

$$F = -k_s r$$

where k_s is a proportionality constant for the force. The equation of motion of the oscillating mass is given by

$$\frac{d^2 r}{dt^2} + \left(\frac{k_s}{m_{red}}\right) r = 0.$$

Solutions of this equation are of form $r = r_0 \exp(-i\omega_{HO} t)$, where r_0 is the amplitude of the oscillation and the oscillation frequency ω_{HO} is given by

$$\omega_{HO} = \left(\frac{k_s}{m_{red}}\right)^{1/2}.$$

The equation of motion written as

$$\frac{d^2 r}{dt^2} + \omega_{HO}^2 \, r = 0$$

is used in Chapter 4 to show that electric field $\mathbf{E}(t)$ variations in a black-body cavity have quantised energies given by the quantum solutions of the one-dimensional harmonic oscillator.

The restoring force of the classical harmonic oscillator is related to the potential energy U_p of the oscillator by $F = -dU_p/dr$ so that the potential energy is given by

$$U_p = \frac{1}{2} k_s r^2 = \frac{1}{2} m_{red} \, \omega_{HO}^2 \, r^2.$$

The Schrodinger equation for the one-dimensional harmonic oscillator is consequently given by

$$-\frac{\hbar^2}{2m_{red}} \frac{d^2 \psi_p}{dr^2} + \frac{1}{2} m_{red} \, \omega_{HO}^2 \, r^2 \psi_p = E_p \psi_p \tag{A.22}$$

where ψ_p are appropriate wavefunctions and E_p are the energy eigenvalues. Multiplying throughout by $2/\hbar\omega_{HO}$, we have

$$\left[-\frac{\hbar}{m_{red}\,\omega_{HO}} \frac{d^2}{dr^2} + \frac{m_{red}\,\omega_{HO}}{\hbar} r^2\right] \psi_p = \frac{2E_p}{\hbar\omega_{HO}} \psi_p.$$

Introducing the variables

$$y = \left(\frac{m_{red}\,\omega_{HO}}{\hbar}\right)^{1/2} r,$$

$$E_p^* = \frac{E_p}{\hbar\omega_{HO}},$$

the Schrodinger equation becomes

$$\left(\frac{d^2}{dy^2} - y^2\right) \psi_p = -2E_p^* \psi_p. \tag{A.23}$$

Multiplication of the terms in the left-hand side of the equal sign shows that

$$\left(\frac{d}{dy}+y\right)\left(\frac{d}{dy}-y\right)\psi_p=\left(\frac{d^2}{dy^2}-y^2-1\right)\psi_p,$$

$$\left(\frac{d}{dy}-y\right)\left(\frac{d}{dy}+y\right)\psi_p=\left(\frac{d^2}{dy^2}-y^2+1\right)\psi_p.$$

If we multiply Equation A.23 by $(d/dy-y)$, then

$$\left(\frac{d}{dy}-y\right)\left(\frac{d}{dy}+y\right)\left(\frac{d}{dy}-y\right)\psi_p=(-2E_p^*-1)\left(\frac{d}{dy}-y\right)\psi_p. \tag{A.24}$$

For this relationship to be valid, we must have that

$$\left(\frac{d}{dy}-y\right)\psi_p=0 \tag{A.25}$$

or that

$$\left(\frac{d}{dy}-y\right)\psi_p=\psi_{p+1}, \tag{A.26}$$

where ψ_{p+1} is another wavefunction solution of the Schrodinger equation. Equation A.25 has wavefunction solutions which diverge at large r so can be discounted. Using Equation A.26, Equation A.24 becomes

$$\left(\frac{d}{dy}-y\right)\left(\frac{d}{dy}+y\right)\psi_{p+1}=(-2E_p^*-1)\psi_{p+1},$$

which can be re-written as

$$\left(\frac{d^2}{dy^2}-y^2\right)\psi_{p+1}=(-2E_p^*-2)\psi_{p+1}. \tag{A.27}$$

Equation A.27 shows that the energy eigenvalues E_{p+1}^* of the wavefunction ψ_{p+1} are given by

$$E_{p+1}^*=E_p^*+1. \tag{A.28}$$

In absolute terms the energy eigenvalues are given by

$$E_{p+1}=(E_p+1)\,\hbar\omega_{HO}. \tag{A.29}$$

Equation A.23 can be directly solved to determine the ground state solution of the Schrodinger equation. If we substitute $\psi_0=\exp(-y^2/2)$ into Equation A.23, we can verify this wavefunction as a solution of the harmonic oscillator Schrodinger equation and find an energy eigenvalue. We have that

$$E_0^*=\frac{1}{2}$$

which must be the ground state (minimum) energy as it is not possible to satisfy Equation A.28 with a lower positive energy. Combining this result with Equation A.29 means that the harmonic oscillator energies are given by

$$E_p = \left(p + \frac{1}{2}\right) \hbar\omega_{HO} \tag{A.30}$$

where $p = 0, 1, 2,...$ is an integer. This energy spacing is used in the derivation of the Planck black-body radiation expression and in our treatment of the vibration energies of molecules. In the Planck black-body expression, the ground state energy is known as the zero-point energy.

A.4 The Thomas–Reiche–Kuhn Sum Rule

The Thomas-Reiche-Kuhn sum rule provides an upper limit to an oscillator strength f_{1j} for a radiative transition. Due to the proportionality between oscillator strength and transition probability (see Equation 4.29), the rule also gives an upper limit for a radiative transition probability and hence the opacity (and rate of stimulated emission) of a transition (see Equation 6.29). The rule is stated as Equation 10.29 in Chapter 10 in the form:

$$\sum_j f_{1j} = \frac{2m_0}{\hbar^2}\left(\frac{1}{3}\right)\sum_j (E_j - E_1)\left|\int \psi_1^* \mathbf{r} \psi_j dV\right|^2 = 1. \tag{A.31}$$

This version of the Thomas-Reiche-Kuhn sum rule includes a factor $1/3$ as the integral over the wavefunctions is angle averaged to allow for random orientations of atoms or ions to the direction of incoming light in the absorption process. We use the symbol j for a different quantum state in order to avoid confusion with the imaginary number $i = \sqrt{-1}$. The rule shows that a summation of oscillator strengths from a quantum state 1 to all other possible states j for an atom or ion is equal to one.

The Thomas-Reiche-Kuhn sum rule can be proved for an atomic or ionic quantum system with energy eigenvalues given by the Hamiltonian operator

$$\hat{H}_0 = \frac{\hat{p}^2}{2m_0} + V(r)$$

where $\hat{p}$ is the momentum operator and $V(r)$ is the potential energy in the field of the nuclear charge. We need to consider the commutation of operators. The commutator of two operators $\hat{A}$ and $\hat{B}$ is defined to be

$$[\hat{A}, \hat{B}] = \hat{A}\hat{B} - \hat{B}\hat{A}.$$

Operators in quantum mechanics do not always commute, which means that

$$[\hat{A}, \hat{B}] \neq 0.$$

For example, the commutation of the operator for position $\hat{x}$ in a particular direction with the commutation of $\hat{x}$ and the Hamiltonian $\hat{H}_0$ is given by

$$[\hat{x}, [\hat{x}, H_0]] = \left[\hat{x}, \left[\hat{x}, \frac{\hat{p}^2}{2m_0} + V(r)\right]\right] = \left[\hat{x}, \left[\hat{x}, \frac{\hat{p}^2}{2m_0}\right]\right] = \frac{1}{2m_0}[\hat{x}, 2i\hbar\hat{p}] = -\frac{\hbar^2}{m_0} \tag{A.32}$$

as $[\hat{x}, \hat{p}] = i\hbar$.

The sum rule for all operators $\hat{f}$ classed as Hermitian, where

$$\int \psi^* \hat{f} \psi dV = \int \psi \hat{f} \psi^* dV$$

have the property that

$$\frac{1}{2} \int \psi_1^* [\hat{f}, [\hat{f}, H_0]] \psi_1 dV = -\sum_j (E_j - E_1) \left| \int \psi_1^* \hat{f} \psi_j dV \right|^2 . \tag{A.33}$$

A proof of this relationship for Hermitian operators can be found in, for example, Wang [120]. From Equation A.32, we have that

$$\int \psi_1^* [\hat{x}, [\hat{x}, H_0]] \psi_1 dV = -\frac{\hbar^2}{m_0} \int \psi_1^* \psi_1 dV = -\frac{\hbar^2}{m_0}. \tag{A.34}$$

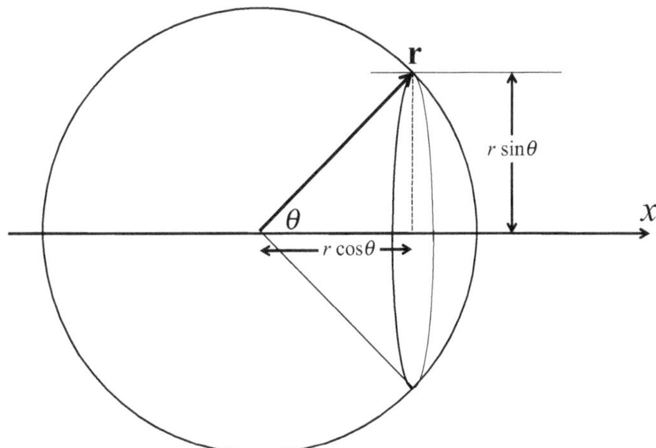

Figure A.1 The geometry required for an integration of all possible values of the square of the component of a vector **r** along a direction x. The locus of the vector at an angle θ to x is illustrated.

Using the position operator so that $\hat{f} = \hat{x} = x$ in Equation A.33 and equating this to Equation A.34 gives

$$\sum_j (E_j - E_1) \left| \int \psi_1^* x \psi_j dV \right|^2 = \frac{\hbar^2}{2m_0}. \quad \text{(A.35)}$$

The version of the sum rule given by Equation A.35 implicitly assumes a fixed atomic orientation to the electric field of the electromagnetic radiation (in the x-direction). Taking account of the random orientation of atoms to an incoming radiation field in the integration over the wavefunctions introduces the factor $1/3$ in Equation A.31. If the component x in the direction of the electric field of a position vector $\mathbf{r}$ is given by $x = r\cos\theta$, then the average value of x^2 for an ensemble of atoms at random angles θ to the electric field is given by integration over the surface of a sphere of radius r (see Figure A.1). We have for the average of x^2:

$$x^2 = r^2 \frac{\int_0^\pi (\cos^2\theta)\, 2\pi \sin\theta \, d\theta}{\int_0^\pi 2\pi \sin\theta \, d\theta} = \frac{1}{3} r^2.$$

References

[1] Ackermann, W., Asova, G., Ayvazyan, et al. 2007. Operation of a free-electron laser from the extreme ultraviolet to the water window. *Nat. Photonics*, **1**(6), 336–342.

[2] Alexander, D. R., and Ferguson, J. W. 1994. Low-temperature Rosseland opacities. *Astrophys. J.*, **437**, 879–891.

[3] Aslanyan, V., and Tallents, G. J. 2014. Local thermodynamic equilibrium in rapidly heated high energy density plasmas. *Phys. Plasmas*, **21**(6), 062702.

[4] Attwood, D. 2000. *Soft X-rays and extreme ultraviolet radiation: principles and applications*. Cambridge University Press, Cambridge, UK.

[5] Atzeni, S., and Meyer-ter-Vehn, J. 2009. *The physics of inertial fusion*. Oxford University Press, Oxford, UK.

[6] Bailey, J. E., Nagayama, T., Loisel, G. P. et al. 2015. A higher-than-predicted measurement of iron opacity at solar interior temperatures. *Nature*, **517**(7532), 56–59.

[7] Bar-Shalom, A., Klapisch, M., and Oreg, J. 2001. HULLAC, an integrated computer package for atomic processes in plasmas. *J. Quant. Spect. Rad. Trans.*, **71**, 169–188.

[8] Bates, D. R., Kingston, A. E., and McWhirter, R. W. P. 1962. Recombination between electrons and atomic ions 1. Optically thin plasmas. *Proc. R. Soc. A*, **267**, 297–312.

[9] Bernstein, J., and Dyson, F. 2003. Opacity bounds. *Publ. Astron. Soc. Pac.*, **115**(814), 1383–1387.

[10] Blitz, L., and Spergel, D. N. 1991. The shape of the galaxy. *Astrophys. J.*, **370**, 205–224.

[11] Boiko, V. A., Faenov, A. Y., and Pikuz, S. A. 1978. X-ray spectroscopy of multiply-charged ions from laser plasmas. *J. Quant. Spect. Rad. Trans.*, **19**, 11–50.

[12] Boiko, V. A., Pikuz, S. A., and Faenov, A. Y. 1979. The determination of laser plasma electron density by K spectra of multicharged ions. *J. Phys. B.*, **12**, 1889–1910.

[13] Bombarda, F., Giannella, R., Kallne, et al. 1988. Observations and comparisons with theory of the heliumlike and hydrogenlike resonance lines and satellites of nickel from the JET tokamak. *Phys. Rev. A*, **37**, 504–522.

[14] Burgess, A., and Tully, J. A. 1978. On the Bethe approximation. *J. Phys. B*, **11**, 4271–4282.

[15] Campbell, G., Conn, R. W., and Shoji, T. 1991 (Feb. 5). High density plasma deposition and etching apparatus. US Patent 4,990,229.

[16] Chandrasekhar, S. 1930. The ionization formula and the new statistics. *Phil. Mag.*, **9**, 292–299.

[17] Chen, F. F. 1984. *Plasma physics and controlled fusion*. Plenum, New York, US.

[18] Chung, H. K., Chen, M. H., Morgan, W. L., Ralchenko, Y., and Lee, R. W. 2005. FLYCHK: Generalized population kinetics and spectral model for rapid spectroscopic analysis for all elements. *High Energ. Dens. Phys.*, **1**, 3–12.

[19] Ciricosta, O., Vinko, S. M., Barbrel, et al. 2016. Measurements of continuum lowering in solid-density plasmas created from elements and compounds. *Nat. Commun.*, **7**, 11713.

[20] Colvin, J., and Larsen, J. 2014. *Extreme physics: properties and behavior of matter at extreme conditions*. Cambridge University Press, Cambridge, UK.

[21] Crowley, B. J. B. 2014. Continuum lowering – a new perspective. *High Energ. Dens. Phys.*, **13**, 84–102.

[22] Crowley, B. J. B., and Harris, J. W. 2001. Modelling of plasmas in an average-atom local density approximation: the CASSANDRA code. *J. Quant. Spec. Rad. Trans.*, **71**, 257–272.

[23] Dendy, R. O. 1990. *Plasma dynamics*. Oxford Science Publications, Oxford, UK.

[24] Dere, K. P., Landi, E., Mason, H. E., Fossi, B. C. M., and Young, P. R. 1997. CHIANTI – an atomic database for emission lines I. Wavelengths greater than 50 Angstrom. *Astron. Astrophys. Suppl. Ser.*, **125**, 149–173.

[25] Dicke, R. H. 1953. The effect of collisions upon the Doppler width of spectral lines. *Phys. Rev.*, **89**, 472–473.

[26] Dirac, P. A. M. 1948. *The principles of quantum mechanics*. Clarendon Press, Oxford, UK.

[27] Djaoui, A., and Rose, S. J. 1992. Calculation of the time-dependent excitation and ionization in a laser-produced plasma. *J. Phys. B*, **25**, 2745–2762.

[28] Drake, R. P. 2006. *High-energy-density physics*. Springer, Berlin, Germany.

[29] El-Naschie, M. S. 2014. Casimir-like energy as a double Eigenvalues of quantumly entangled system leading to the missing dark energy density of the cosmos. *Int. J. High Energy Phys.*, **1**(5), 55–63.

[30] Emma, P., Akre, R., Arthur, et al. 2010. First lasing and operation of an angstrom-wavelength free-electron laser. *Nat. Photonics*, **4**, 641–647.

[31] Ferland, G. J., Korista, K. T., Verner, D. A., Ferguson, J. W., Kingdon, J. B., and Verner, E. M. 1998. CLOUDY 90: numerical simulation of plasmas and their spectra. *Publ. Astron. Soc. Pac.*, **110**, 761–778.

[32] Fletcher, L. B., Kritcher, A. L., Pak, A. et al. 2014. Observations of continuum depression in warm dense matter with X-ray Thomson scattering. *Phys. Rev. Lett.*, **112**, 145004.

[33] Florescu-Mitchella, A. I., and Mitchel, J. B. A. 2006. Dissociative recombination. *Phys. Rep.*, **430**, 277.

[34] Forslund, D. W., Kindel, J. M., Lee, K., Lindman, E. L., and Morse, R. L. 1975. Theory and simulation of resonant absorption in a hot plasma. *Phys. Rev. A*, **11**(Feb.), 679–683.

[35] Freidberg, J. P. 2007. *Plasma physics and fusion energy*. Cambridge University Press, Cambridge, UK.

[36] Fridman, A. 2008. *Plasma chemistry*. Cambridge University Press, Cambridge, UK.

[37] Gabriel, A. H. 1972. Dielectronic satellite spectra for highly-charged helium-like lines. *Monthly Not. R. Astron. Soc.*, **160**, 99–119.

[38] Griem, H. R. 1997. *Principles of plasma spectroscopy*. Cambridge University Press, Cambridge, UK.

[39] Gu, M. F. 2008. The flexible atomic code. *Can. J. Phys.*, **86**, 675–689.

[40] Guillot, T. 1999. Interiors of giant planets inside and outside the solar system. *Science*, **286**, 72–77.

[41] Guzman, F., O'Mullane, M., and Summers, H. P. 2013. ADAS tools for collisional-radiative modelling of molecules. *J. Nucl. Mater.*, **438**, S585.

[42] Haan, S. W., Lindl, J. D., Callahan, D. A. et al. 2011. Point design targets, specifications, and requirements for the 2010 ignition campaign on the National Ignition Facility. *Phys. Plasmas*, **18**, 051001.

[43] Haken, H., and Wolf, H. C. 1994. *The physics of atoms and quanta*. Berlin, Heidelberg: Springer Berlin Heidelberg.

[44] Hammer, J. H., and Rosen, M. D. 2003. A consistent approach to solving the radiation diffusion equation. *Phys. Plasmas*, **10**, 1829–1845.

[45] Hewish, A., Bell, S. J., Pilkington, J. D. H., Scott, P. F., and Scott, R. A. 1968. Observation of a rapidly pulsating radio source. *Nature*, **217**, 709–713.

[46] Hill, E. G., and Rose, S. J. 2012. Modelling of Silicon in inertial confinement fusion conditions. *High Energ. Dens. Phys.*, **8**, 307–312.

[47] Hirata, C. M. 1992. Wouthuysen-Field coupling strength and application to high-redshift 21-cm radiation. *Mon. Note. R. Astron. Soc.*, **367**, 259–274.

[48] Hoarty, D. J., Allan, P., James, S. F. et al. 2013. Observations of the effect of ionization-potential depression in hot dense plasma. *Phys. Rev. Lett.*, **110**(26), 265003.

[49] Hughes, T. P. 1975. *Plasma and laser light*. Institute of Physics, Bristol, UK.

[50] Hurricane, O. A., Callahan, D. A., Casey, D. T. et al. 2014. Fuel gain exceeding unity in an inertially confined fusion implosion. *Nature*, **506**, 343–348.

[51] Hutchinson, I. H. 2002. *Principles of plasma diagnostics*. Cambridge University Press, Cambridge, UK.

[52] Ichimaru, S. 1982. Strongly coupled plasmas – high density classical plasmas and degenerate electron liquids. *Rev. Mod. Phys.*, **54**(4), 1017–1059.

[53] Iglesias, C. A. 2015. Enigmatic photon absorption in plasmas near solar interior conditions. *High Energ. Dens. Phys.*, **15**(Jun), 4–7.

[54] Irons, F. E. 1979. The escape factor in plasma spectroscopy I. The escape factor defined and evaluated. *J. Quant. Spect. Rad. Trans.*, **22**, 1–20.

[55] Ishikawa, T., Aoyagi, H., Asaka, T. et al. 2012. A compact X-ray free-electron laser emitting in the sub-angstrom region. *Nat. Photonics*, **6**(8), 540–544.

[56] Janicki, C. 1990. A computer program for the free-free and bound-free Gaunt factors of Rydberg systems. *Comput. Phys. Commun.*, **60**, 281.

[57] Karzas, W. J., and Latter, R. 1961. Electron radiative transitions in a coulomb field. *Astrophys. J. Suppl. V*, **55**, 167.

[58] Keldysh, L. V. 1965. Ionization in field of a strong electromagentic wave. *Sov. Phys. JETP*, **20**, 1307.

[59] Killiana, T. C., Pattard, T., Pohl, T. et al. 2007. Ultracold neutral plasmas. *Phys. Rep.*, **449**, 77–130.

[60] Kim, Y., and Rudd, M. E. 1994. Binary encounter dipole model for electron-impact ionization. *Phys. Rev. A.*, **50**, 3954–3967.

[61] Kohn, W. 1999. Nobel lecture: electronic structure of matterwave functions and density functionals. *Rev. Mod. Phys.*, **71**, 1253–1266.

[62] Kramers, H. A. 1923. On the theory of X-ray absorption and of the continuous X-ray spectrum. *Phil. Mag.*, **46**, 836–871.

[63] Kramida, A., Ralchenko, Y., Reader, J., and NIST ASD Team. 2015. NIST Atomic Spectra Database (ver. 5.3). Available: http://physics.nist.gov/asd (accessed 13 April 2017). National Institute of Standards and Technology, Gaithersburg, MD.

[64] Lamoreaux, S. K. 1997. Demonstration of the Casimir force in the 0.6 to 6 mm Range. *Phys. Rev. Lett.*, **78**, 5–8.

[65] Larsen, J. 2017. *Foundations of high energy density physics*. Cambridge University Press, Cambridge, UK.

[66] Lindl, J. D. 1995. Development of the indirect-drive approach to inertial confinement fusion and the target physics basis for ignition and gain. *Phys. Plasmas*, **2**, 3933–4024.

[67] Lindl, J. D., Amendt, P., Berger, R. L. et al. 2004. The physics basis for ignition using indirect-drive targets on the National Ignition Facility. *Phys. Plasmas*, **11**, 339–491.

[68] Lotz, W. 1967. An empirical formula for the electron-impact ionization cross-section. *Z. Physik*, **206**, 205–211.

[69] Loudon, R. 1983. *The quantum theory of light*. Oxford University Press, Oxford, UK.

[70] Marchand, E. W. 1978. *Gradient index optics*. Academic Press, New York, US.

[71] Marjoribanks, R. S., Richardson, M. C., Jaanimagi, P. A., and Epstein, R. 1992. Electron-temperature measurement in laser-produced plasmas by the ratio of isoelectronic line intensities. *Phys. Rev. A.*, **46**, R1747–R1750.

[72] Massey, H. S. W., and Burhop, E. H. S. 1952. *Electronic and ionic impact phenomena*. Clarendon Press, Oxford, UK.

[73] Menzel, D. H., and Pekeris, C. L. 1935. Absorption coefficients and hydrogen line intensities. *Monthly Not. R. Astron. Soc.*, **96**(1), 0077–0111.

[74] Morales, M. F., and Wyithe, J. S. B. 2010. Reionization and cosmology with 21-cm fluctuations. *Ann. Rev. Astron. Astrophys.*, **48**, 121–171.

[75] Mott, N.F., and Massey, H.S.W. 1949. *The theory of atomic collisions*. Clarendon Press, Oxford, UK.

[76] Nagler, B., Zastrau, U., Faeustlin, R. R. et al. 2009. Turning solid aluminium transparent by intense soft X-ray photoionization. *Nat. Phys.*, **5**(9), 693–696.

[77] Nikiforov, A. F., Novikov, V. G., and Uvarov, V. B. 2005. *Quantum statistical models of hot dense matter*. Birkhauser, Basel, Switzerland.

[78] Pal'chikov, V. G. 1998. Relativistic transition probabilities and oscillator strengths in hydrogen-like atoms. *Phys. Scr.*, **57**, 581–593.

[79] Parail, V., Belo, P., Boerner, P. et al. 2009. Integrated modelling of ITER reference scenarios. *Nuclear Fusion*, **49**(7), 075030.

[80] Paris, A., and Davies, E. 2015. Hydrogen clouds from comets 266/P Christensen and P/2008 Y2 (Gibbs) are candidates for the source of the 1977 WOW signal. *Washington Acad. Sci.*, 25–31.

[81] Peacock, N. J., Robinson, D. C., Forrest, M. J., Wilcock, P. D., and Sannikov, V. V. 1969. Measurement of the Electron Temperature by Thomson Scattering in Tokamak T3. *Nature*, **224**, 488–490.

[82] Pert, G. J. 1978. The analytic theory of linear resonant absorption. *Plasma Phys.*, **20**, 175–188.

[83] Pert, G. J. 1990. Models of collisional-radiative recombination. *J. Phys. B.*, **23**, 619–650.

[84] Pert, G. J. 2013. *Introductory fluid mechanics for physicists and mathematicians*. Wiley, Oxford, UK.

[85] Phillips, K. J. H. 2004. The solar flare 3.8–10 keV X-ray spectrum. *Astrophys. J.*, **605**, 921–930.

[86] Pradhan, A. K., and Nahar, S. N. 2011. *Atomic astrophysics and spectroscopy*. Cambridge University Press, Cambridge, UK.

[87] Purcell, E. M. 1985. *Electricity and magnetism*. McGraw-Hill, New York, US.

[88] McWhirter, R. W. P. 1965. *Plasma diagnostic techniques*. Edited by Huddlestone, R. H., and Leonard, S. L. Academic Press, New York, US.

[89] Ralchenko, Y. 2016. *Modern methods in collisional-radiative modeling of plasmas*. Springer, Berlin, Germany.

[90] Randewich, A., and Danson, C. 2014. High energy density physics at the Atomic Weapons Establishment. *High Power Laser Sci. Eng.*, **2**, e40.

[91] Rocca, J. J. 1999. Table-top soft X-ray lasers. *Rev. Sci. Instrum.*, **70**(10), 3799–3827.

[92] Rutherford, E. 1911. The scattering of α and β particles by matter and the structure of the atom. *Phil. Mag.*, **21**, 669.

[93] Rybicki, G. B., and Lightman, A. P. 1979. *Radiative processes in astrophysics*. Wiley-Interscience, New York, US.

[94] Sagan, C., Sagan, L. S., and Drake, F. 1972. A message from Earth. *Science*, **175**, 881–884.

[95] Salzmann, D. 1998. *Atomic physics in hot plasmas*. Oxford University Press, Oxford, UK.

[96] Sampson, D. H., and Zhang, H. L. 1992. Use of the van Regemorter formula for collision strengths or cross sections. *Phys. Rev.*, **A45**, 1556.

[97] Samukawa, S., Hori, M., Rauf, S. et al. 2012. The 2012 Plasma Roadmap. *J. Phys. D.*, **45**, 253001.

[98] Schawlow, A. L. 1984. Lasers in historical perspective. *IEEE J. Quant. Electron.*, **QE-20**, 558.

[99] Sheffield, J., Froula, D., Glenzer, S. H., and Luhmann, N. C. 2011. *Plasma scattering of electromagnetic radiation: theory and measurement techniques*. Academic Press, Amsterdam, The Netherlands.

[100] Smith, R., Tallents, G. J., Zhang, J. et al. 1999. Saturation behavior of two X-ray lasing transitions in Ni-like Dy. *Phys. Rev. A*, **59**(1), R47–R50.

[101] Smith, R. K., Brickhouse, N. S., Liedahl, D. A, and Raymond, J. C. 2001. Collisional plasma models with APEC/APED: emission-line diagnostics of hydrogen-like and helium-like ions. *Astrophys. J.*, **556**, L91–L95.

[102] Sobelman, I. I., and Vainshtein, L. A. 1998. *Excitation of atoms and broadening of spectral lines*. Springer, Berlin, Germany.

[103] Stenzel, R. L. 1999. Whistler waves in space and laboratory plasma. *J. Geophys. Res.*, **104**, 14379–14396.

[104] Stewart, J. C., and Pyatt, K. D. 1966. Lowering of ionization potentials in plasmas. *Astrophys. J.*, **144**, 1203.

[105] Tallents, G., Wagenaars, E., and Pert, G. 2010. Optical lithography: lithography at EUV wavelengths. *Nat. Photonics*, **4**(12), 809–811.

[106] Tallents, G. J. 1980. An experimental study of recombination in a laser-produced plasma. *Plasma Phys.*, **22**, 709–718.

[107] Tallents, G. J. 1984. The relative intensities of hydrogen-like fine structure. *J. Phys. B.*, **17**, 3677–3691.

[108] Tallents, G. J. 2003. The physics of soft X-ray lasers pumped by electron collisions in laser plasmas. *J. Phys. D.*, **366**, R259–R276.

[109] Tallents, G. J. 2016. Free electron degeneracy effects on collisional excitation, ionization, de-excitation and three-body recombination. *High Energ. Dens. Phys.*, **20**(9), 9–16.

[110] Tallents, G. J., Wilson, S. A., West, A., Aslanyan, V., Lolley, J., and Rossall, A. K. 2017. The creation of radiation dominated plasmas using laboratory extreme ultra-violet lasers. *High Energ. Dens. Phys.*, **23**(3), 129–132.

[111] Tennyson, J. 2011. *Astronomical spectroscopy: an introduction to the atomic and molecular physics of astronomical spectra*. World Scientific, Singapore.

[112] Trumper, J., Poetscj, W., Reppin, C., Voges, W., Staubert, R., and Kendziorra, E. 1978. Evidence for strong cyclotron line emission in hard X-ray spectrum of Hercules X1. *Astrophys. J.*, **219**(3), L105–L110.

[113] Tseng, W. L., Johnson, R. E., Thomsen, M. F., Cassidy, T. A., and Elrod, M. K. 2011. Neutral H_2 and H_2^+ ions in the Saturnian magnetosphere. *J. Geophys. Res.*, **116**, A03209.

[114] Abels-van Maanen, A. E. P. M. 1985. A package for non-coronal impurity data. *JET-DN-T* (85)29.

[115] van Regemorter, H. 1962. Rate of collisional excitation in stellar atmospheres. *Astrophys. J.*, **A132**, 906.

[116] Vinko, S. M., Ciricosta, O., Cho, B. I. et al. 2012. Creation and diagnosis of a solid-density plasma with an X-ray free-electron laser. *Nature*, **482**(7383), 59–62.

[117] Vinko, S. M., Ciricosta, O., and Wark, J. S. 2014. Density functional theory calculations of continuum lowering in strongly coupled plasmas. *Nat. Commun.*, **5**, 3533.

[118] von Frisch, K. 1967. *The dance language and orientation of bees*. Harvard University Press, Cambridge, MA, US.

[119] Walter, F., Brinks, E., de Blok, W. J. G. et al. 2008. THINGS: the H1 nearby galaxy survey. *Astron. J.*, **136**, 2563–2647.

[120] Wang, W. 1999. Generalization of the Thomas-Rieche-Kuhn and the Bethe sum rules. *Phys. Rev. A*, **60**, 262–266.

[121] NASA. Voyager: the interstellar mission, http://voyager.jpl.nasa.gov/spacecraft/goldenrec1.html (accessed 8 March 2017).

[122] Weinert, F. 1995. Wrong theory-right experiment: the significance of the Stern–Gerlach experiments. *Studies in History and Philosophy Mod. Phys.*, **26**, 75–86.

[123] Wing, W. H., Ruff, G. A., Lamb, W. E., and Spezeski, J. J. 1976. Observation of the infrared spectrum of the hydrogen molecular ion HD^+. *Phys. Rev. Lett.*, **36**, 1488–1491.

[124] Zel'dovich, Ya. B., and Raizer, Yu. P. 1967. *Physics of shock waves and high temperature hydrodynamic phenomena*. Academic, New York, US.

[125] Zhang, J., MacPhee, A. G., Lin, J. et al. 1997. A saturated X-ray laser beam at 7 nanometers. *Science*, **276**(5315), 1097–1100.

Index

www.ingramcontent.com/pod-product-compliance
Lightning Source LLC
LaVergne TN
LVHW082020150826
845671LV00005B/214

* 9 7 8 1 1 0 8 4 1 9 5 4 3 *